Finite Element Methods and Navier-Stokes Equations

Mathematics and Its Applications

Finite Element Methods and Navier-Stokes Equations

by

C. Cuvelier and A. Segal

Department of Mathematics and Informatics,
University of Technology, Delft, The Netherlands

and

A. A. van Steenhoven

Department of Mechanical Engineering,
University of Technology, Eindhoven, The Netherlands

D. Reidel Publishing Company

A MEMBER OF THE KLUWER ACADEMIC PUBLISHERS GROUP

Dordrecht / Boston / Lancaster / Tokyo

Library of Congress Cataloging in Publication Data

CIP

Cuvelier, C. (Cornelius), 1948–
Finite element methods and Navier-Stokes equations.

(Mathematics and its applications)
Bibliography: p.
Includes index.
1. Finite element method. 2. Navier-Stokes equations. I. Segal, A. (August), 1948– . II. Steenhoven, A. A. van (Anton A), 1951– . III. Title. IV. Series: Mathematics and its applications (D. Reidel Publishing Company)
TA347.F5C88 1986 620'.001'515353 86–3240

ISBN-13: 978-1-4020-0309-7 e-ISBN-13: 978-94-010-9333-0

DOI: 10.1007/978-94-010-9333-0

Published by D. Reidel Publishing Company,
P.O. Box 17, 3300 AA Dordrecht, Holland

Sold and distributed in the U.S.A. and Canada
by Kluwer Academic Publishers,
190 Old Derby Street, Hingham, MA 02043, U.S.A.

In all other countries, sold and distributed
by Kluwer Academic Publishers Group,
P.O. Box 322, 3300 AH Dordrecht, Holland

Software and mathematical support available from the authors

Delft University of Technology
Department of Mathematics and Informatics
P.O. Box 356
2600 AJ Delft
The Netherlands

TABLE OF CONTENTS

PREFACE

In present-day technology the engineer is faced with complicated problems which can be solved only by advanced numerical methods using digital computers. He should have mathematical models at his disposal in order to simulate the behavior of physical systems. In many cases, including problems in physics, mechanics, chemistry, biology, civil engineering, electrodynamics, solid mechanics, space engineering, petroleum engineering, weather-forecasting, mass transport, multiphase flow and, last but not least, hydrodynamics, such a physical system can be described mathematically by one or more partial differential equations. Some of these problems require extreme accuracy, while for other problems only the qualitative behavior of the solution need to be studied.

The finite element method (FEM) is one of the most commonly used methods for solving partial differential equations (PDEs). It makes use of the computer and is very general in the sense that it can be applied to both steady-state and transient, linear and nonlinear problems in geometries of arbitrary space dimension. The FEM is in fact a method which transforms a PDE into a system of linear (algebraic) equations.

The following aspects may be identified in the study of a physical phenomenon:

(i) engineering-mathematical sciences to formulate the problem correctly in terms of PDEs

(ii) numerical methods to construct and to solve the system of algebraic equations; applied numerical functional analysis to give error estimates and convergence proofs

(iii) informatics and programming to perform the calculations efficiently on the computer.

In this work we shall pay attention to all three of the aspects, considering in particular the finite element analysis of the Navier-Stokes equations, for which we restrict ourselves to the study of incompressible fluids. Applications include all kinds of internal fluid flows in complex geometries.

Part I (Chapters 1 to 5) serves as an introduction to the finite element analysis of PDEs of elliptic type. In part II (Chapters 6 to 10) three approaches to finite element analysis of the Navier-Stokes equations will be studied, including time dependent terms. In part III (Chapters 11 to 14) a more mathematical approach will be followed to give error estimates and convergence proofs. Finally in part IV (Chapters 15 to 18) we overview some current areas of research.

This work is a written extension of two series of lectures. The first series was given at Eindhoven University of Technology (The Netherlands) for graduate students in mechanical engineering. The second was a series of postdoctoral lectures given at Delft University of Technology (The Netherlands) for research workers whose backgrounds covered almost all areas of engineering. As a consequence of the engineering background of the audience we attempted to eliminate any use of functional analysis. However, in response to the needs of mathematical engineers, we have incorporated in this work some chapters of a more mathematical nature.

We express our hope that this work may be of use to all finite element analysts of the Navier-Stokes equations and that it elucidates the fact that numerical analysis of PDEs is based on the intersection of engineering sciences, applied mathematics and informatics. The experience of acting in a team has made the writing of this book a satisfying task for the authors.

We wish to thank Mr. M.A. Hulsen (Delft University of Technology) for his valuable comments concerning non-Newtonian fluids and Mr. F.N. Van De Vosse (Eindhoven University of Technology) for many useful discussions concerning the time-dependent aspects. Finally we are deeply indebted to Mrs. Gonny Van Werkhoven (Delft University of Technology) who expertly typed the manuscript and to Mr. G. Van De Akker (Audio-Visual Center, Eindhoven University of Technology) who supplied all the figures.

C. Cuvelier
A. Segal
A.A. Van Steenhoven

Delft, Eindhoven
September 1985.

PART I

INTRODUCTION TO THE FINITE ELEMENT METHOD

Introduction

The aim of Part I is to introduce the basic concepts of the Finite Element Method (FEM) for the numerical solution of elliptic Partial Differential Equations (PDEs).

First of all we shall give, in Chapter 1, some examples of elliptic and parabolic PDEs occurring in the fields of heat transfer, lubrication theory, mass transport, elasticity theory and hydrodynamics.

In Chapter 2 we shall pay some attention to the construction of approximate solutions of PDEs using the finite difference method. Three examples will be treated: the Poisson equation in 1D and 2D (one and two dimensions) with Dirichlet and Neumann boundary conditions and the convection-diffusion equation. These examples will show clearly how the finite difference method transforms a PDE into a system of linear algebraic equations. In particular it will be shown that there are some difficulties in satisfying the boundary conditions, especially when working in complex geometries.

In Chapter 3 we shall see that some elliptic PDEs can be formulated equivalently as extremal (minimization or maximization) problems which amounts to finding the stationary value of a specific integral, known as a functional. Based on these extremal problems we shall construct approximate solutions using basis functions. This method is known as the Ritz method and leads to a system of linear algebraic equations which, in general, has a bad condition. As a particular example of the Ritz method, i.e. a particular choice of the basis functions, we shall study the FEM, which leads to a well-conditioned system of linear equations. The method of Ritz is thus based on finding the stationary value of a functional for which the PDE is just the Euler equation. Unfortunately, it is generally difficult or even impossible to find the correct functional corresponding to a PDE. Therefore we shall discuss an other integral (or variational) method based on the PDE itself. This method is known as the Galerkin method and is also applicable to those PDEs for which no equivalent optimization formulation exists. Just like the Ritz method, the Galerkin method defines a procedure to transform the PDE into a system of linear equations. The FEM will be introduced as a particular example of the Galerkin method. Using the FEM we shall explain how to construct basis functions based on the Ritz or the Galerkin method.

In Chapter 4 we shall introduce various types of finite elements: 1D elements, triangular and quadrilateral elements in 2D and tetrahedral and hexahedral elements in 3D. For the construction of higher-order basis functions we introduce the barycentric coordinates.

In Chapter 5 we shall discuss and illustrate the practical aspects of the FEM, such as the assembly of the final system of linear equations and the use of barycentric coordinates to simplify the computation of integrals. Furthermore, some numerical integration rules will be given as well as some rules of the thumb for error estimates. For more precise mathematical results we refer to Part III. Since the application of the FEM leads to the solution of a system of linear equations, it is worthwhile to pay attention to solution methods for systems of (linear) equations which account "optimally" for the systems properties. This will be the subject of Section 5.6.

Chapter 1

Examples of Partial Differential Equations

1.1 Classification of PDEs

Let us consider the n dimensional Euclidean space $\mathbb{R}^n$ in which the independent variables are denoted by $x_1, x_2, \ldots, x_n$. In practical (time-independent) applications n is equal to 1, 2 or 3. For time-dependent problems one of the independent variables denotes the time t. In practice n is then equal to 2, 3 or 4 with $x_n = t$. A variable u which depends on $\underline{x}=(x_1,\ldots,x_n)$ is written as $u = u(\underline{x})$. Partial derivatives of the dependent variable u are written as follows: $\frac{\partial u}{\partial x_i}$, $i = 1,2,\ldots,n$, $\frac{\partial^2 u}{\partial x_i \partial x_j}$, $i,j = 1,2,\ldots,n$, etc.

Differential equations are equations for which the unknowns are functions of one or more independent variables and which contain these functions and their derivatives as well. If the unknowns are functions of more than one variable, then the equation is called a PDE. A PDE is said to be of order p if it contains at least one derivative of order p and does not contain derivatives of higher order.

The general form of a PDE of order p in an unknown function u is:

$$\phi(x_1,\ldots,x_n,u,\frac{\partial u}{\partial x_1},\ldots,\frac{\partial u}{\partial x_n},\frac{\partial^2 u}{\partial x_1^2},\frac{\partial^2 u}{\partial x_1 \partial x_2},\ldots,\frac{\partial^p u}{\partial x_n^p}) = 0 \qquad (1.1.1)$$

where ϕ is a function of the indicated quantities. Equation (1.1.1) is in general a nonlinear relation.

In this work we shall mainly be concerned with second-order PDEs (p=2), i.e. equations of the following type:

$$\mathcal{L}u \equiv -\sum_{i,j=1}^{n} a_{ij}\frac{\partial^2 u}{\partial x_i \partial x_j} + \psi(x_1,\dots,x_n,u,\frac{\partial u}{\partial x_1},\dots,\frac{\partial u}{\partial x_n}) = 0 \qquad (1.1.2)$$

where ψ is a function of the indicated quantities, and where the coefficients a_{ij} may depend on $x_1,\dots,x_n,u,\frac{\partial u}{\partial x_1},\dots,\frac{\partial u}{\partial x_n},\frac{\partial^2 u}{\partial x_1^2},\frac{\partial^2 u}{\partial x_1 \partial x_2},\dots,\frac{\partial^2 u}{\partial x_n^2}$.

The PDE (1.1.2) is called quasilinear if the coefficients a_{ij} depend only on $x_1,\dots,x_n,u,\frac{\partial u}{\partial x_1},\dots,\frac{\partial u}{\partial x_n}$. The equation (1.1.2) is called linear if it can be written in the following form

$$-\sum_{i,j=1}^{n} a_{ij}\frac{\partial^2 u}{\partial x_i \partial x_j} + \sum_{i=1}^{n} a_i \frac{\partial u}{\partial x_i} + a_0 u = f \qquad (1.1.3)$$

where a_{ij}, $i,j=1,\dots,n$, a_i, $i=1,\dots,n$, a_0 and f depend on $x_1,\dots,x_n$ only. In all other cases equation (1.1.2) is called a nonlinear PDE. As is well-known, second-order PDEs can be classified into three types: elliptic, hyperbolic and parabolic. In order to give criteria according to which the PDE is of a particular type, we consider the n×n matrix $A = (a_{ij})$ composed of the coefficients of the second-order derivatives. Because $\frac{\partial^2 u}{\partial x_i \partial x_j} = \frac{\partial^2 u}{\partial x_j \partial x_i}$, we can change the coefficients a_{ij} into $\frac{1}{2}(a_{ij}+a_{ji})$ so that the matrix can be taken as symmetric. Since the coefficients a_{ij} depend on $\underline{x} = (x_1,\dots,x_n)$, the eigenvalues of A depend also on $\underline{x}$. Let us denote by $n^+(\underline{x})$ the number of positive eigenvalues of A at the point $\underline{x} \in \mathbb{R}^n$; the number of negative eigenvalues at $\underline{x}$ will be denoted by $n^-(\underline{x})$ and $n^0(\underline{x})$ is the number of zero-valued eigenvalues at $\underline{x}$. Notice that, since A is symmetric, all the eigenvalues are real (see Strang (1980)) and that $n = n^+ + n^- + n^0$.

Equation (1.1.2) is called a PDE of elliptic type in a region $\Omega \subset \mathbb{R}^n$ if $n^+(\underline{x}) = n$ for all $\underline{x} \in \Omega$ or $n^-(\underline{x}) = n$ for all

$\underline{x} \in \Omega$. An example of an elliptic PDE is the Poisson equation in $\mathbb{R}^n$:

$$-\left(\frac{\partial^2 u}{\partial x_1^2} + \dots + \frac{\partial^2 u}{\partial x_n^2}\right) = f \ .$$

All the eigenvalues of A are equal to one, so that $n^+ = n$ and $n^- = n^0 = 0$.

Equation (1.1.2) is an equation of hyperbolic type in the region $\Omega \subset \mathbb{R}^n$ if $n^+(\underline{x}) = n-1$ and $n^-(\underline{x}) = 1$ for all $\underline{x} \in \Omega$ or $n^+(\underline{x}) = 1$ and $n^-(\underline{x}) = n-1$ for all $\underline{x} \in \Omega$. An example of a hyperbolic equation in $\mathbb{R}^n$ is:

$$\frac{\partial^2 u}{\partial x_n^2} - \left(\frac{\partial^2 u}{\partial x_1^2} + \dots + \frac{\partial^2 u}{\partial x_{n-1}^2}\right) = f \ .$$

The eigenvalues are equal to -1 and 1 ($n-1$ times), so that $n^+ = n-1$ and $n^- = 1$. In applications we usually have $x_n = t$. Equation (1.1.2) is termed ultra-hyperbolic if $n^0(\underline{x}) = 0$ and $1 < n^+(\underline{x}) < n-1$ for all $\underline{x} \in \Omega$.

Equation (1.1.2) is designated parabolic in $\Omega \subset \mathbb{R}^n$ if $n^0(\underline{x}) > 0$ for all $\underline{x} \in \Omega$. An example of a parabolic equation is

$$\frac{\partial u}{\partial x_n} - \left(\frac{\partial^2 u}{\partial x_1^2} + \dots + \frac{\partial^2 u}{\partial x_{n-1}^2}\right) = f$$

where, often, $x_n = t$.

In 2D (i.e. $n=2$) the above mentioned criteria reduce to

(i) $a_{12}a_{21} - a_{11}a_{22} < 0$: PDE is elliptic,

(ii) $a_{12}a_{21} - a_{11}a_{22} > 0$: PDE is hyperbolic,

(iii) $a_{12}a_{21} - a_{11}a_{22} = 0$: PDE is parabolic.

Notice that $a_{12} = a_{21}$.

In general there are infinitely many functions u satisfying a given PDE. A problem for an unknown function u can only be called physically realistic if the problem is well-posed, which means that the unknown function u exists (that is, there is at least one solution), that the function u is unique (that is, there is at most one solution) and that the function u depends continuously on the data of the problem (like coefficients, boundary conditions, geometry of the region). This usually means that elliptic PDEs are associated with boundary conditions, while parabolic and hyperbolic PDEs are associated with both initial and boundary conditions. This point will become clear in the examples.

1.2 Laplace and Poisson Equation

Many steady-state (i.e. time independent) processes in physics and mechanics are described by elliptic PDEs, as exemplified by problems in heat conduction, electrostatics, lubrication theory, elasticity theory and hydrodynamics.

Let us consider the problem of heat conduction in an n-dimensional open region $\Omega \subset \mathbb{R}^n$ the boundary of which is denoted by Γ. The vector $\underline{n}$ denotes the unit outer normal on Γ (see Fig. 1.1).

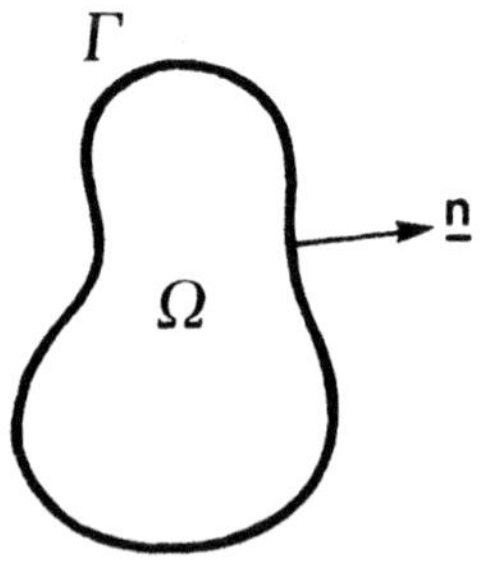

Fig. 1.1 Region Ω with boundary Γ and unit outer normal $\underline{n}$.

The heat transfer in isotropic media is described by the following constitutive law of Fourier: the heat flow per unit area (= heat flux vector) $\underline{q}$ is proportional to the gradient of the temperature T:

$$\underline{q} = -k \text{ grad } T \tag{1.2.1}$$

where the parameter $k > 0$ is the thermal conductivity of the fluid. When we assume the medium to be at rest and the temperature distribution to be constant in time, the energy equation (see Section 1.6) reads:

$$\text{div } \underline{q} = \rho \, f_T \tag{1.2.2}$$

where f_T is the rate of heat production per unit mass (=heat source) and ρ denotes the density of the fluid. Combination of (1.2.1) and (1.2.2) leads to the following equation for the temperature T:

$$-\text{ div}(k \text{ grad } T) = \rho \, f_T. \tag{1.2.3}$$

In general the thermal conductivity of the fluid k depends on the spatial coordinates $\underline{x}$: $k = k(\underline{x})$. When k is uniform throughout the material, equation (1.2.3) reduces to:

$$-\, k \text{ div grad } T = \rho \, f_T. \tag{1.2.4}$$

Equation (1.2.3) as well as equation (1.2.4) is a linear second order PDE of elliptic type since $k(\underline{x}) > 0$ in $\Omega \cup \Gamma$. Equation (1.2.4) is called the Poisson equation and the operator $\mathcal{L}$ (see (1.1.2)) is given by

$$\mathcal{L}T \equiv -\, k \text{ div grad } T - \rho \, f_T \equiv -\, k\left(\frac{\partial^2 T}{\partial x_1^2} + \ldots + \frac{\partial^2 T}{\partial x_n^2}\right) - \rho \, f_T.$$

If there are no heat sources, then $f_T \equiv 0$ and (1.2.4) can be written as

$$-\, \Delta \, T = 0 \tag{1.2.5}$$

with

$$\Delta = \frac{\partial^2}{\partial x_1^2} + \ldots + \frac{\partial^2}{\partial x_n^2}$$

which is known as the Laplace equation for T.

In order to determine the temperature distribution in the region Ω, we need boundary conditions on Γ. We have among others the following possibilities:

(i) prescribed temperature T on Γ, called Dirichlet type boundary conditions:

$$T(\underline{x}) = g_0(\underline{x}) \qquad \underline{x} \in \Gamma$$

(ii) prescribed heat flux on Γ, called Neumann type boundary conditions:

$$(\underline{q}\cdot\underline{n})(\underline{x}) \equiv -(k \text{ grad } T.\underline{n})(\underline{x}) \equiv -(k \frac{\partial T}{\partial n})(\underline{x}) = g_1(\underline{x}), \; \underline{x} \in \Gamma$$

(iii) a relation between temperature and heat flux on Γ, known as the Newton's law of cooling, called Robbins type boundary conditions:

$$h\, T(\underline{x}) + k \frac{\partial T}{\partial n} (\underline{x}) = g_2(\underline{x}) \;, \qquad \underline{x} \in \Gamma$$

where h denotes the heat transfer coefficient.

(iv) a combination of (i), (ii) and (iii).

The problem can now be formulated as follows

$$\left| \begin{array}{l} \text{Find } T = T(\underline{x}),\; \underline{x} \in \Omega \cup \Gamma, \text{ such that} \\ -\text{ div } (k \text{ grad } T) = \rho\, f_T \quad \text{in } \Omega \\ T = g_0 \quad \text{on } \Gamma_0 \\ -k \dfrac{\partial T}{\partial n} = g_1 \quad \text{on } \Gamma_1 \\ h\, T + k \dfrac{\partial T}{\partial n} = g_2 \quad \text{on } \Gamma_2 \end{array} \right. \qquad (1.2.6)$$

where Γ_0, Γ_1 and Γ_2 are disjoint parts of Γ such that $\Gamma_0 \cup \Gamma_1 \cup \Gamma_2 = \Gamma$ (see Fig. 1.2). Later on we shall solve this problem numerically with the possibility of vanishing Γ_0, Γ_1 or Γ_2.

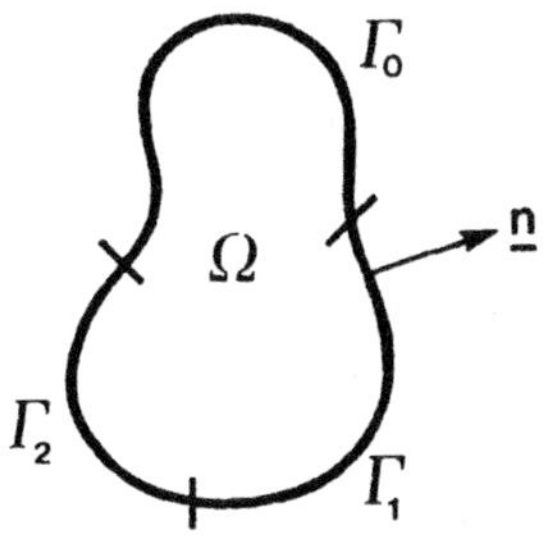

Fig. 1.2 Region Ω with $\Gamma = \Gamma_0 \cup \Gamma_1 \cup \Gamma_2$.

A problem similar to the preceding one is that of diffusion of mass in a binary fluid with constant mass density of the solution (see Bird, Stewart, Lightfoot (1960)). Fourier's law is now replaced by Fick's law of diffusion:

$$\underline{q} = - k \text{ grad } C \tag{1.2.7}$$

where $\underline{q}$ denotes the mass flux vector relative to the mass average velocity, k is the coefficient of diffusion and $C = C(\underline{x})$ the mass concentration.

Assuming the fluid to be at rest and the concentration distribution to be constant in time, the diffusion equation reads:

$$- \text{div } (k \text{ grad } C) = f_C \tag{1.2.8}$$

with f_C the rate of production of mass per unit volume. When k is constant, (1.2.8) is the Poisson equation and if, moreover, there are no sources, then C satisfies the Laplace equation. On the boundary Γ of Ω a condition must be specified of the Dirichlet, Neumann or Robbins type or a combination of these.

Another physical phenomenon which leads to the Poisson equation is the description of an electrostatic field in a dielectric medium Ω. This problem is governed by the Maxwell equations:

$$\operatorname{curl} \underline{E} = 0 \tag{1.2.9}$$

$$\operatorname{div} (\varepsilon \underline{E}) = \rho_{free} \tag{1.2.10}$$

where $\underline{E}$ is the electric field vector, $\varepsilon = \varepsilon(\underline{x}) > 0$ is the permittivity of the dielectricum and $\rho_{free} = \rho_{free}(\underline{x})$ denotes the density of the electric charges put on conductors or at known places in space. Since curl $\underline{E} = 0$, $\underline{E}$ can be written as the gradient of a potential:

$$\underline{E} = - \operatorname{grad} \phi \tag{1.2.11}$$

where ϕ is known as the electric potential.
Substituting (1.2.11) into (1.2.10), we obtain the following elliptic PDE

$$- \operatorname{div} (\varepsilon \operatorname{grad} \phi) = \rho_{free} \quad \text{in } \Omega$$

which reduces for a uniform medium to the Poisson equation

$$- \Delta \phi = \frac{\rho_{free}}{\varepsilon} .$$

If we assume that the boundary Γ is a conductor, then the component of the electric field $\underline{E}$ in the tangent plane must vanish

$$\underline{E} - (\underline{E} \cdot \underline{n}) \underline{n} = \underline{0} \quad \text{on } \Gamma$$

which amounts to saying that

$$\phi = \text{constant on } \Gamma. \tag{1.2.12}$$

According to its definition, the electric potential ϕ is defined up to an additive constant which can be fixed by specifying ϕ in one point.
Other possible boundary conditions are (i) potential given on Γ (Dirichlet type), (ii) normal derivative of the potential, i.e. the electric field given on Γ (Neumann type), or (iii) a combination of (i) and (ii).

Similar equations are found for the magnetic vector potential in the case of steady currents (see Panofsky, Philips

(1962), Landau, Lifshitz (1969)).

In the following example we consider the potential flow of an incompressible fluid in a region Ω. For an incompressible fluid, the mass-conservation equation (1.6.1) takes the simple form:

$$\text{div}\ \underline{u} = 0 \tag{1.2.13}$$

where $\underline{u}$ denotes the velocity vector.
If, moreover, the velocity field is irrotational, i.e. curl $\underline{u} = \underline{0}$, there exists a velocity potential ϕ such that

$$\underline{u} = \text{grad}\ \phi\ . \tag{1.2.14}$$

Substitution of (1.2.14) into (1.2.13) yields a PDE governing the velocity potential:

$$-\Delta\phi = 0\ . \tag{1.2.15}$$

The boundary conditions for ϕ are obtained from (1.2.14). When the normal component of the velocity u_n is known on Γ, then it follows from (1.2.14) that

$$\frac{\partial\phi}{\partial n} = \text{grad}\ \phi\cdot\underline{n} = \underline{u}\cdot\underline{n} = u_n \quad \text{on} \quad \Gamma\ . \tag{1.2.16}$$

The velocity potential ϕ is thus defined by the Laplace equation (1.2.15) together with the Neumann boundary condition (1.2.16) and must be fixed at an arbitrary constant in one point.

Besides, a more general description can be given for an incompressible fluid.
Since div $\underline{u} = 0$, there exists a vector potential $\underline{\Psi}$ such that

$$\underline{u} = \text{curl}\ \underline{\Psi} \tag{1.2.17}$$

In 2D $\underline{\Psi}$ is equal to

$$\underline{\Psi} = \begin{vmatrix} 0 \\ 0 \\ \psi \end{vmatrix}$$

where ψ is known as the streamfunction. Thus

$$u_1 = \frac{\partial\psi}{\partial x_2} \qquad u_2 = -\frac{\partial\psi}{\partial x_1} \,. \tag{1.2.18}$$

Introducing the local vorticity of the fluid to be $\underline{\omega}$ = curl $\underline{u}$ we find

$$\text{curl } \underline{u} = \text{curl curl } \underline{\Psi} = \text{curl} \begin{vmatrix} \frac{\partial\psi}{\partial x_2} \\ -\frac{\partial\psi}{\partial x_1} \\ 0 \end{vmatrix} = \begin{vmatrix} 0 \\ 0 \\ -\Delta\psi \end{vmatrix} = \underline{\omega} = \begin{vmatrix} 0 \\ 0 \\ \omega \end{vmatrix} .$$

Hence

$$-\Delta\psi = \omega \quad \text{in } \Omega \,.$$

When the vorticity ω is only dependent on spatial coordinates $\underline{x}$, we arrive to the Poisson equation, while for the irrotational case the Laplace equation is found for ψ.

Boundary conditions for ψ are obtained from the relation (1.2.17). When on a part of the boundary the flow vector is parallel to it (or equal to zero), then ψ is constant on that part. For boundary parts with mass flux, we find that

$$\frac{\partial\psi}{\partial t} = \underline{u}\cdot\underline{n} = u_n$$

with u_n the normal component of the velocity on that part. In general we can write also

$$\frac{\partial\psi}{\partial n} = \text{grad } \psi\cdot\underline{n} = -u_2n_1 + u_1n_2 = -u_t$$

where u_t denotes the tangential component of the velocity and $\underline{t} = (-n_2,n_1)$ the unit tangent. A value of ψ can be set arbitrarily, since ψ is defined only to within an arbitrary constant.

So, it turns out that the streamfunction for a 2D flow of an incompressible fluid with constant vorticity is governed by the Poisson equation with, for example, mixed Dirichlet-Neumann boundary conditions. See also Section 6.3.

Analogous relations can be found for axi-symmetric flow situations (see Batchelor (1967)).

1.3 Steady state convection-diffusion equation

In Section 1.2 an equation was derived for the temperature distribution in a medium which is at rest. If the medium in which the heat flow is considered is moving with given velocity $\underline{v} = (v_1,\dots,v_n)$, then there is an additional contribution to the equation of energy as used in the preceding section. If the effects of compression and viscous dissipation are neglected (see Section 1.6) the following relation for the steady state can be derived:

$$- \text{div } (k \text{ grad } T) + \rho \, c_p \, \underline{v}\cdot\text{grad } T = \rho \, f_T \qquad (1.3.1)$$

where c_p is the specific heat per unit mass at constant pressure. In addition to this equation, T has to satisfy boundary conditions.

Consider for instance the problem of heat transport by a Poiseuille flow through a straight cylindrical tube, see Fig. 1.3.

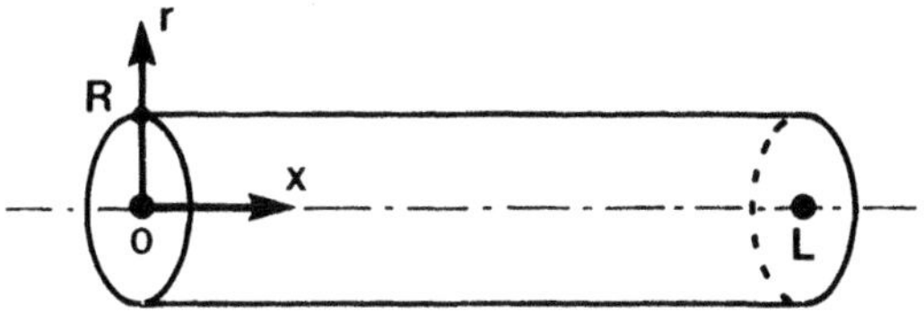

Fig. 1.3 Straight cylindrical tube, $0 \leqslant r \leqslant R$, $0 \leqslant x \leqslant L$.

At the inlet (x=0) we assume the temperature equal to T_0, while the tube is kept at temperature T_1.
If we make the following assumptions: (i) the velocity distribution $\underline{v}$ of the fluid depends only on r and is in the

direction of the x-axis, (ii) the problem is axi-symmetric, which implies that the temperature depends on r and x only, (iii) the physical quantities k, ρ and c_p are constant, then T satisfies the following elliptic PDE

$$- k \Delta_r T + \rho c_p v \frac{\partial T}{\partial x} = \rho f_T \qquad 0 < r < R , \quad 0 < x < L$$

with

$$\Delta_r = \frac{\partial^2}{\partial x^2} + \frac{\partial^2}{\partial r^2} + \frac{1}{r}\frac{\partial}{\partial r}$$

the Laplace operator in cylindrical coordinates and $v = v(r)$ the x-component of $\underline{v}$.
The boundary conditions are

$$T = T_0 \quad \text{at} \quad x = 0 \qquad 0 \leq r \leq R$$

$$T = T_1 \quad \text{at} \quad r = R \qquad 0 < x < L$$

$$\frac{\partial T}{\partial r} = 0 \quad \text{at} \quad r = 0 \qquad 0 < x < L$$

$$\frac{\partial T}{\partial x} = 0 \quad \text{at} \quad x = L \qquad 0 \leq r \leq R.$$

The last boundary condition is justified for large values of L: $L > 0.68 \text{ Pe } R$, where Pe denotes the Peclet number $\text{Pe} = \frac{v_{max} R}{\kappa}$ with v_{max} the velocity at $r = 0$, R the tube radius and $\kappa = \frac{k}{\rho c_p}$ the thermal diffusivity of the fluid (see Bird, Stewart, Lightfoot (1980)).

So we see that in the case of a moving medium too, the temperature is governed by an elliptic PDE with mixed Dirichlet and Neumann boundary conditions.

Generally, equations of the type

$$- \Delta T + \underline{v}.\text{grad } T = f \qquad (1.3.2)$$

are called convection-diffusion equations.

An equation similar to (1.3.2) is satisfied by the concentration C in the case of mass diffusion in a medium which is moving with a given velocity distribution (cf. Section 1.2).

1.4 Time dependent convection-diffusion equation

For a description of time dependent conduction of heat in an isotropic medium $\Omega \subset \mathbb{R}^n$, which is moving with velocity $\underline{v} = \underline{v}(\underline{x})$, we assume that the temperature distribution at time $t = 0$ is known, as well as the heat sources and sinks, if any. The problem is to determine the temperature distribution at later times.

When the effects of viscous dissipation are neglected and the assumption is made that the fluid density ρ is independent of the temperature and pressure changes within the fluid, the equation of energy reads (see Section 1.6):

$$\rho c_p \frac{DT}{Dt} \equiv \rho c_p \frac{\partial T}{\partial t} + \rho c_p \underline{v}\cdot \text{grad } T = \text{div}(k \text{ grad } T) + \rho f_T \quad (1.4.1)$$

in which the notation

$$\frac{D}{Dt} = \frac{\partial}{\partial t} + \underline{v}\cdot\text{grad} \quad (1.4.2)$$

for the material derivative of a field variable is introduced. When ρ, c_p and k are uniform, (1.4.1) reduces to

$$\frac{\partial T}{\partial t} + \underline{v}\cdot\text{grad } T = \kappa \, \Delta \, T + \frac{1}{c_p} f_T \, . \quad (1.4.3)$$

For a fluid at rest we finally find the diffusion equation:

$$\frac{\partial T}{\partial t} = \kappa \, \Delta \, T + \frac{1}{c_p} f_T \quad (1.4.4)$$

The equations (1.4.3) and (1.4.4) are of parabolic type in the variables $x_1,\ldots,x_n,t$, and for the well-posedness of this problem we have to prescribe initial conditions:

$$T(\underline{x},0) = T^0(\underline{x}) \qquad \underline{x} \in \Omega \cup \Gamma$$

and boundary conditions of which we only mention the following two possibilities:

Dirichlet type: $T(\underline{x},t)$ prescribed for $\underline{x} \in \Gamma$, $t > 0$

and

Neumann type: $- k \frac{\partial T}{\partial n} (\underline{x},t)$ prescribed for $\underline{x} \in \Gamma$, $t > 0$

A similar equation is found for the time-dependent mass diffusion in a binary system in motion (see Bird, Stewart, Lightfoot (1960)).

1.5 Reynolds equation

This example is taken from the theory of lubrication (see Pinkus, Sternlight (1961), Cameron (1966)). Here the unknown is the pressure distribution p in a film of lubricant, with viscosity μ, contained in the narrow clearance between two cylinders C_1 (with radius r_1) and C_2 (with radius r_2) of equal length L, open at both sides to atmospheric pressure p_a. The axis of C_1 and C_2 are parallel though distinct and the motion of C_2 is a rotation around its axis with constant angular velocity ω. The situation is depicted in Fig. 1.4.

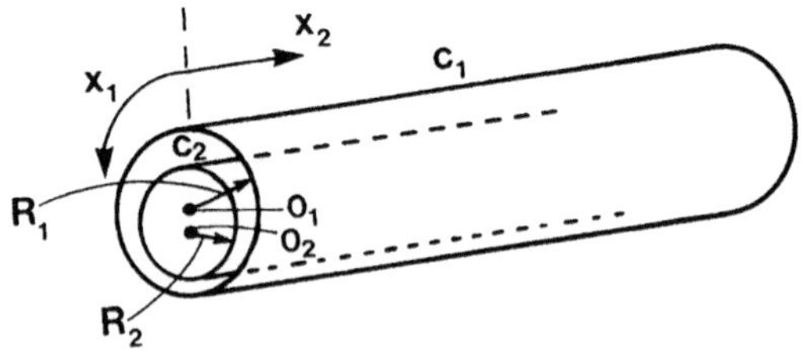

Fig. 1.4 Journal bearing.

We assume the spacing between the two cylinders so small that the pressure in the lubricant can be considered as constant in the radial direction, which implies that $p = p(x_1,x_2)$.
The thickness of the film is given by

$$h(x_1,x_2) = (1 + \varepsilon \cos \frac{x_1}{2\pi r_1})c$$

with ε = eccentricity $= \frac{e}{c}$, $0 \leqslant \varepsilon < 1$, $e = |0_1-0_2|$ = distance between the central axis, $c = r_1-r_2$.

It can be proved that the general equations describing the dynamics of the lubricant (i.e. conservation of mass, momentum and energy) reduce to the so-called Reynolds equation for the pressure:

$$- \text{div } (h^3 \text{grad } p) = -6 \mu \omega r_1 \frac{\partial h}{\partial x_1} \quad \begin{cases} 0 < x_1 < 2\pi r_1 \\ 0 < x_2 < L \end{cases} \tag{1.5.1}$$

which is a PDE of elliptic type, since $h > 0$.
The boundary conditions for p are

(i) periodicity in x_1: $p(0,x_2) = p(2\pi r_1,x_2) \qquad 0 \leqslant x_2 \leqslant L$

$$\frac{\partial p}{\partial x_1}(0,x_2) = \frac{\partial p}{\partial x_1}(2\pi r_1,x_2) \qquad 0 \leqslant x_2 \leqslant L$$

(ii) $p(x_1,0) = p(x_1,L) = p_a \qquad 0 \leqslant x_1 \leqslant 2\pi r_1$

1.6 Equations of fluid dynamics; Navier-Stokes equations

The equations describing the dynamics of fluid flow will be introduced in this section. The equations will be derived from the basic principles of conservation of mass, momentum and energy. For a complete derivation of these equations we refer to Batchelor (1967) or Landau, Lifshitz (1971); we give here only a brief outline.
We shall use the nabla-notation

$$\nabla = \begin{vmatrix} \frac{\partial}{\partial x_1} \\ \vdots \\ \frac{\partial}{\partial x_n} \end{vmatrix} .$$

For a scalar function ϕ, $\nabla\phi$ denotes grad ϕ; for a vector $\underline{u}$, $\nabla\cdot\underline{u}$ denotes div $\underline{u}$; for a tensor $\underline{\underline{\sigma}}$ with components σ_{ij}, $i,j=1,\ldots,n$, $\nabla\cdot\underline{\underline{\sigma}}$ denotes the vector with components $(\nabla\cdot\underline{\underline{\sigma}})_i = \sum_{j=1}^{n} \frac{\partial}{\partial x_j} \sigma_{ij}$.

The first principle is the conservation of mass which can be expressed by

$$\frac{\partial\rho}{\partial t} + \nabla\cdot(\rho\underline{u}) = 0 \qquad (1.6.1)$$

where $\rho = \rho(\underline{x},t)$ is the density of the fluid, t the time and $\underline{u} = \underline{u}(\underline{x},t)$ the velocity vector.

From the balance of momentum we obtain

$$\rho(\frac{\partial\underline{u}}{\partial t} + (\underline{u}\cdot\nabla)\underline{u}) = \nabla\cdot\underline{\underline{\sigma}} + \rho\,\underline{f} \qquad (1.6.2)$$

where $\underline{\underline{\sigma}}$ is the stress tensor and $\underline{f} = \underline{f}(\underline{x},t)$ denotes the body force per unit mass.

For a fluid it is convenient to regard the stress tensor $\underline{\underline{\sigma}}$ as the sum of an isotropic part $-p\underline{\underline{I}}$, having the same form as the stress tensor for a fluid at rest with a hydrostatic pressure p, and a remaining non-isotropic part $\underline{\underline{d}}$, termed the deviatoric stress tensor, caused entirely by the fluid motion:

$$\underline{\underline{\sigma}} = -p\underline{\underline{I}} + \underline{\underline{d}} \quad \text{with } \underline{\underline{I}} = \text{unit tensor}. \qquad (1.6.3)$$

To deduce the dependence of the deviatoric stress tensor on the local velocity gradients, it is assumed that d_{ij} is a linear function of the various components of the velocity gradients and that the fluid is isotropic. Such a fluid is called a Newtonian fluid and its constitutive relation reads:

$$\underline{\underline{d}} = 2\mu(\underline{\underline{e}} - \frac{1}{3}(\nabla.\underline{u})\underline{\underline{I}}) \quad \text{with} \quad e_{ij} = \frac{1}{2}(\frac{\partial u_i}{\partial x_j} + \frac{\partial u_j}{\partial x_i}) \tag{1.6.4}$$

in which $\underline{\underline{e}}$ is the symmetrical part of the velocity gradient tensor, known as the rate of strain tensor, and μ is the fluid viscosity depending on the temperature $T(\underline{x},t)$. Substitution into the equation of motion (1.6.2) then gives for the velocity

$$\rho(\frac{\partial \underline{u}}{\partial t} + (\underline{u}.\nabla)\underline{u}) = \rho\ \underline{f} - \nabla\ p + \nabla.[2\mu(\underline{\underline{e}} - \frac{1}{3}(\nabla.\underline{u})\underline{\underline{I}})]. \tag{1.6.5}$$

These equations are known as the Navier-Stokes equations for fluid motion.

For the case the differences in temperature are small enough for μ to be taken as uniform over the fluid, the Navier-Stokes equations reduce to:

$$\rho(\frac{\partial \underline{u}}{\partial t} + (\underline{u}.\nabla)\underline{u}) = \rho\ \underline{f} - \nabla\ p + \mu[\Delta\underline{u} + \frac{1}{3}\nabla(\nabla.\underline{u})]\ . \tag{1.6.6}$$

In the case of an incompressible fluid, these relations further reduce to

$$\rho(\frac{\partial \underline{u}}{\partial t} + \underline{u}.\nabla\underline{u}) = \rho\underline{f} - \nabla p + \mu\Delta\underline{u}\ . \tag{1.6.7}$$

Neglecting the convective term yields the Stokes equations for such a fluid:

$$\rho\ \frac{\partial \underline{u}}{\partial t} = \rho\underline{f} - \nabla p + \mu\Delta\underline{u} \tag{1.6.8}$$

The general equation of energy for a mass of homogeneous fluid is found from the change in the internal energy of the fluid E due to the performance of work on the fluid, by a gain of heat through conduction and the presence of heat sources f_T, all defined per unit mass. Calculating the internal energy balance for such a mass of homogeneous fluid, yields the next expression:

$$\frac{DE}{Dt} = \frac{-p}{\rho}\ \nabla.\underline{u} + \frac{2\mu}{\rho}\ (\underline{\underline{e}}:\underline{\underline{e}} - \frac{1}{3}\ (\nabla.\underline{u})^2) + \frac{1}{\rho}\ \nabla.(k\nabla T) + f_T \tag{1.6.9}$$

with

$$\underline{\underline{e}}:\underline{\underline{e}} = \sum_{i,j=1}^{n} e_{ij}e_{ij} \ . \tag{1.6.10}$$

This relation shows the separate contributions to the work done associated with the compression and shearing of the fluid, the heat conduction and heat sources, respectively. The second term on the right-hand side of the equation is non-negative and leads to an irreversible addition of heat, which is a consequence of the effect of viscosity on the fluid. Therefore this term is also denoted as the rate of dissipation of mechanical energy per unit mass.

Finally the internal energy can be rewritten in terms of the rate of change of entropy S per unit mass using the relation

$$\frac{DE}{Dt} = T\,\frac{DS}{Dt} - p\,\frac{D(\frac{1}{\rho})}{Dt} = c_p\,\frac{DT}{Dt} - \frac{\beta T}{\rho}\,\frac{Dp}{Dt} - p\,\frac{D(\frac{1}{\rho})}{Dt} \tag{1.6.11}$$

with β the coefficient of thermal expansion.

The basic equations can be simplified by assuming that $\beta=0$ and that the fluid is incompressible, which means that the density ρ is constant and by assuming that the rate of dissipation of mechanical energy term in the energy equation is negligible.

With these assumptions the equations (1.6.1), (1.6.8) and (1.6.9) reduce to the following equations of motion

$$\left|\begin{array}{l} \nabla.\underline{u} = 0 \\[2ex] \rho(\dfrac{\partial \underline{u}}{\partial t} + (\underline{u}.\nabla)\underline{u}) - \mu\Delta\underline{u} + \nabla p = \rho\underline{f} \\[2ex] \rho\ c_p(\dfrac{\partial T}{\partial t} + (\underline{u}.\nabla)T) - \nabla.(k\nabla T) = \rho f_T \end{array}\right. \tag{1.6.12}$$

To this system of PDEs, which is parabolic both in $\underline{u}$ and in T, we add initial conditions

$$\underline{u}(\underline{x},0) = \underline{u}^0(\underline{x}), \quad T(\underline{x},0) = T^0(\underline{x}) \qquad \underline{x} \in \Omega \cup \Gamma \tag{1.6.13}$$

and boundary conditions of which we only mention the Dirichlet and Neumann type in 2D.

For $\underline{u}$ we can prescribe one or two velocity components on Γ (Dirichlet boundary conditions):

$$u_i \quad \text{prescribed on} \quad \Gamma, \quad t > 0$$

or one or two Neumann conditions which are conditions for the normal and/or tangential components of the stress (see also Section 11.5):

$$\left| \begin{array}{l} \sigma_n \equiv (\underline{\underline{\sigma}}\cdot\underline{n})\cdot\underline{n} = \sum_{i,j=1}^{2} \sigma_{ij} n_i n_j = \text{normal stress}, \\ \text{prescribed on } \Gamma,\ t > 0 \end{array} \right. \qquad (1.6.14)$$

$$\left| \begin{array}{l} \sigma_t \equiv (\underline{\underline{\sigma}}\cdot\underline{n})\cdot\underline{t} = \sum_{ij,=1}^{2} \sigma_{ij} n_i t_j = \text{tangential stress}, \\ \text{prescribed on } \Gamma,\ t > 0 \end{array} \right. \qquad (1.6.15)$$

where $\underline{n}$ is the unit normal and $\underline{t}$ is the unit tangent on Γ (see Fig. 1.5).

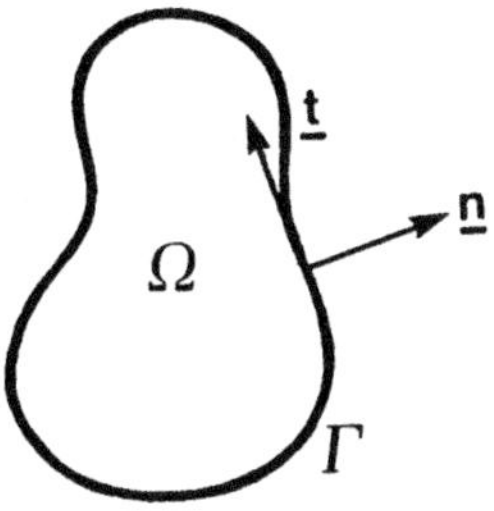

Fig. 1.5. Unit normal $\underline{n}$ and unit tangent $\underline{t}$ on Γ.

Concerning the boundary conditions for the temperature we only give two examples:

Dirichlet : T prescribed on Γ, $t > 0$

Neumann : $-\, k \dfrac{\partial T}{\partial n}$ prescribed on Γ, $t > 0$

When the problem is time independent, the terms $\frac{\partial \underline{u}}{\partial t}$ and $\frac{\partial T}{\partial t}$ can be neglected as well as the initial conditions. The resulting system is a quasi-linear system of elliptic PDEs.

If the fluid can be considered isothermal, then the steady-state motion for constant μ is determined by the following problem:

Find $\underline{u} = \underline{u}(\underline{x})$ and $p = p(\underline{x})$, $\underline{x} \in \Omega \cup \Gamma$, such that

$$\nabla \cdot \underline{u} = 0 \tag{1.6.16}$$

$$-\mu\Delta\underline{u} + \rho(\underline{u}\cdot\nabla)\underline{u} + \nabla p = \rho\underline{f} \tag{1.6.17}$$

with boundary conditions for $\underline{u}$ on Γ.

This system of equations will be studied in part II.

The time dependent version of the Navier-Stokes equations for an incompressible fluid is

Find $\underline{u} = \underline{u}(\underline{x},t)$ and $p = p(\underline{x},t)$, $\underline{x} \in \Omega \cup \Gamma$, $t > 0$

such that

$$\nabla \cdot \underline{u} = 0 \tag{1.6.18}$$

$$\rho\left(\frac{\partial \underline{u}}{\partial t} + (\underline{u}\cdot\nabla)\underline{u}\right) - \mu\Delta\underline{u} + \nabla p = \rho\underline{f} \tag{1.6.19}$$

with boundary conditions for $\underline{u}$ on Γ for $t > 0$ and initial conditions for $\underline{u}$ in $\Omega \cup \Gamma$ at $t = 0$.

This time dependent problem will be the subject of Chapter 10.

Finally we remark that if the curvature of the boundary Γ is negligible then, in 2D, σ_n and σ_t satisfy

$$\sigma_n = -p + 2\mu \frac{\partial u_n}{\partial n} \tag{1.6.20}$$

$$\sigma_t = \mu\left(\frac{\partial u_n}{\partial t} + \frac{\partial u_t}{\partial n}\right) \tag{1.6.21}$$

with $u_n = \underline{u}\cdot\underline{n}$ and $u_t = \underline{u}\cdot\underline{t}$.

In the general n dimensional case the Neumann boundary conditions for the Navier-Stokes equations can be written as

$$\underline{\underline{\sigma}}\cdot\underline{n} \quad \text{prescribed on } \Gamma \tag{1.6.22}$$

where the vector $\underline{\underline{\sigma}}\cdot\underline{n}$ represents physically the surface force on Γ. Written out in normal and tangential components, (1.6.22) is equivalent to

$$\sigma_n \equiv (\underline{\underline{\sigma}}\cdot\underline{n})\cdot\underline{n} = \text{surface force acting in normal direction} \tag{1.6.23}$$

and

$$\underline{\sigma}_t \equiv \underline{\underline{\sigma}}\cdot\underline{n} - \sigma_n\underline{n} = \text{surface force acting in tangent plane} \tag{1.6.24}$$

prescribed on Γ.

1.7 Equations of linear elasticity

The example of this section is taken from the theory of small displacements of an elastic body (see Landau, Lifshitz (1967), Sokolnikov (1956)). In fact this topic exceeds the scope of this work. Nevertheless, we still give the equations since they are similar to those obtained by a penalty (or perturbation) approach to the Navier-Stokes equations (see Chapter 8).

Let Ω be a region in $\mathbb{R}^n$ (in practice n=2 or 3) with boundary Γ. We consider small displacements of a body which occupies the region $\Omega \cup \Gamma$ in its natural state. We denote by $\underline{u} = \underline{u}(\underline{x})$ the displacement vector of the points of $\Omega \cup \Gamma$ under influence of a given body force per unit volume $\underline{f}$ and stress vector $\underline{g}$ distributed over the boundary.

The principle of conservation of momentum gives

$$-\nabla\cdot\underline{\underline{\sigma}} = \underline{f} \quad \text{in } \Omega \tag{1.7.1}$$

where $\underline{\underline{\sigma}}$ denotes the stress tensor. In linear elasticity the relation between the stress tensor $\underline{\underline{\sigma}}$ and the infinitesimal strain tensor $\underline{\underline{\varepsilon}}$ is given by the generalized Hooke's law which reads for isotropic materials

$$\underline{\underline{\sigma}} = \lambda \; tr(\underline{\underline{\varepsilon}})\underline{\underline{I}} + 2\mu\underline{\underline{\varepsilon}} \; , \quad tr(\underline{\underline{\varepsilon}}) = trace(\underline{\underline{\varepsilon}}) \tag{1.7.2}$$

where

$$\varepsilon_{ij} = \frac{1}{2}\left(\frac{\partial u_i}{\partial x_j} + \frac{\partial u_j}{\partial x_i}\right) .$$

The parameters λ and μ are positive constants, called the Lamé coefficients.

Combination of (1.7.1) and (1.7.2) leads to the following PDE for $\underline{u}$

$$- \mu\Delta\underline{u} - (\lambda+\mu)\, \nabla(\nabla.\underline{u}) = \underline{f} \quad \text{in} \quad \Omega \; .$$

Let Γ_0 and Γ_1 be two disjoint parts of Γ such that $\Gamma = \Gamma_0 \cup \Gamma_1$. We assume that the body can not move along Γ_0 and that on Γ_1 the surface force is active.

We formulate the problem as follows

Find $\underline{u} = \underline{u}(\underline{x})$, $\underline{x} \in \Omega \cup \Gamma$, such that

$$-\mu\Delta\underline{u} - (\lambda+\mu)\, \nabla(\nabla.\underline{u}) = \underline{f} \quad \text{in} \quad \Omega \tag{1.7.3}$$

$$\underline{u} = \underline{0} \quad \text{on} \quad \Gamma_0 \tag{1.7.4}$$

$$\underline{\underline{\sigma}}.\underline{n} = g_1 \quad \text{on} \quad \Gamma_1 \tag{1.7.5}$$

Just as in Section 1.6 the Neumann condition (1.7.5) can be split up into a component σ_n in the normal direction on Γ and a component $\underline{\sigma}_t$ in the tangent plane.

In the 2D case with negligible curvature of Γ, one has

$$\sigma_n = \lambda \; \text{div} \; \underline{u} + 2\mu \frac{\partial u_n}{\partial n}$$

$$\sigma_t = \mu\left(\frac{\partial u_n}{\partial t} + \frac{\partial u_t}{\partial n}\right)$$

with $u_n = \underline{u}.\underline{n}$ and $u_t = \underline{u}.\underline{t}$.

Finally let us remark that equations of the type

$$- \mu_1 \Delta \underline{u} - \mu_2 \nabla(\nabla . \underline{u}) = \rho \underline{f} \qquad (1.7.6)$$

are elliptic for any positive value of μ_1 and μ_2.

Introducing a quantity $p = p(\underline{x})$ by

$$p = - \mu_2 \nabla . \underline{u}$$

(1.7.6) can be written as

$$\left| \begin{array}{l} - \mu_1 \Delta \underline{u} + \nabla p = \rho \underline{f} \\ \\ \nabla . \underline{u} = - \dfrac{1}{\mu_2} p \end{array} \right. \qquad (1.7.7)$$

which, for large values of μ_2, can be considered as a perturbation of the stationary Stokes equations (1.6.8). However, some caution should be exercised for large values of μ_2. We shall return to this perturbation aspect of the (Navier-) Stokes equations in Chapter 8.

1.8 Comments

In this chapter we have treated some examples of elliptic PDEs with various types of boundary conditions and parabolic PDEs with boundary conditions and initial conditions.

For an elliptic PDE of the second order or a system of second-order elliptic PDEs there must be given at every point of Γ exactly one boundary condition for each of the dependent variables. An elliptic PDE with its boundary conditions is often called a boundary value problem (BVP). Concerning the different types of linear boundary conditions we mention once more those who occur most frequently. See also Chapter 11.

(i) Dirichlet type boundary conditions. The value of the dependent variable is prescribed.

Examples (i)' heat conduction, T = temperature:

$$T(\underline{x}) = g_0(\underline{x}) \quad \underline{x} \in \Gamma$$

(i)" Navier-Stokes equations, $\underline{u}$ = velocity:

$$\underline{u}(\underline{x}) = \underline{g}_0(\underline{x}) \quad \underline{x} \in \Gamma$$

(ii) Neumann type boundary conditions. The "normal derivative" of the dependent variable is prescribed.

Examples (ii)' T = temperature: $- k \frac{\partial T}{\partial n}(\underline{x}) = g_1(\underline{x}) \quad \underline{x} \in \Gamma$

(ii)" $\underline{u}$ = velocity, p = pressure:

$$\underline{\underline{\sigma}} \cdot \underline{n}(\underline{x}) = \underline{g}_1(\underline{x}) \quad \underline{x} \in \Gamma$$

(iii) Robbins type boundary conditions. A combination of Dirichlet and Neumann conditions is prescribed

Examples (iii)' T = temperature: $hT(\underline{x}) + k \frac{\partial T}{\partial n}(\underline{x}) = g_2(\underline{x})$

$$\underline{x} \in \Gamma$$

(iii)" $\underline{u}$ = velocity, p = pressure:

$$h\underline{u}(x) + \underline{\underline{\sigma}} \cdot \underline{n}(\underline{x}) = \underline{g}_2(\underline{x}) \quad \underline{x} \in \Gamma$$

When a function g is equal to zero we call the corresponding boundary condition homogeneous, otherwise it is called non-homogeneous.

The fact that in each point of Γ a boundary condition must be given for each dependent variable, does not guarantee that the corresponding problem is well-posed (i.e. has precisely one solution which depends continuously on the data). Consider for example the Poisson equation for the temperature T with right hand side f_T in $\Omega \subset \mathbb{R}^n$. The boundary conditions of Dirichlet or Robbins type for T on the whole boundary determine the solution uniquely. However, the same Poisson equation with Neumann type boundary conditions

$$\left| \begin{array}{ll} -\Delta T = f_T & \text{in } \Omega \\ \\ -\frac{\partial T}{\partial n} = g_1 & \text{on } \Gamma \end{array} \right. \qquad (1.8.1)$$

has in general no solution. A necessary and sufficient

condition for the solvability of (1.8.1) is the following so-called compatibility condition:

$$\int_\Omega f_T \, d\Omega - \int_\Gamma g_1 \, d\Gamma = 0 . \qquad (1.8.2)$$

When this condition is satisfied, a solution exists and is unique up to an additive constant.

The relation (1.8.2) is obtained by integration of the Poisson equation over Ω, using the gradient theorem (or Green's formula) and substitution of the boundary conditions:

$$\int_\Omega f_T \, d\Omega = \int_\Omega (-\Delta T) d\Omega = - \int_\Gamma \frac{\partial T}{\partial n} d\Gamma = \int_\Gamma g_1 \, d\Gamma .$$

Physically this compatibility condition means that the heat production in the region Ω must be equal to the net heat flux through Γ, otherwise the process can not be stationary.

A similar compatibility condition must be satisfied in the case of an incompressible fluid (i.e. $\nabla \cdot \underline{u} = 0$, $\underline{u}$ = velocity) with Dirichlet type boundary conditions ($\underline{u}(\underline{x}) = \underline{g}_0(\underline{x})$, $\underline{x} \in \Gamma$):

$$0 = \int_\Omega \nabla \cdot \underline{u} \, d\Omega = \int_\Gamma \underline{u} \cdot \underline{n} \, d\Gamma = \int_\Gamma \underline{g}_0 \cdot \underline{n} \, d\Gamma$$

which means that the net mass transport through the boundary Γ is zero.

We may say that the determination of the correct boundary conditions corresponding to the physical problem is a fundamental step in the formulation of the mathematical model.

A necessary condition for the well-posedness of problems of parabolic type is that the solution must be given at a certain time, say $t = 0$, and must satisfy boundary conditions on Γ for $t > 0$. A parabolic PDE together with its initial and boundary conditions belongs to the class of initial-boundary value problems (IBVP). This kind of problems will be treated in Chapter 10.

Chapter 2

Finite Difference Schemes for Poisson equation and Convection-Diffusion equation

Introduction

In this chapter we shall construct a finite difference scheme for the Poisson equation in 1D and 2D and for the convection-diffusion equation in 1D. The aim of this chapter is to see how the finite difference method (FDM) transforms a BVP (i.e. a PDE with boundary conditions) into a system of (linear) algebraic equations. Of course we do not present here a thorough treatment of the FDMs; for this we refer to Forsythe, Wasow (1960), Richtmeyer, Morton (1967), Mitchell, Griffith (1980), Samarski, Andreev (1976), Collatz (1960), Marchuk (1977), Isaacson, Keller (1966), Godounov, Riabenki (1973).

The FDM is a largely used method for the numerical solution of PDEs. The basic concepts are as follows. The region $\Omega \subset \mathrm{IR}^n$ on which the PDE is defined is first subdivided by a grid (mesh, or net) with a finite number of mesh points. All the functions, which have continuous varying arguments in the region Ω, are now replaced by functions whose arguments are defined on the grid points. These functions are called discrete functions. Next the derivatives occurring in the PDE and in the boundary conditions are replaced by finite difference approximations, which are linear combinations of values taken by the discrete functions on the grid points. In this way the PDE with its boundary conditions is transformed into a system of algebraic equations (called finite difference scheme) which is linear if the PDE and the boundary conditions are linear.

This (finite difference) procedure seems to be quite simple. However the following two questions are hard to answer in general, especially when working on irregular regions Ω or with complicated boundary conditions:

(i) How do we choose the grid?

(ii) Which finite difference approximation do we choose?

2.1 1D Poisson equation with Dirichlet boundary conditions

The derivative $\frac{du}{dx}$ of a function $u = u(x)$ is defined as follows:

$$\frac{du}{dx}(x) = \lim_{h\to 0} \frac{u(x+h)-u(x)}{h} =$$

$$= \lim_{h\to 0} \frac{u(x)-u(x-h)}{h} =$$

$$= \lim_{h\to 0} \frac{u(x+\frac{1}{2}h)-u(x-\frac{1}{2}h)}{h}$$

and can be approximated in various ways. We shall use the Taylor series expansion, which for $u(x+h)$ can be written, at the point x, as

$$u(x+h) = u(x) + h\frac{du}{dx}(x) + \frac{h^2}{2!}\frac{d^2u}{dx^2}(x) + \frac{h^3}{3!}\frac{d^3u}{dx^3}(x) + \ldots$$

According to the mean-value theorem of classical analysis, there exists $\xi \in [x,x+h]$ such that

$$u(x+h) = u(x) + h\frac{du}{dx}(x) + \frac{h^2}{2!}\frac{d^2u}{dx^2}(x) + \frac{h^3}{3!}\frac{d^3u}{dx^3}(\xi). \quad (2.1.1)$$

Likewise we have

$$u(x-h) = u(x) - h\frac{du}{dx}(x) + \frac{h^2}{2!}\frac{d^2u}{dx^2}(x) - \frac{h^3}{3!}\frac{d^3u}{dx^3}(\xi) \quad (2.1.2)$$

for some $\xi \in [x-h,x]$.

Combination of (2.1.1) and (2.1.2) gives:

$$\frac{du}{dx}(x) = \frac{u(x+h)-u(x)}{h} + O(h) \qquad \text{(forward finite difference)}$$

$$= \frac{u(x)-u(x-h)}{h} + O(h) \qquad \text{(backward finite difference)}$$

$$= \frac{u(x+\frac{1}{2}h)-u(x-\frac{1}{2}h)}{h} + O(h^2) \quad \text{(central finite difference)}$$

where the local truncation error $O(h^p)$, $p = 1,2$, is a term of order h^p which means that, for sufficiently small h, it is in absolute value smaller than ch^p, c a constant. Notice that for the central finite difference the local truncation error is one order of h higher than for the forward or backward finite difference.

For the second order derivative of a function $u = u(x)$ we write

$$\frac{d^2u}{dx^2} = \frac{d}{dx}\left(\frac{du}{dx}\right).$$

Applying central finite differences, we find

$$\frac{d^2u}{dx^2} = \lim_{h\to 0} \frac{\frac{du}{dx}(x+\frac{1}{2}h) - \frac{du}{dx}(x-\frac{1}{2}h)}{h} = \lim_{h\to 0} \frac{\frac{u(x+h)-u(x)}{h} - \frac{u(x)-u(x-h)}{h}}{h} =$$

$$= \lim_{h\to 0} \frac{u(x-h) - 2u(x) + u(x+h)}{h^2}.$$

The local truncation error for small, non zero, values of h can be found by developing the terms $u(x+h)$ and $u(x-h)$ into Taylor series up to the fourth order term. This gives the second finite difference

$$\frac{d^2u}{dx^2} = \frac{u(x-h) - 2u(x) + u(x+h)}{h^2} + O(h^2). \qquad (2.1.3)$$

Let us now consider the Poisson equation for u defined on the interval (0,1)

$$-\frac{d^2u}{dx^2} = f \tag{2.1.4}$$

with homogeneous Dirichlet boundary conditions:

$$u(0) = u(1) = 0. \tag{2.1.5}$$

The second order derivative in (2.1.4) will now be replaced by the expression (2.1.3) to get

$$-\frac{u(x-h) - 2u(x) + u(x+h)}{h^2} + O(h^2) = f(x) \tag{2.1.6}$$

In order to reduce this equation to a system of linear equations we subdivide the interval (0,1) into N intervals of equal length $h = \frac{1}{N}$, see Fig. 2.1.

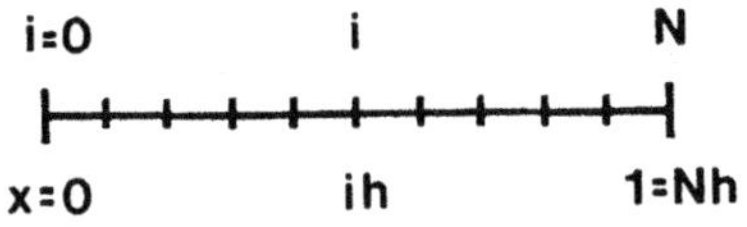

Fig. 2.1 Discretization of the interval (0,1) into N subintervals of equal length $h = \frac{1}{N}$.

On this interval we define for i=0,1,...,N

$$x_i = ih, \; u(x=ih) = u(x_i) = u_i \;, \quad f(x=ih) = f(x_i) = f_i$$

and the Poisson equation (2.1.6) for $x = x_i$ becomes:

$$-\left(\frac{u_{i-1} - 2u_i + u_{i+1}}{h^2}\right) + O(h^2) = f_i \; . \tag{2.1.7}$$

Next we introduce a grid-function $\tilde{u}$, i.e. a function defined on the points x_i, $i = 0,1,\ldots,N$, by

$$\left|\begin{array}{l} -\left(\dfrac{\tilde{u}_{i-1} - 2\tilde{u}_i + \tilde{u}_{i+1}}{h^2}\right) = f_i \qquad \text{for } i = 1,\dots,N-1 \\ \\ \tilde{u}_0 = \tilde{u}_N = 0 \end{array}\right. \tag{2.1.8}$$

Since the term $O(h^2)$ in (2.1.7) is assumed to be small, the grid-function $\tilde{u}$ can be considered as an approximation of u.

Notice that (2.1.8) is a system of linear equations for the unknowns $\tilde{u}_i$, $i = 0,\dots,N$. We shall write down explicitly this system for $i = 1,\dots,N-1$ and substitute the boundary conditions for $\tilde{u}_0$ and $\tilde{u}_N$:

$$i=1: \quad \frac{1}{h^2}(-\not{\tilde{u}_0}+2\tilde{u}_1-\tilde{u}_2) = f_1$$

$$i=2: \quad \frac{1}{h^2}(-\tilde{u}_1+2\tilde{u}_2-\tilde{u}_3) = f_2$$

$$i=3: \quad \frac{1}{h^2}(-\tilde{u}_2+2\tilde{u}_3-\tilde{u}_4) = f_3$$

$$\cdot \quad \cdot \quad \cdot \quad \cdot$$

$$\cdot \quad \cdot \quad \cdot \quad \cdot$$

$$i=N-1: \quad \frac{1}{h^2}(-\tilde{u}_{N-2}+2\tilde{u}_{N-1}-\not{\tilde{u}_N}) = f_{N-1}$$

Introducing the following vectors:

$$\underline{\tilde{u}} = \begin{vmatrix} \tilde{u}_1 \\ \tilde{u}_2 \\ \cdot \\ \cdot \\ \cdot \\ \tilde{u}_{N-1} \end{vmatrix} \qquad \underline{F} = \begin{vmatrix} f_1 \\ f_2 \\ \cdot \\ \cdot \\ \cdot \\ f_{N-1} \end{vmatrix}$$

the system of equations for $\tilde{u}_1, \tilde{u}_2, \ldots, \tilde{u}_{N-1}$ can be written in matrix notation:

$$A\,\underline{\tilde{u}} = \underline{F} \qquad (2.1.9)$$

where the tri-diagonal $(N-1)\times(N-1)$ matrix A is given by

$$A = \frac{1}{h^2}\begin{vmatrix} 2 & -1 & & & & & \\ -1 & 2 & -1 & & & & \\ & -1 & 2 & -1 & & & \\ & & \cdot & \cdot & \cdot & & \\ & & & \cdot & \cdot & \cdot & \\ & & & & -1 & 2 & -1 \\ & & & & & -1 & 2 \end{vmatrix} .$$

System (2.1.9) consists of N-1 linear equations with N-1 unknowns. The matrix A is positive definite, which means that (i) A is symmetric : $A = A^T$ and (ii) for any vector $\underline{v} \in \mathrm{IR}^{N-1}$ with $\underline{v} \neq \underline{0}$: $\underline{v}^T A \underline{v} > 0$. In fact we have:

$$\underline{v}^T A \underline{v} = \frac{1}{h^2}\left\{v_1^2 + v_{N-1}^2 + \sum_{i=2}^{N-1}(v_i - v_{i-1})^2\right\} > 0 \quad \text{if} \quad \underline{v} \neq \underline{0},$$

where the superscript T denotes the transposed. Consequently system (2.1.9) is invertible and can be solved using, for instance, an LU-decomposition or Cholesky's algorithm (see also Section 5.6).

Thus we have seen that the FDM transforms a PDE into a system of linear algebraic equations which, due to nice properties of the system can be solved numerically. Moreover, it can be proved that the approximate solution $\tilde{u}$ converges to the solution u of the original problem. When $u \in C^4(0,1)$, i.e. u is four times continuously differentiable on (0,1), the following estimate for the difference $u-\tilde{u}$ holds

$$\max_{1\leq i\leq N-1} |u_i - \tilde{u}_i| \leq \frac{1}{96} \sup_{0\leq x\leq 1} \left|\frac{d^4u}{dx^4}(x)\right| h^2 \qquad (2.1.10)$$

which means that for $N \to \infty$ (or $h \to 0$) $\tilde{u}$ converges to u.

2.2 1D Poisson equation with other type of boundary conditions

2.2.1 Mixed homogeneous Dirichlet-Neumann boundary conditions

Let us consider the 1D Poisson equation for u on the interval (0,1) with mixed homogeneous Dirichlet-Neumann boundary conditions:

$$\left| \begin{array}{l} -\dfrac{d^2u}{dx^2} = f \quad \text{on} \quad (0,1) \\[2ex] u(0) = \dfrac{du}{dx}(1) = 0 \end{array} \right. \tag{2.2.1}$$

The interval (0,1) is subdivided into N subintervals of equal length $h = \frac{1}{N}$. Using the finite difference expression (2.1.3) for the second order derivative in (2.2.1) we find that the approximate solution $\tilde{u}$ verifies:

$$\frac{1}{h^2}(-\tilde{u}_{i-1}+2\tilde{u}_i-\tilde{u}_{i+1}) = f_i \qquad \text{for } i = 1,\dots,N\,. \tag{2.2.2}$$

The Dirichlet boundary condition at $x = 0$ implies

$$\tilde{u}_0 = 0\,.$$

The Neumann boundary condition at $x = 1$ can be written as

$$\frac{du}{dx}(1) = 0 \quad \Rightarrow \quad \frac{u_{N+1}-u_{N-1}}{2h} + O(h^2) = 0$$

which implies the following relation for $\tilde{u}$

$$\tilde{u}_{N+1} = \tilde{u}_{N-1} \tag{2.2.3}$$

where a point $x_{N+1} = 1+h$ is added to the grid.
Equation (2.2.2) for $i = N$ reads

$$\frac{1}{h^2}(-\tilde{u}_{N-1}+2\tilde{u}_N-\tilde{u}_{N+1}) = f_N$$

which, after substitution of (2.2.3), reduces to

$$\frac{1}{h^2}(-\tilde{u}_{N-1}+\tilde{u}_N) = \tfrac{1}{2} f_N.$$

The explicit form of the system of linear equations is now

$$i=1: \quad \frac{1}{h^2}(-\not{\tilde{u}}_0+2\tilde{u}_1-\tilde{u}_2) = f_1$$

$$i=2: \quad \frac{1}{h^2}(-\tilde{u}_1+2\tilde{u}_2-\tilde{u}_3) = f_2$$

$$\cdot \quad \cdot \quad \cdot \quad \cdot$$

$$\cdot \quad \cdot \quad \cdot \quad \cdot$$

$$i=N-1: \quad \frac{1}{h^2}(-\tilde{u}_{N-2}+2\tilde{u}_{N-1}-\tilde{u}_N) = f_{N-1}$$

$$i=N: \quad \frac{1}{h^2}(-\tilde{u}_{N-1}+\tilde{u}_N) = \tfrac{1}{2} f_N$$

or in matrix form

$$A\,\tilde{\underline{u}} = \underline{F}$$

with

$$\tilde{\underline{u}} = \begin{vmatrix} \tilde{u}_1 \\ \tilde{u}_2 \\ \cdot \\ \cdot \\ \cdot \\ \tilde{u}_N \end{vmatrix} \qquad \underline{F} = \begin{vmatrix} f_1 \\ f_2 \\ \cdot \\ \cdot \\ \cdot \\ f_N \end{vmatrix} \qquad A = \frac{1}{h^2}\begin{vmatrix} 2 & -1 & & & \\ -1 & 2 & -1 & & \\ & \cdot & \cdot & \cdot & \\ & & \cdot & \cdot & \cdot \\ & & -1 & 2 & -1 \\ & & & -1 & 1 \end{vmatrix}$$

The tri-diagonal matrix A is of order N × N and positive definite, since

$$\underline{v}^T A \underline{v} = \frac{1}{h^2}\{v_1^2 + \sum_{i=2}^{N}(v_i-v_{i-1})^2\} > 0 \quad \text{for any } \underline{v} \in \mathbb{R}^N,\ \underline{v} \neq \underline{0}.$$

Let us remark that the matrix A is almost the same as the one for the homogeneous Dirichlet problem of Section 2.1, only one equation is added for the unknown $\tilde{u}_N$.

2.2.2 Non-homogeneous Dirichlet boundary conditions

Consider the same problem (2.2.1) but now with a non-homogeneous Dirichlet boundary condition at x = 0:

$$u(0) = g_0.$$

This gives the following boundary condition for the approximate solution $\tilde{u}$:

$$\tilde{u}_0 = g_0.$$

Substitution of this relation into the explicit form of the equations will change only the right hand side of the first equation, which becomes:

$$\frac{1}{h^2}(2\tilde{u}_1 - \tilde{u}_2) = f_1 + \frac{1}{h^2} g_0.$$

The matrix A does not change.

2.2.3 Non-homogeneous Neumann boundary conditions

Consider problem (2.2.1) with a non-homogeneous Neumann condition at x = 1:

$$\frac{du}{dx}(1) = g_1$$

which implies for $\tilde{u}$

$$\frac{1}{2h}(\tilde{u}_{N+1} - \tilde{u}_{N-1}) = g_1.$$

Substitution into (2.2.2) for i = N gives only a change in the right hand side of the last equation:

$$\frac{1}{h^2}(-\tilde{u}_{N-1}+\tilde{u}_N) = \tfrac{1}{2} f_N + \frac{1}{h} g_1 .$$

The matrix A remains unchanged.

2.2.4 Non-homogeneous Robbins boundary conditions

At x = 1 we take now

$$\frac{du}{dx}(1) + \alpha u(1) = g_2$$

which gives for $\tilde{u}$:

$$\tilde{u}_{N+1} = \tilde{u}_{N-1} - 2h\alpha\tilde{u}_N + 2g_2 h .$$

The last equation of the system becomes now:

$$\frac{1}{h^2}(-\tilde{u}_{N-1}+(1+\alpha h)\tilde{u}_N) = \tfrac{1}{2} f_N + \frac{1}{h} g_2 .$$

The consequence of this non-homogeneous Robbins boundary condition is that the last row of the matrix changes as well as the last component of the right hand side vector. The matrix remains positive definite because

$$\underline{v}^T A \underline{v} = \frac{1}{h^2}\{v_1^2+\alpha h v_N^2+\sum_{i=2}^{N}(v_i-v_{i-1})^2\} > 0 \quad \text{for any } \underline{v} \in \mathbb{R}^N,\ \underline{v} \neq \underline{0}.$$

In this section we have shown how the FDM works for various types of boundary conditions. In all the cases a system of linear equations is obtained which can be solved uniquely. In the following section we shall apply the FDM to a problem in 2D.

2.3 2D Poisson equation with Dirichlet boundary conditions

Let us denote by Ω the 2D unit square $(0,1) \times (0,1)$ as shown in

Fig. 2.2 and by Γ the boundary of Ω.

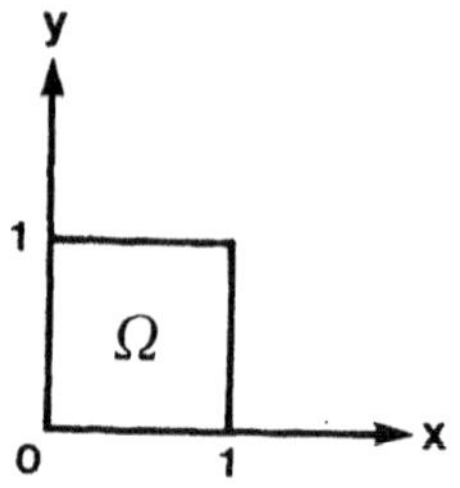

Fig. 2.2 $\Omega = (0,1) \times (0,1)$.

Consider the following Poisson equation for $u = u(x,y)$ with homogeneous Dirichlet boundary conditions:

$$\begin{cases} -\Delta u = f & \text{in } \Omega \quad , \quad \Delta = \dfrac{\partial^2}{\partial x^2} + \dfrac{\partial^2}{\partial y^2} \\ u = 0 & \text{on } \Gamma \end{cases} \tag{2.3.1}$$

Just as in the 1D case, the FDM is a procedure to define an approximation of the solution u of (2.3.1) in a finite number of points in the region Ω.

The second order derivatives in (2.3.1) will be replaced by their corresponding finite difference expressions. For simplicity we take equal spacing h in both directions:

$$\frac{\partial^2 u}{\partial x^2}(x,y) = \frac{u(x-h,y)-2u(x,y)+u(x+h,y)}{h^2} + O(h^2)$$

$$\frac{\partial^2 u}{\partial y^2}(x,y) = \frac{u(x,y-h)-2u(x,y)+u(x,y+h)}{h^2} + O(h^2)$$

The finite difference scheme for the Poisson equation is:

$$-\frac{u(x-h,y)+u(x,y-h)-4u(x,y)+u(x+h,y)+u(x,y+h)}{h^2} + O(h^2) = f(x,y) \tag{2.3.2}$$

We now introduce a uniform mesh in Ω consisting of the nodal points $(x_i,y_i) = (ih,jh)$ with $i = 0,1,\ldots,N$, $j = 0,1,\ldots,N$ and $N = \frac{1}{h}$ (see Fig. 2.3)

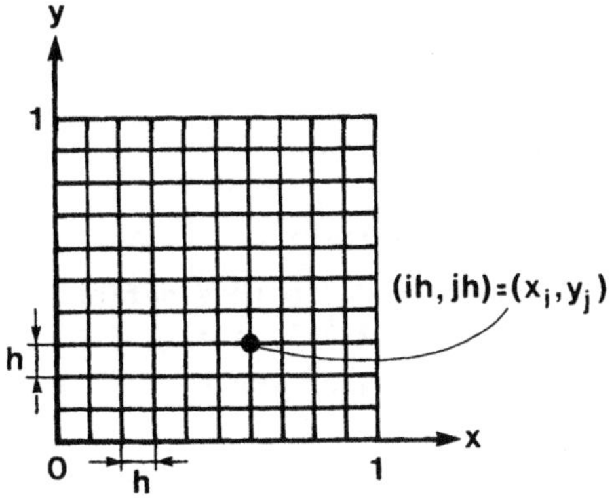

Fig. 2.3 Mesh on Ω.

We shall use the following notations

$$(x_i,y_j) = (ih,jh)$$

$$u(x_i,y_j) = u_{i,j}$$

$$f(x_i,y_j) = f_{i,j}$$

For $(x,y) = (ih,jh)$, equation (2.3.2) becomes

$$-\frac{u_{i-1,j}+u_{i,j-1}-4u_{i,j}+u_{i+1,j}+u_{i,j+1}}{h^2} + O(h^2) = f_{i,j}.$$

Next, we define the approximate solution $\tilde{u}$, defined on the points (x_i,y_j), $i = 0,1,\ldots,N$, $j = 0,1,\ldots,N$, by:

$$-\frac{1}{h^2}(\tilde{u}_{i-1,j}+\tilde{u}_{i,j-1}-4\tilde{u}_{i,j}+\tilde{u}_{i+1,j}+\tilde{u}_{i,j+1}) = f_{i,j} \qquad (2.3.3)$$

for $i = 1,2,\ldots,N-1$, $j = 1,2,\ldots,N-1$.

For nodal points (x_i,y_j) on the boundary Γ we have:

$$u_{i,j} = 0 \qquad (x_i,y_j) \in \Gamma$$

which gives for the approximate solution

$$\tilde{u}_{i,j} = 0 \qquad (x_i,y_j) \in \Gamma$$

i.e.

$$\tilde{u}_{i,0} = \tilde{u}_{i,N} = \tilde{u}_{0,j} = \tilde{u}_{N,j} = 0 \qquad (2.3.4)$$

for $i = 0,\ldots,N$, $j = 0,\ldots,N$.

The approximate solution $\tilde{u}$ is thus defined by the system (or finite difference scheme) (2.3.3) and (2.3.4).
When we introduce the following vectors

$$\underline{\tilde{u}} = (\tilde{u}_{1,1}\tilde{u}_{2,1}\ldots\tilde{u}_{N-1,1}\tilde{u}_{1,2}\ldots\tilde{u}_{N-1,2}\tilde{u}_{1,3}\ldots\tilde{u}_{1,N-1}\ldots\tilde{u}_{N-1,N-1})^T$$

$$\underline{F} = (f_{1,1}f_{2,1}\ldots f_{N-1,1}f_{1,2}\ldots f_{N-1,2}f_{1,3}\ldots f_{1,N-1}\ldots f_{N-1,N-1})^T$$

system (2.3.3), (2.3.4) can be written in matrix notation

$$A\,\underline{\tilde{u}} = \underline{F} \qquad (2.3.5)$$

with

$$A = \begin{vmatrix} D & -I & & & & \\ -I & D & -I & & & \\ & \cdot & \cdot & \cdot & & \\ & & \cdot & \cdot & \cdot & \\ & & & \cdot & \cdot & \cdot \\ & & & -I & D & -I \\ & & & & -I & D \end{vmatrix}$$

where I is the unit $(N-1) \times (N-1)$ matrix and D the $(N-1) \times (N-1)$ matrix given by

$$D = \begin{vmatrix} 4 & -1 & & & & & \\ -1 & 4 & -1 & & & & \\ & \cdot & \cdot & \cdot & & & \\ & & \cdot & \cdot & \cdot & & \\ & & & \cdot & \cdot & \cdot & \\ & & & & 1 & 4 & -1 \\ & & & & & -1 & 4 \end{vmatrix} .$$

A is an $(N-1)^2 \times (N-1)^2$ matrix and is positive definite, consequently (2.3.5) is uniquely solvable for the $(N-1)^2$ unknowns $\tilde{u}$. Let us remark that the matrix A is sparse and that each row has at most five non-zero elements. The particular (row by row or natural) numbering of the nodal points adopted here, i.e.:

(11)(21)(31)...(N-1 1) (12) (22)...(N-1 2) (13)...(1 N-1)...(N-1 N-1)

gives rise to a block tri-diagonal matrix A as can be seen from its structure. Each diagonal matrix D corresponds to the set of equations (2.3.3) associated with one horizontal line in the mesh. The only non-zero elements for the equation corresponding to the point (x_i, y_j) are the elements in the diagonal block corresponding to the unknowns $\tilde{u}_{i-1,j}, \tilde{u}_{i+1,j}$ and the elements in the off-diagonal blocks corresponding to the unknowns $\tilde{u}_{i,j-1}$ and $\tilde{u}_{i,j+1}$.

Let us emphasize the fact that this penta-diagonal structure of A is a direct consequence of the particular numbering of the nodal points. For an abitrary numbering the matrix still remains sparse but will not have any particular structure.

What we have shown by this example is that, in the 2D case too, the FDM transforms a BVP into a system of linear algebraic equations for an approximate solution $\tilde{u}$.

2.4 Boundary conditions, geometry and variable coefficients in 2D

2.4.1 Boundary conditions

The treatment of various types of boundary conditions in the 2D case is similar to the situation in 1D.

(i) When homogeneous Neumann boundary conditions are given on, let us say, boundary x = 1 (see the example of Section 2.3), then

$$\frac{\partial u}{\partial x}(1,y) = 0 \Rightarrow \frac{\tilde{u}_{N+1,j}-\tilde{u}_{N-1,j}}{2h} = 0 \Rightarrow \tilde{u}_{N+1,j} = \tilde{u}_{N-1,j}.$$

The consequence of this type of boundary condition is that N-1 equations for the unknowns $\tilde{u}_{N,j}$, $j=1,\dots,N-1$ are added to the system (2.3.5). The matrix A changes, but remains positive definite (cf. Section 2.2).

(ii) Non-homogeneous Dirichlet or Neumann boundary conditions at x = 1 only change a number of components of the right hand side vector of the system for the corresponding homogeneous boundary conditions (cf. Section 2.2).

$$u(1,y) = g_0(y) \Rightarrow \tilde{u}_{N,j} = g_0(jh)$$

$$\frac{\partial u}{\partial x}(1,y) = g_1(y) \Rightarrow \tilde{u}_{N+1,j} = u_{N-1,j} + 2hg_1(jh).$$

(iii) When non-homogeneous Robbins boundary conditions are given at x = 1, we have

$$\frac{\partial u}{\partial x}(1,y)+\alpha u(1,y) = g_2(y) \Rightarrow \frac{\tilde{u}_{N+1,j}-\tilde{u}_{N-1,j}}{2h} + \alpha\tilde{u}_{N,j} = g_2(jh)$$

$$\Rightarrow \tilde{u}_{N+1,j} = \tilde{u}_{N-1,j} - 2h\alpha\tilde{u}_{N,j} + 2hg_2(jh).$$

In this case the matrix as well as the right hand side vector will change (cf. Section 2.2).

2.4.2 Geometry

When the region Ω is not a rectangle (or a union of rectangles) with sides parallel to the coordinate axis, nodal points adjacent to the boundary require special treatment. In general, interpolation techniques are used to take the boundary conditions into account. We shall explain this from the following example:

Consider the geometry of Fig. 2.4 and the following Dirichlet problem for the Poisson equation:

$$\left| \begin{array}{ll} -\Delta u = f & \text{in } \Omega \\ u = g_0 & \text{on } \Gamma \end{array} \right. \qquad (2.4.1)$$

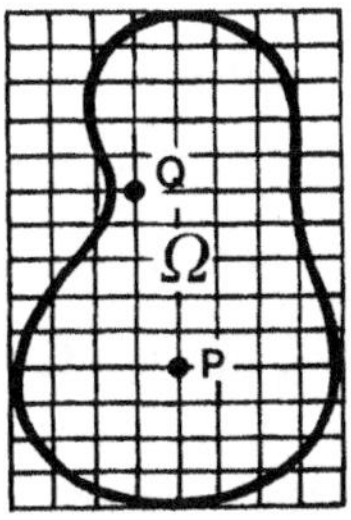

Fig. 2.4 Region Ω and mesh with uniform spacing h.

When P = (ih,jh) is a nodal point in Ω for which the neighbour points ((i-1)h,jh), ((i+1)h,jh), (ih,(j-1)h), (ih,(j+1)h) also belong to Ω (see Fig. 2.4), then the equation associated to the point P is (cf. (2.3.3)):

$$-(\tilde{u}_{i-1,j}+\tilde{u}_{i,j-1}-4\tilde{u}_{i,j}+\tilde{u}_{i+1,j}+\tilde{u}_{i,j+1}) = h^2 f_{ij}$$

However for a point Q (see Fig. 2.4) the situation is different, since one of the neighbours lies outside Ω. To simplify the notation let us renumber the neighbour points of Q as indicated in Fig. 2.5.

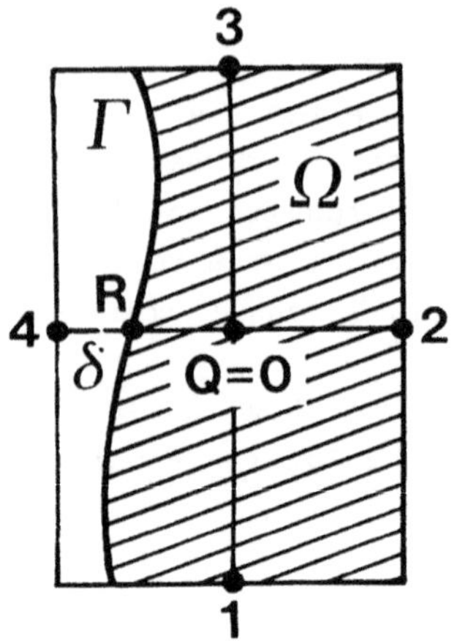

Fig. 2.5 Close-up of point Q.

For the point Q (i.e. point number 0), we have the following equation

$$-(\tilde{u}_1+\tilde{u}_2-4\tilde{u}_0+\tilde{u}_3+\tilde{u}_4) = h^2 f_0 \qquad (2.4.2)$$

The boundary condition "$u = g_0$ on Γ" is given for point R and not for point 4 which lies at a distance δ from R (see Fig. 2.5). In order to eliminate the unknown $\tilde{u}_4$ from the equation we use a linear interpolation on the interval [point 4, point 0]. This gives:

$$\tilde{u}(R) = \tilde{u}_4 \frac{h-\delta}{h} + \tilde{u}_0 \frac{\delta}{h} = g_0(R)$$

from which we obtain an expression for $\tilde{u}_4$:

$$\tilde{u}_4 = - \tilde{u}_0 \frac{\delta}{h-\delta} + g_0(R) \frac{h}{h-\delta} \cdot$$

Substitution into (2.4.2) yields the following equation for u_0:

$$-(\tilde{u}_1+\tilde{u}_2-(4+ \frac{\delta}{h-\delta})\tilde{u}_0+\tilde{u}_3) = h^2 f_0 + g_0(R) \frac{h}{h-\delta} \cdot$$

It will be clear that boundary conditions in complex geometries present considerable additional difficulties, especially for

Neumann and Robbins boundary conditions, which have to do with normal derivatives.

2.4.3 Variable coefficients

Until now we restricted ourselves to the Poisson equation, i.e. an equation with constant coefficients. A slightly more complicated equation is the Reynolds equation (see Section 1.5). In x and y variables, the equation reads:

$$-\frac{\partial}{\partial x}\left(a^3 \frac{\partial u}{\partial x}\right) - \frac{\partial}{\partial y}\left(a^3 \frac{\partial u}{\partial y}\right) = -\ 6\mu\omega r_1 \frac{\partial a}{\partial x} \tag{2.4.3}$$

where, in general, $a = a(x,y)$.

To construct the finite difference approximation of (2.4.3), we first apply a central finite difference to $a^3 \frac{\partial u}{\partial x}$, next a central finite difference to $\frac{\partial u}{\partial x}$. We proceed similarly for the y variable.

On the region Ω we define a uniform mesh with grid spacing h. Using the following notations:

$$u_{i,j} = u(ih,jh) \qquad a_{i+\frac{1}{2},j} = a((i+\tfrac{1}{2})h,jh)$$

we find

$$\frac{\partial}{\partial x}\left(a^3 \frac{\partial u}{\partial x}\right) \Rightarrow \frac{\left(a^3 \frac{\partial \tilde{u}}{\partial x}\right)_{i+\frac{1}{2},j} - \left(a^3 \frac{\partial \tilde{u}}{\partial x}\right)_{i-\frac{1}{2},j}}{h}$$

$$\Rightarrow \frac{a^3_{i+\frac{1}{2},j}\left(\frac{\tilde{u}_{i+1,j}-\tilde{u}_{i,j}}{h}\right) - a^3_{i-\frac{1}{2},j}\left(\frac{\tilde{u}_{i,j}-\tilde{u}_{i-1,j}}{h}\right)}{h} =$$

$$\Rightarrow \frac{a^3_{i-\frac{1}{2},j}\tilde{u}_{i-1,j} - (a^3_{i-\frac{1}{2},j} + a^3_{i+\frac{1}{2},j})\tilde{u}_{i,j} + a^3_{i+\frac{1}{2},j}\tilde{u}_{i+1,j}}{h^2}. \tag{2.4.4}$$

Likewise we obtain

$$\frac{\partial}{\partial y}\left(a^3 \frac{\partial u}{\partial y}\right) \Rightarrow \frac{a^3_{i,j-\frac{1}{2}}\tilde{u}_{i,j-1} - (a^3_{i,j-\frac{1}{2}} + a^3_{i,j+\frac{1}{2}})\tilde{u}_{i,j} + a^3_{i,j+\frac{1}{2}}\tilde{u}_{i,j+1}}{h^2}. \tag{2.4.5}$$

Substitution of (2.4.4) and (2.4.5) into (2.4.3) leads to the equation associated to the unknown $\tilde{u}_{i,j}$.

It will be clear that, because of the variable coefficients, the final system of linear algebraic equations becomes more complicated.

2.5 Comments

It emerges from the preceding discussion that the FDM is directly applicable to the PDE and its boundary conditions leading to a system of linear algebraic equations in a simple straightforward way. Complications in the applicability of the FDM are due to the treatment of boundary conditions on non-rectangular regions and the fact that complicated equations involve some calculations. Another disadvantage is that a local refinement is not possible, even not if a variable grid spacing is used. See for instance Fig. 2.6.

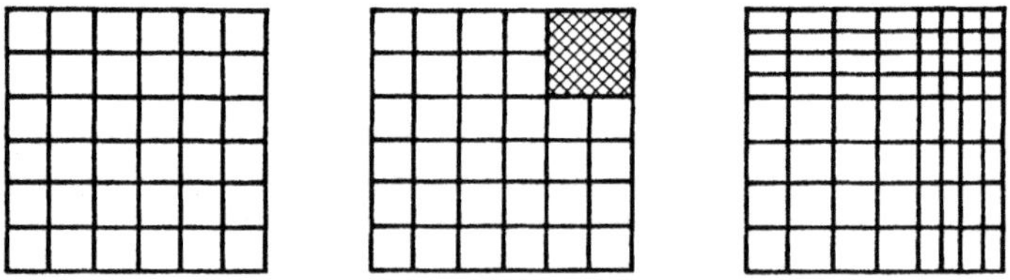

Fig.2.6 Course grid — Desired local refinement in the region, however, FDM is not applicable. — Refinement, FDM applicable.

Recent developments in FDMs are the use of body fitted coordinates and finite volume techniques. Body fitted coordinates are used to transform the independent and dependent variables in order to get a rectangular computational grid in a curvilinear physical domain. For a survey see Thompson (1982). Finite volume techniques use an integral formulation (as opposed to a differential formulation) of the problem. They reduce the problem of deriving boundary conditions and are very useful for the Navier-Stokes equations. See Patanker (1980).

Two important notions in the theory of FDMs applied to elliptic problems are: consistency and convergence. In the derivation of the finite difference approximation to the differential equation and boundary conditions, we neglected terms of $O(h)$ and $O(h^2)$. These terms together are called the truncation errors. Consistency of the finite difference approximation ensures that the truncation errors go to zero if the grid spacing tends to zero. In other words, the finite difference scheme approximates the desired BVP and not any other BVP. Consistency of a finite difference scheme does not imply however that its solution $\tilde{u}$ converges to the solution u of the BVP as the grid spacing tends to zero. This property of the finite difference scheme is called convergence. However consistency is a necessary condition for convergence. We will come back to the concepts of consistency and convergence in the more theoretical context of Chapter 12.

2.6 Convection-Diffusion equation

Let us consider the 1D convection-diffusion equation

$$-\frac{d^2u}{dx^2} + v\,\frac{du}{dx} = 0 \qquad \text{on} \quad (0,1) \tag{2.6.1}$$

where, for instance, u represents the temperature and v denotes the ratio of velocity and thermal diffusivity (see Section

1.3). We assume that $v > 0$. The solution u of (2.6.1) is assumed to satisfy the following non-homogeneous Dirichlet boundary conditions:

$$u(0) = 0 \quad u(1) = 1 \tag{2.6.2}$$

System (2.6.1), (2.6.2) will be used as a model to show that some non-physical phenomena can occur in the approximate solution when the first order term dominates the second order term in (2.6.1), i.e. for large values of v. These problems are encountered by using both FDMs as FEMs in one as well as in more dimensional cases, and are concerned with the approximation of the first order terms.

When we discretize, for $v = 40$ and $h = \frac{1}{10}$, the first order term by a central or a backward finite difference (see Section 2.1), we observe the behavior of the solution as depicted in Fig. 2.7.

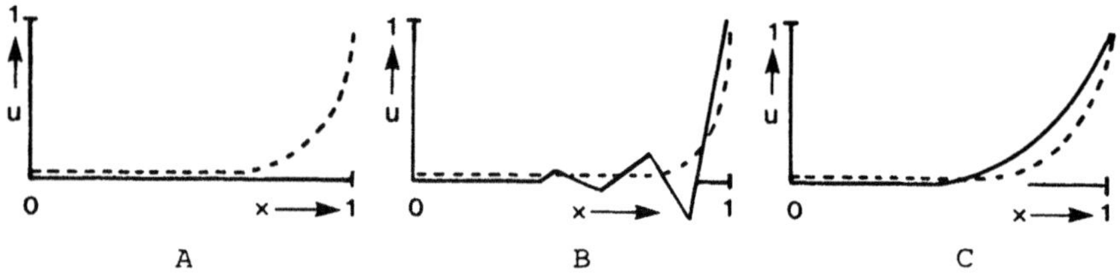

Fig. 2.7

A. exact solution

B. approximation by central finite difference, $h = \frac{1}{10}$, $v=40$.

C. approximation by backward finite difference.

For the central difference case, the solution shows non-physical oscillations (also called wiggles), while for a backwardly discretized first order term an important error is found in the approximate solution, but no oscillations occur. When, however, in the case of a central finite difference, we choose $h = \frac{1}{40}$, the oscillations disappear.

This kind of phenomena will be explained in this section.

Central finite difference

We subdivide the interval (0,1) into N subintervals of equal length $h = \frac{1}{N}$ and we define

$$x_i = ih \qquad u(x_i) = u_i.$$

We now replace in (2.6.1) the second order term by

$$-\frac{d^2u}{dx^2} = -\frac{u_{i-1}-2u_i+u_{i+1}}{h^2} + O(h^2)$$

and the first order term by (v is assumed to be constant):

$$\frac{du}{dx} = v\,\frac{u_{i+1}-u_{i-1}}{2h} + O(h^2).$$

We obtain

$$-\frac{u_{i-1}-2u_i+u_{i+1}}{h^2} + v\left(\frac{u_{i+1}-u_{i-1}}{2h}\right) + O(h^2) = 0 \quad i = 1,2,\ldots,N-1$$

and

$$u_0 = 0 \qquad u_N = 1.$$

Next we define the approximate solution $\tilde{u}$ by neglecting the term $O(h^2)$:

$$\left|\begin{array}{ll} \frac{1}{h^2}\left[(-1-\frac{vh}{2})\tilde{u}_{i-1}+2\tilde{u}_i+(-1+\frac{vh}{2})\tilde{u}_{i+1}\right] = 0 & i = 1,2,\ldots,N-1 \\ \tilde{u}_0 = 0 \qquad \tilde{u}_N = 1 & \end{array}\right. \tag{2.6.3}$$

System (2.6.3) can equivalently be written in matrix notation:

$$A\,\underline{\tilde{u}} = \underline{F}$$

with

$$A = \frac{1}{h^2} \begin{vmatrix} b & c & & & & & 0 \\ a & b & c & & & & \\ & a & b & c & & & \\ & & \cdot & \cdot & \cdot & & \\ & & & \cdot & \cdot & \cdot & \\ & & & & a & b & c \\ 0 & & & & & a & b \end{vmatrix} , \quad \underline{\tilde{u}} = \begin{vmatrix} \tilde{u}_1 \\ \cdot \\ \cdot \\ \cdot \\ \tilde{u}_{N-1} \end{vmatrix} , \quad \underline{F} = \begin{vmatrix} 0 \\ \cdot \\ \cdot \\ \cdot \\ 0 \\ \frac{-c}{h^2} \end{vmatrix}$$

$$a = -1 - \frac{vh}{2} , \quad b = 2 , \quad c = -1 + \frac{vh}{2} .$$

Notice that in this case the matrix A is not symmetric and consequently not positive-definite.

In order to explain the observed oscillations in the approximate solution, we consider the following system of difference equations:

$$\left| \begin{array}{ll} a\tilde{u}_{i-1} + b\tilde{u}_i + c\tilde{u}_{i+1} = 0 & i = 1,2,\ldots,N-1 \\ \tilde{u}_0 = 0 \quad \tilde{u}_N = 1 & \end{array} \right. \tag{2.6.4}$$

with a, b and c satisfying

$$b > 0 \quad a+b+c = 0 \quad a < 0.$$

This system can be solved exactly writing the approximate solution as

$$\tilde{u}_i = \alpha + \beta r^i .$$

Substitution of this expression into (2.6.4) gives

$$ar^{i-1} + br^i + cr^{i+1} = 0 .$$

Since

$$a+b+c = 0 \Rightarrow b = -a-c$$

we find

$$cr^2 + (-a-c)r + a = 0$$

which is satisfied for

$$r = \frac{a}{c} .$$

Thus

$$\tilde{u}_i = \alpha + \beta\left(\frac{a}{c}\right)^i .$$

The constants α and β are determined by the boundary conditions. The solution of (2.6.4) is given by

$$\tilde{u}_i = \frac{1-\left(\frac{a}{c}\right)^i}{1-\left(\frac{a}{c}\right)^N} \qquad (2.6.5)$$

We distinguish between the following three cases:

(i) $c < 0$.

Since $a < 0$, we find $\frac{a}{c} > 0$.
If $\frac{a}{c} > 1$ then both the nominator as the denominator in (2.6.5) are negative, which implies $\tilde{u}_i$ positive.
If $0 < \frac{a}{c} < 1$, $\tilde{u}_i$ is positive since nominator and denominator are positive.

(ii) $c > 0$.

Since $a < 0$, we have $\frac{a}{c} < 0$. On the other hand $a+c = -b$ with $b > 0$, so that $c < -a$ which gives $\frac{a}{c} < -1$.
When $\frac{a}{c} < -1$, then $\tilde{u}_i$ is alternatively positive and negative: if $\tilde{u}_i > 0$ then $\tilde{u}_{i+1} < 0$, if $\tilde{u}_i < 0$ then $\tilde{u}_{i+1} > 0$. The approximate solution shows an oscillatory behavior as depicted in Fig. 2.7.b.

(iii) $c = 0$.

In this case the difference equations reduce to $a\tilde{u}_{i-1} + b\tilde{u}_i = 0$ implying that $\tilde{u}_i$ depends only on $\tilde{u}_{i-1}$ and not on $\tilde{u}_{i+1}$. The solution will be identical to zero except in the last point where $\tilde{u}_N = 1$.

In summary we can say that

(i) If $c < 0$ then $\tilde{u}_i > 0$; no oscillations occur.

(ii) If $c > 0$ then $\tilde{u}_i$ is alternatively positive and negative; oscillations occur.

(iii) If $c = 0$ then $\tilde{u}_i$ vanishes everywhere except in point $x_N=1$; no oscillations occur but the approximation is quite inaccurate.

This means that the following condition must be satisfied by the grid spacing h in order to suppress the oscillations

$$c = -1 + \frac{vh}{2} < 0$$

or equivalently

$$hv < 2 \tag{2.6.6}$$

For the numerical solution of the convection-diffusion equation with non-homogeneous Dirichlet boundary conditions by a central finite difference method, the grid spacing h must satisfy condition (2.6.6) which, for large values of v, implies a severe condition on h. Condition (2.6.6) is called the cell Reynolds number condition and the expression hv is called the cell Péclet or, on the analogy of the Navier-Stokes equations (see Chapter 6), the cell Reynolds number. We note that when v becomes large, equation (2.6.1) degenerates to a first order equation for which a prescription of two Dirichlet boundary conditions is inadequate. This also explains the boundary layer behavior of the solution.

A condition similar to (2.6.6) is found if we use a forward finite difference approximation in the discretization of the first order term in (2.6.1). The difference equation becomes:

$$-\frac{\tilde{u}_{i-1}-2\tilde{u}_i+\tilde{u}_{i+1}}{h^2} + v\,\frac{\tilde{u}_{i+1}-\tilde{u}_i}{h} = 0$$

which can be written as

$$-\tilde{u}_{i-1} + (2-vh)\tilde{u}_i + (-1+vh)\tilde{u}_{i+1} = 0\ . \tag{2.6.7}$$

In this case no oscillations occur if the following cell Reynolds number condition holds:

$$c = -1 + vh < 0$$

which means

$$hv < 1 \tag{2.6.8}$$

This condition is more severe than (2.6.6) while, moreover, the truncation error is O(h) because the forward finite difference is a first order accurate approximation of the first order derivative in (2.6.1).

Backward finite difference

When we use the backward finite difference approximation for the convection term in (2.6.1), we obtain the following scheme:

$$-\frac{\tilde{u}_{i-1}-2\tilde{u}_i+\tilde{u}_{i+1}}{h^2} + v\,\frac{\tilde{u}_i-\tilde{u}_{i-1}}{h} = 0 \tag{2.6.9}$$

or equivalently

$$(-1-vh)\tilde{u}_{i-1} + (2+vh)\tilde{u}_i - \tilde{u}_{i+1} = 0\ . \tag{2.6.10}$$

In this situation the cell Reynolds number condition $c < 0$ is satisfied for any value of h and $v > 0$. No oscillations occur, but the solution is rather inaccurate which of course is due to the first order accurate approximation of the convection term.

Let us calculate the local truncation error for this case to find out for what differential equation the finite difference scheme (2.6.9) has a local truncation error of $O(h^2)$.
We consider (2.6.9) and develop $\tilde{u}_{i-1}$ and $\tilde{u}_{i+1}$ into Taylor series at the point $x_i = ih$:

$$\tilde{u}_{i+1} = \tilde{u}_i + h\,\frac{d\tilde{u}}{dx} + \frac{h^2}{2}\,\frac{d^2\tilde{u}}{dx^2} + \frac{h^3}{6}\,\frac{d^2\tilde{u}}{dx^3} + O(h^4)$$

$$\tilde{u}_{i-1} = \tilde{u}_i - h\,\frac{d\tilde{u}}{dx} + \frac{h^2}{2}\,\frac{d^2\tilde{u}}{dx^2} - \frac{h^3}{6}\,\frac{d^3\tilde{u}}{dx^3} + O(h^4)\ .$$

Equation (2.6.9) becomes:

$$- \frac{\tilde{u}_{i-1} - 2\tilde{u}_i + \tilde{u}_{i+1}}{h^2} + v\left(\frac{\tilde{u}_i - \tilde{u}_{i-1}}{h}\right) =$$

$$= - \frac{h^2 \frac{d^2\tilde{u}}{dx^2} + O(h^4)}{h^2} + v\left(\frac{h \frac{d\tilde{u}}{dx} - \frac{h^2}{2} \frac{d^2\tilde{u}}{dx^2} + \frac{h^3}{6} \frac{d^3\tilde{u}}{dx^3} + O(h^4)}{h}\right)$$

$$= - \frac{d^2\tilde{u}}{dx^2} + v \frac{d\tilde{u}}{dx} - \frac{vh}{2} \frac{d^2\tilde{u}}{dx^2} + O(h^2) \Rightarrow$$

$$- \left(1 + \frac{vh}{2}\right) \frac{d^2 u}{dx^2} + v \frac{du}{dx} = 0 \ . \tag{2.6.11}$$

From this we deduce that scheme (2.6.9) is a second order accurate approximation of (2.6.11) which differs a term $\frac{vh}{2} \frac{d^2 u}{dx^2}$ from the original equation (2.6.1). The quantity $\frac{vh}{2}$ is called artificial viscosity and causes that the approximate solution is less steep than the exact solution. In particular, the differences are essential for large values of v.

The results we have obtained for the 1D convection-diffusion equation can be extrapolated to second order PDEs with dominating first order derivative with $v > 0$ and for which the solution changes strongly in a boundary layer due to the non-homogeneous Dirichlet boundary conditions.
The approximate solution shows the following behavior for $v > 0$.

(i) for a central finite difference
 - oscillations can occur for $hv > 2$
 - no oscillations for $hv < 2$

(ii) for a forward finite difference
 - possibility for oscillations
 - inaccurate solution

(iii) for a backward finite difference
- no oscillations
- inaccurate solution.

When v is a negative constant, then one should read in (i) $h|v|$ in stead of hv and the conclusions of (ii) and (iii) must be interchanged. An approximation of the type: backward for positive v and forward for negative v is called upwind differencing.

In fact these oscillatory phenomena are strongly determined by the fact that the solution changes drastically in a boundary layer near $x = 1$ due to the non-homogeneous Dirichlet conditions. The oscillations can largely be suppressed by choosing, for instance, a non-uniform grid with sufficiently small spacing in the boundary layer and, if possible, by taking a Neumann instead of a Dirichlet boundary condition. An example will be given in Chapter 9.

The appearance of oscillations is essentially due to the non diagonal dominance of the matrix A. Since $a < 0$ and $b = -a-c$ we have

$$|b| = |a| + |c| \qquad \text{A diagonal dominant}$$

for $c \leq 0$, and

$$|b| < |a| + |c| \qquad \text{A non-diagonal dominant}$$

for $c > 0$, which cases are exactly distinguished by the cell Reynolds number.

One should not read here that finite difference equations with non-diagonal dominant matrices will have oscillatory solutions. This depends also on the boundary conditions and on the PDE itself, but the possibility for oscillations is present.

Another possibility for the appearance of oscillations is that one is solving a physically unrealistic problem. Consider, for example, the convection-diffusion equation for the heat-transport in a Poiseuille flow through a straight cylindrical tube with radius R, as discussed in Section 1.3. The initial fluid temperature is T = 0 and the wall temperature changes from T = 0 to T = 1 at x = 0 (see Fig. 2.8).

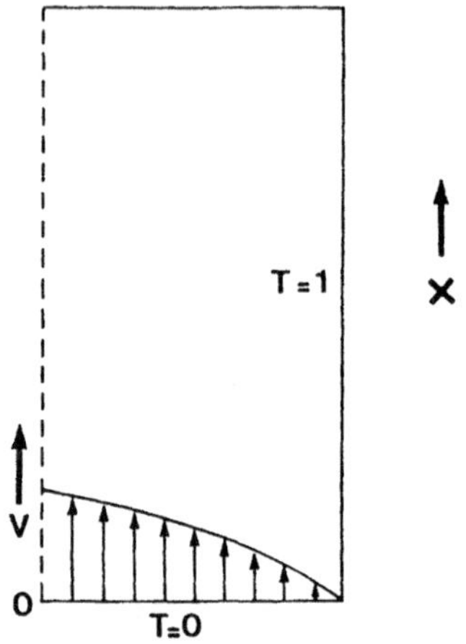

Fig. 2.8

Let us assume the boundary condition T = 1 at x = L, $0 \leqslant r \leqslant R$. If the length L is not large enough (for instance $L < 0.68$ Pe.R with Pe the Peclet number) then this Dirichlet boundary condition at the outlet is not realistic and gives rise to an oscillatory numerical solution as is shown in Fig. 2.9 (i). The oscillations disappear whenever L is large enough (see Fig. 2.9 (ii)). When the Neumann boundary condition $\frac{\partial T}{\partial x} = 0$ is taken at the outlet, no oscillations occur, even not for unrealistic small values of L (see Fig. 2.9 (iii)).

Since oscillations are primarily a consequence of the matrix structure and since, as we shall see later on, the FEM leads to a matrix A having the same structure as the matrix obtained using the FDM, identical phenomena will occur when applying a FEM.

For all those cases we can reduce the oscillatory behavior by formulating the problem with physically realistic conditions and/or using a non-uniform grid with sufficiently small spacing in the region close to the boundary and/or replacing Dirichlet by Neumann boundary conditions.

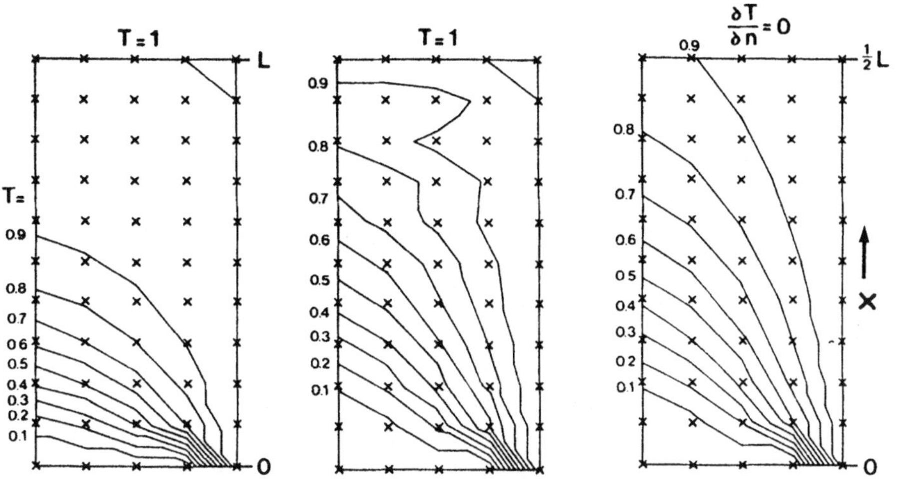

Fig. 2.9 (i) (ii) (iii)

Chapter 3

The Finite Element Method

Introduction

A PDE describing a physical phenomenon can often be formulated equivalently as a variational problem. As we have seen, a (partial) differential formulation of a problem is to find an unknown function which satisfies a relation between the function itself and its (partial) derivatives and which is, moreover, subject to given boundary conditions. The relation is called PDE and the relation together with the boundary conditions is termed a boundary value problem BVP.
A variational (or extremal) formulation is a problem of the type: Find a function that minimizes, maximizes or makes stationary a functional subject to given boundary conditions. The two formulations are called equivalent if both problems admit the same unique solution. By a functional we mean a mapping from a set of admissible functions into the real numbers. The set of admissible functions consists of those functions for which the functional takes finite values and which satisfy imposed boundary conditions.

It follows from the calculus of variations that a necessary condition for a functional to be extremized or made stationary is that the Euler-Lagrange equation is satisfied for the sought solution together with boundary conditions. This Euler-Lagrange equation is precisely the PDE of the problem. The variational formulation often has advantages over the (partial) differential formulation, especially from the numerical point of view. Firstly the functional contains derivatives of lower order than the differential problem. A consequence is that an approximate solution can be sought in a larger class of functions. Secondly, the variational formulation treats in a simple way complicated boundary conditions, such as, for instance, Dirichlet, Neumann or Robbins boundary conditions in non-rectangular regions.

In this chapter we shall first derive the Euler-Lagrange differential equation, which is a necessary condition for an admissible function to extremize the functional. Examples will be given for the Poisson equation in 1D and 2D for various types of boundary conditions.

The situation is then as follows. For the numerical solution of a PDE we can equivalently solve the associated variational formulation (if it exists). One general method for obtaining an approximate solution of a variational problem is the Ritz method. The FEM will be introduced as a special case of the Ritz method. Unfortunately, it is sometimes impossible to find the correct functional corresponding to a PDE. Therefore we shall discuss an other variational formulation based on the PDE itself. This method is known as the Galerkin method and is also applicable to those PDEs for which no equivalent extremal problem exists. The FEM will be discussed as a particular example of the Galerkin method.

3.1 Extremal problem; Euler-Lagrange equation

In this section we shall derive the Euler-Lagrange equation for a 1D problem. A simple problem in terms of one independent variable is to determine the optimum of the integral

$$J(w) = \int_{x_0}^{x_1} \phi(x,w,\frac{dw}{dx})\, dx \tag{3.1.1}$$

with $w = w(x)$ and $x_0 \leqslant x \leqslant x_1$. We assume that the function $w(.)$ as well as the function $\phi(.,.,.)$ are continuous with continuous (partial) first and second order derivatives. The problem of the calculus of variations is now to find a function u so as to optimize the functional J. As boundary conditions we impose

$$u(x_0) = g_0 \tag{3.1.2}$$

where g_0 is given.
The class of admissible functions on which J is defined is thus the set of twice continuously differentiable functions on $[x_0,x_1]$, denoted by $C^2[x_0,x_1]$, verifying (3.1.2).
Assume that the optimization problem is in fact a minimization problem and let u be its solution, then

$$\left| \begin{array}{ll} J(u) \leq J(w) & \text{for all admissible } w \\ u(x_0) = g_0 & \end{array} \right. \tag{3.1.3}$$

Let v be an arbitrary function in $C^2[x_0,x_1]$ which vanishes at $x = x_0$: $v(x_0) = 0$. If λ is a parameter, then

$$w(x) = u(x) + \lambda v(x) \tag{3.1.4}$$

represents a family of admissible functions (see Fig. 3.1.)

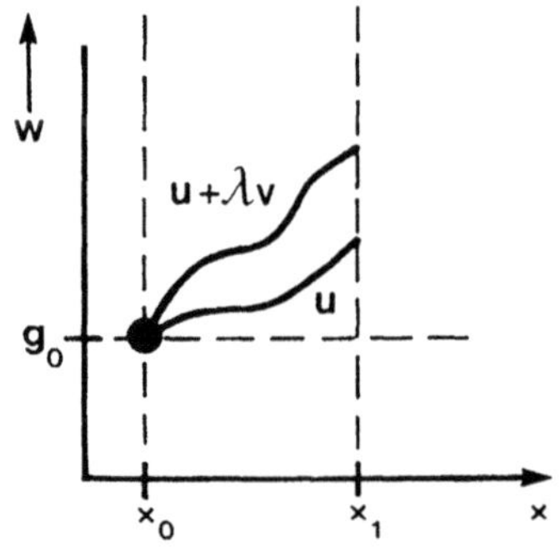

Fig. 3.1 Solution u and family of admissible functions $u+\lambda v$.

Substituting (3.1.4) into (3.1.1) leads to an integral as a function of λ

$$J(w) = \int_{x_0}^{x_1} \phi(x,u+\lambda v,\frac{du}{dx} + \lambda \frac{dv}{dx})\, dx \equiv I(\lambda) \ .$$

For $\lambda = 0$, (3.1.4) yields $w(x) = u(x)$ and since u minimizes the functional, we conclude that I must have a minimum for $\lambda = 0$. A necessary condition for this is that

$$\frac{dI}{d\lambda} = 0 \quad \text{for } \lambda = 0 \quad \text{and for all } v \in C^2[x_0,x_1] \quad \text{with } v(x_0) = 0. \tag{3.1.5}$$

Thus we obtain

$$\frac{d}{d\lambda} \int_{x_0}^{x_1} \phi(x,u+\lambda v,\frac{du}{dx} + \lambda \frac{dv}{dx})dx = 0 \text{ for } \lambda = 0 \text{ and for all } v \in C^2[x_0,x_1] \text{ with } v(x_0) = 0. \tag{3.1.6}$$

Let us introduce the following notation

$$p = \frac{dw}{dx} = \frac{du}{dx} + \lambda \frac{dv}{dx}$$

then

$$\phi = \phi(x,w,p) \text{ and } \frac{d\phi}{d\lambda} = \frac{\partial\phi}{\partial x}\frac{\partial x}{\partial\lambda} + \frac{\partial\phi}{\partial w}\frac{\partial w}{\partial\lambda} + \frac{\partial\phi}{\partial p}\frac{\partial p}{\partial\lambda}$$

and (3.1.6) becomes, as λ does not depend on x:

$$\int_{x_0}^{x_1} \{\frac{\partial\phi}{\partial w} v + \frac{\partial\phi}{\partial p}\frac{dv}{dx}\}dx = 0 \ .$$

Partial integration of the second term gives

$$\int_{x_0}^{x_1} \{\frac{\partial\phi}{\partial w} - \frac{d}{dx}(\frac{\partial\phi}{\partial p})\} v\, dx + [\frac{\partial\phi}{\partial p} v]_{x_0}^{x_1} = 0 \quad \text{for } w = u \ .$$

Using now the boundary condition for v : $v(x_0) = 0$, we deduce

$$\int_{x_0}^{x_1} \{\frac{\partial\psi}{\partial w} - \frac{d}{dx}(\frac{\partial\psi}{\partial p})\} v \, dx + [\frac{\partial\psi}{\partial p} v](x_1) = 0 \quad \text{for } w = u \quad (3.1.7)$$

This relation must be satisfied for all $v \in C^2[x_0,x_1]$ with $v(x_0) = 0$, thus certainly for all $v \in C^2[x_0,x_1]$ with $v(x_0) = v(x_1) = 0$. For this special subclass of functions v (3.1.7) reduces to

$$\int_{x_0}^{x_1} \{\frac{\partial\psi}{\partial w} - \frac{d}{dx}(\frac{\partial\psi}{\partial p})\} v \, dx = 0 \quad \text{for } w = u \text{ and for all}$$

$$v \in C^2[x_0,x_1], \; v(x_0) = v(x_1) = 0 \; .$$

Using Du Bois Reymond's lemma, which will be proved at the end of this section, we obtain

$$\frac{\partial\psi}{\partial w} - \frac{d}{dx}(\frac{\partial\psi}{\partial p}) = 0 \quad \text{for } w = u$$

or equivalently

$$\frac{\partial\psi}{\partial w} - \frac{d}{dx}(\frac{\partial\psi}{\partial(\frac{dw}{dx})}) = 0 \quad \text{for } w = u \; . \quad (3.1.8)$$

This equation is independent of v, so it can be substituted into relation (3.1.7) to give

$$[\frac{\partial\psi}{\partial p} v](x_1) = 0 \quad \text{for } w = u \text{ and for all } v \in C^2[x_0,x_1]$$

$$\text{with } v(x_0) = 0 \; .$$

Since the value of v at $x = x_1$ is arbitrary, we deduce the following condition:

$$\frac{\partial\psi}{\partial p}(x_1) = 0 \quad \text{for } w = u$$

or equivalently

$$\frac{\partial\psi}{\partial(\frac{dw}{dx})}(x_1) = 0 \quad \text{for } w = u \; . \quad (3.1.9)$$

The result is thus that a necessary condition for an admissible function u to minimize the functional J is that u satisfies the following BVP

$$\left|\begin{array}{ll} \text{(i)} & \dfrac{\partial\phi}{\partial u} - \dfrac{d}{dx}\left(\dfrac{\partial\phi}{\partial(\frac{du}{dx})}\right) = 0 \qquad \phi = \phi(x,u,\frac{du}{dx}) \\ \text{(ii)} & u(x_0) = g_0 \\ \text{(iii)} & \dfrac{\partial\phi}{\partial(\frac{du}{dx})}(x_1) = 0 \end{array}\right. \tag{3.1.10}$$

Equation (3.1.10) (i) is called the Euler-Lagrange equation for the minimization problem; the two conditions (3.1.10) (ii) and (iii) are the associated boundary conditions. The boundary condition (3.1.10) (ii) is called essential since this condition was already imposed on the solution of the minimization problem. The second boundary condition (3.1.10) (iii) is called natural since it is not explicitly imposed on the solution of the minimization problem, but it follows directly from it.

The example we have treated in this section was 1D. The treatment in more dimensional cases is similar, we shall exemplify this in the next section.

In the derivation of the Euler-Lagrange equation reference was made to the following important result in the calculus of variations:

Lemma (of DU BOIS REYMOND) (see Courant (1936)) (3.1.11)
Let G be a continuous function on $[a,b]$: $G \in C[a,b]$, and let

$$\int_a^b G(x)\, v(x)\, dx = 0 \tag{3.1.12}$$

for all $v \in C^2[a,b]$ with $v(a) = v(b) = 0$.

Then

$$G(x) = 0 \qquad \text{for all } x \in [a,b].$$

Proof

Suppose there exists a point $x_0 \in (a,b)$ for which $G(x_0) \neq 0$, say $G(x_0) > 0$. Because G is continuous on $[a,b]$ there exists a neighbourhood $(x_0-\delta, x_0+\delta) \subset (a,b)$ such that

$$G(x) > 0 \qquad \text{for } |x-x_0| < \delta .$$

We choose $v \in C^2[a,b]$ as follows (see Fig. 3.2)

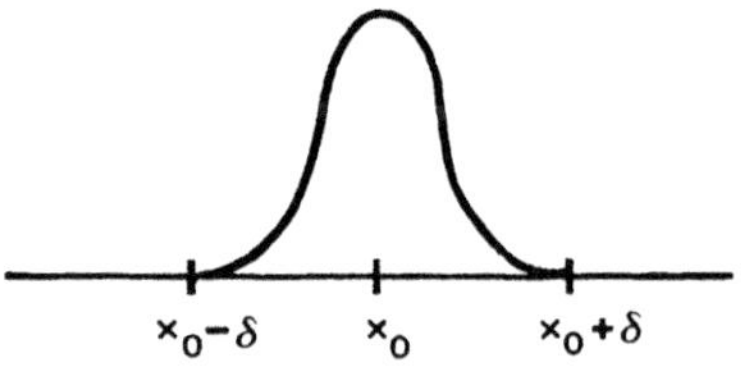

Fig. 3.2 The function v.

$$v(x) = \begin{cases} \exp\{-\dfrac{1}{\delta^2-(x-x_0)^2}\} & \text{if } |x-x_0| < \delta \\ 0 & \text{if } |x-x_0| \geq \delta \end{cases}$$

then

$$\int_a^b G(x)\, v(x)\, dx = \int_{x_0-\delta}^{x_0+\delta} G(x).\exp\{-\frac{1}{\delta^2-(x-x_0)^2}\}dx > 0$$

which contradicts (3.1.12).
Consequently $G(x) = 0$ for all $x \in (a,b)$ and by the continuity of G we obtain

$$G(x) = 0 \qquad \text{for all } x \in [a,b] .$$

Remark An extension of the Du Bois Reymond lemma to more dimensions is straightforward.

3.2 Extremal Formulation of the Poisson Equation

3.2.1 1D case

Consider the following minimization problem

> Find a function $u = u(x)$, $0 \leq x \leq 1$, with $u(0) = g_0$, that minimizes
>
> $$J(w) = \int_0^1 \{\tfrac{1}{2}(\frac{dw}{dx})^2 - fw\}dx \ , \qquad f = f(x) \qquad (3.2.1)$$
>
> over the class of functions which are equal to g_0 at $x = 0$.

We assume that a solution $u \in C^2[0,1]$ exists. Then with the notation of Section 3.1 we have

$$\phi(x,w,\frac{dw}{dx}) = \tfrac{1}{2}(\frac{dw}{dx})^2 - fw \ .$$

Substitution into the Euler-Lagrange equation (3.1.10) with

$$\frac{\partial\phi}{\partial w} = -f \qquad \frac{\partial\phi}{\partial(\frac{\partial w}{\partial x})} = \frac{dw}{dx} \qquad \frac{d}{dx}(\frac{\partial\phi}{\partial(\frac{\partial w}{\partial x})}) = \frac{d^2w}{dx^2}$$

gives

$$-\frac{d^2u}{dx^2} = f \quad \text{on} \quad (0,1)$$

and boundary conditions

$$u(0) = g_0 \qquad \text{(essential)}$$

$$\frac{\partial\phi}{\partial(\frac{du}{dx})}(1) = \frac{du}{dx}(1) = 0 \qquad \text{(natural)}.$$

The Euler-Lagrange equation for problem (3.2.1) is thus the Poisson equation with mixed Dirichlet-Neumann boundary conditions for u:

$$\left| \begin{array}{l} -\dfrac{d^2u}{dx^2} = f \qquad \text{on} \quad (0,1) \\[2ex] u(0) = g_0 \\[2ex] \dfrac{du}{dx}(1) = 0 \end{array} \right. \tag{3.2.2}$$

3.2.2 2D case

Let Ω be a bounded open region in 2D with smooth boundary Γ. Consider the minimization problem

> Find $u = u(\underline{x})$, $\underline{x} \in \bar{\Omega} \equiv \Omega \cup \Gamma$, with $u|_\Gamma = 0$ minimizing the following functional
>
> $$J(w) = \tfrac{1}{2} \int_\Omega |\text{grad } w|^2 \, d\Omega - \int_\Omega f \, w \, d\Omega , \qquad f = f(\underline{x}) \tag{3.2.3}$$
>
> over the class of functions which vanish on Γ.

We assume that a solution $u \in C^2(\Omega) \cap C(\bar{\Omega})$ exists and we define the set of admissible functions as the subset of $C^2(\Omega) \cap C(\bar{\Omega})$ consisting of functions which vanish on Γ.
Let $v \in C^2(\Omega) \cap C(\bar{\Omega})$ be arbitrary with $v|_\Gamma = 0$, then

$$w(\underline{x}) = u(\underline{x}) + \lambda v(\underline{x}) , \qquad \lambda \text{ parameter}$$

is a family of admissible functions. Substitution of w into the expression for J leads to

$$J(w) = \tfrac{1}{2} \int_\Omega |\text{grad}(u+\lambda v)|^2 \, d\Omega - \int_\Omega f(u+\lambda v) d\Omega \equiv I(\lambda).$$

A necessary condition for I to take its minimum value for $\lambda = 0$ is that

$$\frac{d}{d\lambda}I(\lambda) = 0 \text{ for } \lambda = 0 \text{ and for all } v \in C^2(\Omega) \cap C(\bar{\Omega}) \tag{3.2.4}$$

with $v|_\Gamma = 0$.

Using the property

$$\frac{\partial}{\partial\lambda} |\text{grad } w|^2 = \frac{\partial}{\partial\lambda} \left\{\left(\frac{\partial w}{\partial x_1}\right)^2 + \left(\frac{\partial w}{\partial x_2}\right)^2\right\} =$$

$$= 2\left\{\frac{\partial w}{\partial x_1}\frac{\partial v}{\partial x_1} + \frac{\partial w}{\partial x_2}\frac{\partial v}{\partial x_2}\right\} = 2 \text{ grad } w. \text{ grad } v ,$$

we find that (3.2.4) can be rewritten as

$$\int_\Omega \text{grad } u. \text{ grad } v \, d\Omega - \int_\Omega f \, v \, d\Omega = 0.$$

Applying Green's formula to the first term, we obtain

$$\int_\Omega (-\Delta u - f)v \, d\Omega + \int_\Gamma \frac{\partial u}{\partial n} v \, d\Gamma = 0 \tag{3.2.5}$$

where $\underline{n}$ is the outward unit normal on Γ. Since $v|_\Gamma = 0$, relation (3.2.5) further reduces to

$$\int_\Omega (-\Delta u - f)v \, d\Omega = 0$$

which must be satisfied for all $v \in C^2(\Omega) \cap C(\bar{\Omega})$ with $v|_\Gamma = 0$. Next we apply the analogue of Du Bois Reymond's lemma (3.1.11) for 2D and it follows that the Euler-Lagrange equation for (3.2.3) reads:

$$\left| \begin{array}{ll} -\Delta u = f & \text{in } \Omega \\ u = 0 & \text{on } \Gamma \end{array} \right. \tag{3.2.6}$$

3.2.3 Various types of boundary conditions

Consider a bounded open region $\Omega \subset \mathbb{R}^n$ with smooth boundary Γ. Γ is subdivided into four disjoint parts Γ_0, Γ_1, Γ_2 and Γ_3 (see Fig. 3.3).

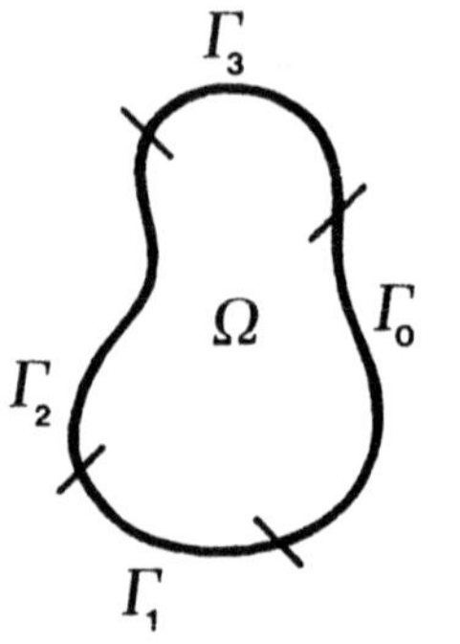

Fig. 3.3 Region Ω with Γ subdivided into four parts.

Consider the following minimization problem

> Find $u = u(\underline{x})$, $\underline{x} \in \Omega \cup \Gamma$, with $u(\underline{x}) = g_0(\underline{x})$ for $\underline{x} \in \Gamma_0$, that minimizes the functional
>
> $$J(w) = \tfrac{1}{2} \int_\Omega |\text{grad } w|^2 d\Omega + \tfrac{1}{2}\alpha \int_{\Gamma_3} |w|^2 d\Gamma - \int_\Omega f\, w\, d\Omega - \int_{\Gamma_1} g_1 w\, d\Gamma \tag{3.2.7}$$
>
> over the class of functions which are equal to g_0 on Γ

for given functions $f = f(\underline{x})$, $\underline{x} \in \Omega$, and $g_i = g_i(\underline{x})$, $\underline{x} \in \Gamma_i$ $i = 0,1$, $\alpha > 0$.

We assume that a solution $u \in C^2(\Omega) \cap C^1(\bar{\Omega})$ exists. The set of admissible functions is defined as the subset of $C^2(\Omega) \cap C^1(\bar{\Omega})$ containing functions that are equal to g_0 on Γ_0. Let $v \in C^2(\Omega) \cap C^1(\bar{\Omega})$ with $v|_{\Gamma_0} = 0$, then we substitute the family of admissible functions

$$w = u + \lambda v$$

with parameter λ into J:

$$J(w) = J(u+\lambda v) \equiv I(\lambda).$$

A necessary condition for minimization is

$$\frac{d}{d\lambda} I(\lambda) = 0 \text{ for } \lambda = 0 \text{ and for all } v \in C^2(\Omega) \cap C^1(\bar{\Omega}), \tag{3.2.8}$$

with $v|_{\Gamma_0} = 0$

which amounts to saying that

$$\int_\Omega \text{grad } u.\text{grad } v \, d\Omega + \alpha \int_{\Gamma_3} u \, v \, d\Gamma - \int_\Omega f \, v \, d\Omega - \int_{\Gamma_1} g_1 v \, d\Gamma = 0 \tag{3.2.9}$$

or, after Green's formula has been applied to the first term:

$$\int_\Omega \{-\Delta u - f\} v \, d\Omega + \alpha \int_{\Gamma_3} u \, v \, d\Gamma + \int_\Gamma \frac{\partial u}{\partial n} v \, d\Gamma - \int_{\Gamma_1} g_1 v \, d\Gamma = 0 \tag{3.2.10}$$

This integral relation must be satisfied for all $v \in C^2(\Omega) \cap C^1(\bar{\Omega})$ with $v|_{\Gamma_0} = 0$, thus in particular for all $v \in C^2(\Omega) \cap C^1(\bar{\Omega})$ with $v|_\Gamma = 0$ for which (3.2.10) reduces to:

$$\int_\Omega \{-\Delta u - f\} v \, d\Omega = 0 \, . \tag{3.2.11}$$

Using Du Bois Reymond's lemma, we find

$$-\Delta u = f \quad \text{in } \Omega \, . \tag{3.2.12}$$

Substitution of (3.2.12) into (3.2.10) gives:

$$\alpha \int_{\Gamma_3} u \, v \, d\Gamma + \int_{\Gamma_1 \cup \Gamma_2 \cup \Gamma_3} \frac{\partial u}{\partial n} v \, d\Gamma - \int_{\Gamma_1} g_1 v \, d\Gamma = 0 \tag{3.2.13}$$

for all $v \in C^2(\Omega) \cap C^1(\bar{\Omega})$ with $v|_{\Gamma_0} = 0$, thus in particular for all $v \in C^2(\Omega) \cap C^1(\bar{\Omega})$ with $v|_{\Gamma_0 \cup \Gamma_1 \cup \Gamma_2} = 0$ for which (3.2.13) can be written as

$$\alpha \int_{\Gamma_3} u \, v \, d\Gamma + \int_{\Gamma_3} \frac{\partial u}{\partial n} v \, d\Gamma = 0 \, .$$

From this relation we deduce

$$\alpha u + \frac{\partial u}{\partial n} = 0 \quad \text{on } \Gamma_3. \tag{3.2.14}$$

After substitution into (3.2.13), this relation further simplifies to

$$\int_{\Gamma_1 \cup \Gamma_2} \frac{\partial u}{\partial n} v \, d\Gamma - \int_{\Gamma_1} g_1 v \, d\Gamma = 0 \tag{3.2.15}$$

for all $v \in C^2(\Omega) \cap C^1(\bar{\Omega})$ with $v|_{\Gamma_0} = 0$, thus in particular for all v with $v|_{\Gamma_0 \cup \Gamma_1} = 0$ for which (3.2.15) becomes

$$\int_{\Gamma_2} \frac{\partial u}{\partial n} v \, d\Gamma = 0$$

which implies

$$\frac{\partial u}{\partial n} = 0 \qquad \text{on } \Gamma_2. \tag{3.2.16}$$

Relation (3.2.15) now simplifies to

$$\int_{\Gamma_1} \left(\frac{\partial u}{\partial n} - g_1\right) v \, d\Gamma = 0$$

for all v with $v|_{\Gamma_0} = 0$, so that

$$\frac{\partial u}{\partial n} = g_1 \qquad \text{on } \Gamma_1. \tag{3.2.17}$$

Hence, the Euler-Lagrange equation for problem (3.2.7) takes the following form:

$$\left| \begin{array}{ll} -\Delta u = f & \text{in } \Omega \\ u = g_0 & \text{on } \Gamma_0 \\ \dfrac{\partial u}{\partial n} = g_1 & \text{on } \Gamma_1 \\ \dfrac{\partial u}{\partial n} = 0 & \text{on } \Gamma_2 \\ \alpha u + \dfrac{\partial u}{\partial n} = 0 & \text{on } \Gamma_3 \end{array} \right. \tag{3.2.18}$$

The non-homogeneous Dirichlet condition on Γ_0 is essential since it was already imposed on the solution of the extremal formulation (3.2.7). On the other hand, the non-homogeneous Neumann condition on Γ_1, the homogeneous Neumann condition on Γ_2 and the homogeneous Robbins condition on Γ_3 are natural since they are automatically satisfied by the variational formulation (3.2.7).

3.3 Comments

In the preceding section we have established a relation between the Poisson equation and a minimization problem. That is to say: given a minimization problem and assuming that it has a unique solution, we have written down a necessary condition, the Poisson equation, for its solution. However, not for all elliptic PDEs there exists an associated minimization problem. A sufficient condition for an elliptic PDE

$$\left| \begin{array}{l} \mathcal{L}u = f \quad \text{in } \Omega \\ u \text{ satisfying homogeneous boundary conditions} \end{array} \right. \tag{3.3.1}$$

to be interpreted as the Euler-Lagrange equation of a minimization problem is that

(i) $\mathcal{L}$ is linear

(ii) $\mathcal{L}$ is self-adjoint (or symmetric):

$$\int_\Omega v \, \mathcal{L} \, u \, d\Omega = \int_\Omega u \, \mathcal{L} \, v \, d\Omega \tag{3.3.2}$$

(iii) $\mathcal{L}$ is positive: $\int_\Omega u \, \mathcal{L} \, u \, d\Omega \geq 0$

for all u and v satisfying the homogeneous boundary conditions. If we assume that (3.3.1) has a unique solution, then u minimizes the following functional

$$J(w) = \tfrac{1}{2} \int_\Omega w \, \mathcal{L} \, w \, d\Omega - \int_\Omega f \, w \, d\Omega$$

over an appropriate set of admissible functions.

When other types of (non-homogeneous) boundary conditions are given, extra terms must be added to the functional or the set of admissible functions must be adjusted.

To illustrate the conditions (3.3.2), let us consider the 1D Poisson equation with mixed homogeneous Neumann-Dirichlet boundary conditions:

$$\left| \begin{aligned} & -\frac{d^2u}{dx^2} = f \qquad \text{on} \quad (0,1) \\ & u(0) = 0 \qquad \frac{du}{dx}(1) = 0 \end{aligned} \right. \tag{3.3.3}$$

The operator $\mathcal{L} = -\frac{d^2}{dx^2}$ is

(i) linear

(ii) self-adjoint: $\int_0^1 v\,\mathcal{L}\,w\,dx = -\int_0^1 v\,\frac{d^2w}{dx^2}\,dx =$

$$\int_0^1 \frac{dv}{dx}\frac{dw}{dx}\,dx - \left[v\,\frac{dw}{dx}\right]_0^1 = \int_0^1 \frac{dv}{dx}\frac{dw}{dx}\,dx =$$

$$-\int_0^1 w\,\frac{d^2v}{dx^2}\,dx + \left[\frac{dv}{dx}\,w\right]_0^1 =$$

$$-\int_0^1 w\,\frac{d^2v}{dx^2}\,dx = \int_0^1 w\,\mathcal{L}\,v\,dx$$

(iii) positive: $\int_0^1 w\,\mathcal{L}\,w\,dx = -\int_0^1 w\,\frac{d^2w}{dx^2}\,dx = \int_0^1 \left(\frac{dw}{dx}\right)^2 dx > 0$

The corresponding functional is (cf. Section 3.2, 1D case)

$$J(w) = \tfrac{1}{2}\int_0^1 \left(\frac{dw}{dx}\right)^2 dx - \int_0^1 f\,w\,dx$$

and the admissible functions must satisfy $w(0) = 0$.

If we take a non-homogeneous Dirichlet condition at $x = 0$: $u(0) = g_0$, then the admissible functions must satisfy $w(0) = g_0$. For a non-homogeneous Neumann condition at $x = 1$: $\frac{du}{dx}(1) = g_1$ a term is added to the functional:

$$J(w) = \frac{1}{2} \int_0^1 \left(\frac{dw}{dx}\right)^2 dx - \int_0^1 f\, w\, dx - g_1 w(1)$$

The term $g_1 w(1)$ is the 1D analogue of the term $\int_{\Gamma_1} g_1 w \, d\Gamma$ in functional (3.2.7).

For many PDEs however the conditions (3.3.2) are not satisfied. In particular there does not exist an equivalent minimization problem for the convection-diffusion equation (1.3.2) and for the Navier-Stokes equations (1.6.16), (1.6.17). For these equations the FEMs which construct a solution of the BVP using the minimization formulation (for instance: the Ritz method in Section 3.4) are not applicable. Later on we shall see that there are FEMs which do not use the minimization problem, but are directly applicable to the BVP (for instance: the Galerkin method in section (3.6)).

Finally let us make a remark concerning the boundary conditions. For an elliptic PDE boundary conditions must be given along the whole boundary. For a minimization problem we only have to prescribe the essential boundary conditions, since the natural conditions are satisfied automatically by the solution. For the three types of boundary conditions we have considered, viz. Dirichlet, Neumann and Robbins type, only the Dirichlet conditions are essential, the other two are natural conditions. From the numerical point of view this is an important advantage of the minimization problem over the PDE formulation since the natural boundary conditions (Neumann, Robbins) are usually the most complicated ones.

3.4 The Ritz method

In this section we shall be concerned with the construction of approximate solutions of a PDE based on its minimization formulation. So we assume that the PDE we are solving admits an equivalent minimization formulation. Moreover, we assume that any admissible function can be written as a linear combination of, in principle, infinitely many basis functions (also called trial functions). When we restrict the number of basis functions to a finite number N, we obtain approximate solutions. This method is called the Ritz method. A classical example of this method is the approximation by Fourier series or power series. The FEM will be introduced as a particular example of the Ritz method.

Consider the following minimization problem defined on the region $\Omega \subset \mathbb{R}^n$ with boundary Γ.

> Find a function $u : \Omega \cup \Gamma \to \mathbb{R}$ with $u(\underline{x}) = g_0(\underline{x})$ for $\underline{x} \in \Gamma$ such that u extremizes
>
> $$J(w) = \int_\Omega \phi(\underline{x}, w, \frac{\partial w}{\partial x_1}, \ldots, \frac{\partial w}{\partial x_n})\, d\Omega \tag{3.4.1}$$
>
> over the class of functions which are equal to g_0 on Γ.

We are going to find a solution u within the class of admissible functions V_{g_0} consisting of $C^1(\Omega) \cap C(\bar{\Omega})$ functions which are equal to g_0 on the boundary Γ. We assume that there exists a basis $\phi_1, \ldots, \phi_N, \ldots$ for the function space V_0 consisting of $C^1(\Omega) \cap C(\bar{\Omega})$ functions which vanish on Γ. Let $\phi_0 \in V_{g_0}$ be arbitrary, then any function $w \in V_{g_0}$ can be written as

$$w(\underline{x}) = \phi_0(\underline{x}) + \sum_{j=1}^{\infty} \alpha_j \phi_j(\underline{x}) \tag{3.4.2}$$

where α_j, $j = 1,2,\ldots$ are coefficients.

An admissible approximate solution $\tilde{u}$ will be written as ϕ_0 plus a finite linear combination of the N basisfunction $\phi_1,\ldots,\phi_N$:

$$\tilde{u}(\underline{x}) = \phi_0(\underline{x}) + \sum_{j=1}^{N} \alpha_j\phi_j(\underline{x}). \tag{3.4.3}$$

We define the following (approximate) problem:

| Find $\tilde{u} \in V_{g_0N}$ which extremizes J on V_{g_0N} (3.4.4)

where V_{g_0N} denotes the class of admissible approximate solutions of the type (3.4.3). Since the coefficients $\alpha_1,\ldots,\alpha_N$ are the unknowns, problem (3.4.4) can be written as

| Find $\{\alpha_1,\ldots,\alpha_N\}$ such that
| $J(\alpha_1,\ldots,\alpha_N)$ takes its extremum value . (3.4.5)

From differential calculus we obtain the following necessary condition:

$$\frac{\partial J}{\partial \alpha_i} = 0 \qquad j = 1,2,\ldots,N. \tag{3.4.6}$$

We have reduced now the original extremal problem (3.4.1) to a system of N equations (3.4.6) for N unknowns $\alpha_1,\ldots,\alpha_N$, which can be solved numerically. The accuracy of the approximate solution depends on the number N of basis functions $\phi_1,\ldots,\phi_N$. Generally, as N increases the accuracy improves. In the following example we shall apply the Ritz method to the 1D Poisson equation.

1D Poisson equation

As we have seen in Section 3.2 the 1D Poisson equation with mixed Dirichlet-Neumann boundary conditions

$$\left|\begin{array}{l} -\dfrac{d^2u}{dx^2} = f \quad \text{on} \quad (0,1) \\ u(0) = 0 \\ \dfrac{du}{dx}(1) = 0 \end{array}\right. \tag{3.4.7}$$

is equivalent to the following minimization problem

$$\left|\begin{array}{l} \text{Find a function } u : [0,1] \to \mathbb{R}, \text{ with } u(0) = 0 \\ \text{such that} \\ J(u) \leq J(w) \text{ for all } w : [0,1] \to \mathbb{R} \\ \text{with } w(0) = 0 \\ J(w) = \frac{1}{2}\int_0^1 (\frac{dw}{dx})^2 dx - \int_0^1 f\, w\, dx. \end{array}\right. \tag{3.4.8}$$

Let us choose the following basis functions:

$$\phi_0 = 0\ ,\quad \phi_1 = x\ ,\quad \phi_2 = x^2\ ,\ \ldots\ ,\quad \phi_N = x^N, \ldots$$

It is a classical result that any function on $[0,1]$ which vanishes at $x = 0$ can be written as a linear combination of these basis functions.

Thus

$$u(x) = \sum_{j=1}^{\infty} \alpha_j \phi_j(x) = \sum_{j=1}^{\infty} \alpha_j x^j$$

and as approximate solution we take functions of the form

$$\tilde{u}(x) = \sum_{j=1}^{N} \alpha_j x^j. \tag{3.4.9}$$

After substitution of expression (3.4.9) into the minimization problem, we find the following approximate problem

$$\left|\begin{array}{l} \text{minimize } J(\tilde{u}) \quad \text{with respect to } \alpha_1, \ldots, \alpha_N \\ J(\tilde{u}) = \frac{1}{2}\int_0^1 (\sum_{j=1}^{N} \alpha_j \frac{d\phi_j}{dx})^2 dx - \int_0^1 f(\sum_{j=1}^{N} \alpha_j \phi_j)\, dx. \end{array}\right. \tag{3.4.10}$$

Since this implies

$$\frac{\partial J}{\partial \alpha_j} = 0 \qquad i = 1,2,\ldots,N$$

we find, using the relation

$$\frac{\partial}{\partial \alpha_i} \left(\sum_{j=1}^{N} \alpha_j \frac{d\phi_j}{dx} \right) = \frac{d\phi_i}{dx} ,$$

that

$$\int_0^1 \left(\sum_{j=1}^{N} \alpha_j \frac{d\phi_j}{dx} \right) \frac{d\phi_i}{dx} \, dx = \int_0^1 f \, \phi_i \, dx \qquad i = 1,2,\ldots,N. \tag{3.4.11}$$

The function ϕ_j being given by $\phi_j(x) = x^j$, (3.4.11) is equivalent to

$$\sum_{j=1}^{N} \alpha_j \int_0^1 j \, i \, x^{j-1} x^{i-1} dx = \int_0^1 f \, x^i \, dx \qquad i = 1,2,\ldots,N . \tag{3.4.12}$$

The integral on the left hand side satisfies

$$\int_0^1 j \, i \, x^{j+i-2} \, dx = \frac{j \, i}{j+i-1}$$

so that (3.4.12) becomes

$$\sum_{j=1}^{N} \alpha_j \frac{j \, i}{j+i-1} = \int_0^1 f \, x^i \, dx \qquad i = 1,2,\ldots,N .$$

The function f is given. Setting now

$$F_i = \int_0^1 f \, x^i \, dx$$

the system of linear equations for $\alpha_1,\ldots,\alpha_N$ can be written as

$$\sum_{j=1}^{N} \alpha_j \frac{j \, i}{j+i-1} = F_i \qquad i = 1,2,\ldots,N$$

or in matrix notation

$$A\,\underline{\alpha} = \underline{F} \tag{3.4.13}$$

with

$$A = \begin{vmatrix} 1 & 1 & 1 & 1 & . & . & 1 \\ 1 & 4/3 & 6/4 & 8/5 & . & . & 2N/N+1 \\ 1 & 6/4 & 9/5 & 12/6 & . & . & 3N/N+2 \\ 1 & 8/5 & 12/6 & 16/7 & . & . & 4N/N+3 \\ . & & & & & & . \\ . & & & & & & . \\ 1 & & & & & & N^2/2N-1 \end{vmatrix},$$

$$\underline{\alpha} = \begin{vmatrix} \alpha_1 \\ \alpha_2 \\ . \\ . \\ \alpha_N \end{vmatrix}, \quad \underline{F} = \begin{vmatrix} F_1 \\ F_2 \\ . \\ . \\ F_N \end{vmatrix}.$$

The matrix A, known as the Hilbert matrix, is positive definite, so that system (3.4.13) has a unique solution $\underline{\alpha}$. From the practical point of view let us remark that the matrix A is a full one and that it has a bad condition, which means that even for moderate values of N the round-off errors become dominant and worsen the accuracy of the approximate solution.

Generally we can say that the Ritz method consists of assuming the form of the solution in terms of basis functions $\phi_1,\ldots,\phi_N$ with unknown parameters $\alpha_1,\ldots,\alpha_N$. The approximate solution is then substituted into the functional. The unknown parameters are determined in such a way that they render the functional stationary. For this we differentiate the functional with respect to each unknown parameter, and the resulting expressions are set equal to zero. This leads to a system of linear equations of the form

$$A \underline{\alpha} = \underline{F}$$

where A : N×N matrix with elements $A_{ij} = \int_0^1 \frac{d\phi_j}{dx} \frac{d\phi_i}{dx} dx$

$\underline{\alpha}$: column matrix with elements $(\underline{\alpha})_i = \alpha_i$

$\underline{F}$: column matrix with elements $(\underline{F})_i = F_i = \int_0^1 f \phi_i \, dx$

The Ritz method has some advantages over the FDM which is based on the PDE itself. The number of basis functions N can be increased easily without many additional computations and a known behavior of the solution can be expressed in the choice of the basis functions. The admissible functions nor the basis functions have to satisfy the natural boundary conditions, while the solution does automatically. The Ritz method seems to be a good approximation method once there is a systematic way of choosing basis functions leading to a sparse system of linear equations for the coefficients. A method which has these properties is the Finite Element Method, which will be introduced in the next section.

3.5 The FEM

3.5.1 Definition

The FEM can be considered as a particular example of the Ritz method, which means that, in this section, we restrict ourselves to the solution of minimization problems. In the next section we shall introduce the FEM for a wider class of problems.

The FEM is a systematic method for generating basis functions on arbitrary regions Ω which will give rise to a sparse matrix A. As we have seen for the 1D Poisson equation the matrix elements of A are given by

$$A_{ij} = \int_0^1 \frac{d\phi_j}{dx} \frac{d\phi_i}{dx} \, dx$$

and it is easy to see that the matrix A will be a sparse one if the basis functions have a "small" support. The support of a function defined on Ω is the closure of the subregion of Ω on which the function differs from zero. It is evident that the matrix element A_{ij} equals zero whenever the supports of ϕ_j and ϕ_i have empty intersection.

The basic idea of the FEM is to subdivide the region $\Omega \subset \mathbb{R}^n$ into subregions. In the 1D case these subregions are intervals, in 2D the subregions are triangles or quadrilaterals and, for instance, tetrahedrons or hexahedrons in 3D.

We require the subdivision of $\bar{\Omega}$ into subregions to satisfy the following properties:

(i) there is a finite number of subregions e_k, $k = 1,2,\ldots,K$

(ii) If e_k and e_ℓ are two subregions, then either $e_k = e_\ell$, or $e_k \cap e_\ell = \emptyset$, or

(ii)' e_k and e_ℓ have a common point when n=1

(ii)'' e_k and e_ℓ have a common edge or a common vertex when n=2 (3.5.1)

(ii)''' e_k and e_ℓ have a common face, a common edge or a common vertex when n=3

(iii) the union of all subregions is precisely the set $\bar{\Omega} = \Omega \cup \Gamma$.

In Fig. 3.4 a subdivision of an interval $\Omega = (0,1)$ into a finite number of subregions is shown. Fig. 3.4 satisfies the properties (3.5.1). In Fig. 3.5 a 2D situation is shown. Fig. 3.5(a) satisfies (3.5.1), while the situation of fig. 3.5(b) does not.

Fig. 3.4 Subdivision of Ω, n=1.

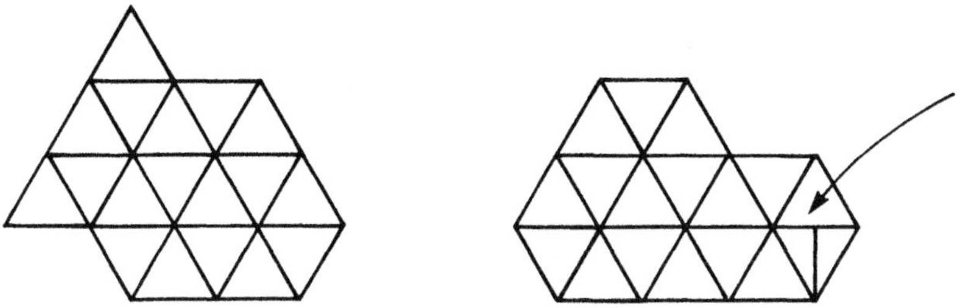

Fig. 3.5 Subdivision of Ω, n=2.

Next we choose in each subregion a finite number of points, preferably on the boundary of the subregion. These points are called nodal points. See Fig. 3.6 for examples in 1D, 2D and 3D. Nodal points will be denoted by $\underline{x}^j$, j=1,2,... with coordinates $(x_1^j,\dots,x_n^j)$.

Finally we define a basis function $\phi_i : \bar{\Omega} \to \mathbb{R}$ for each nodal point $\underline{x}^i$ in $\bar{\Omega}$, i=1,2,...,N. These functions must satisfy the following properties:

(i) ϕ_i takes the value 1 at $\underline{x}^i$ and vanishes at all the other nodal points:

$$\phi_i(\underline{x}^j) = \delta_{ij} = \begin{cases} 1 \text{ if } i=j \\ 0 \text{ if } i\neq j \end{cases} \tag{3.5.2}$$

(ii) ϕ_i has prescribed behavior on each subregion. For example, linear (by which we mean 'affine') or quadratic depending on the number of nodal points on each subregion.

(iii) ϕ_i is continuous on $\bar{\Omega}$.

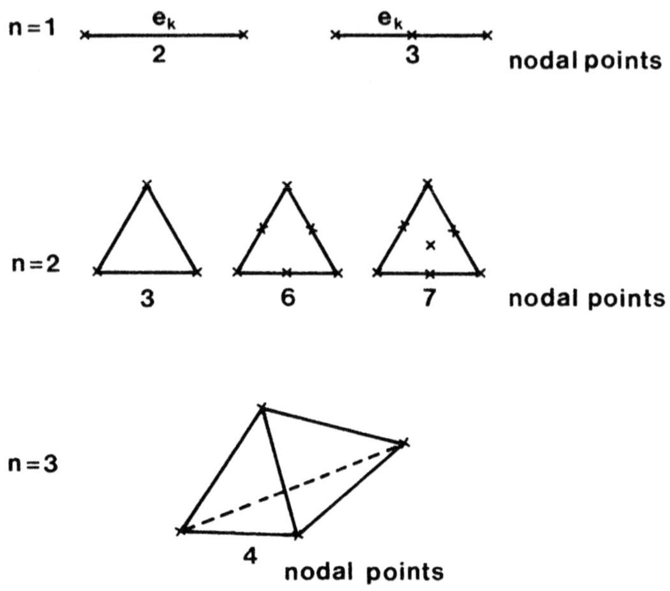

Fig. 3.6 Examples of nodal point distributions.

Using these basis functions ϕ_i an approximate solution $\tilde{u}$ can be constructed. In fact the function

$$\tilde{u}(x) = \sum_{j=1}^{N} \tilde{u}_j \phi_j(\underline{x}) \tag{3.5.3}$$

is a linear combination of basis functions ϕ_j, continuous on $\bar{\Omega}$ with prescribed behavior on the subregions. The parameter $\tilde{u}_i$ in (3.5.3) is precisely the value taken by $\tilde{u}$ at $\underline{x}^i$:

$$\tilde{u}(\underline{x}^i) = \sum_{j=1}^{N} \tilde{u}_j \phi_j(\underline{x}^i) = \sum_{j=1}^{N} \tilde{u}_j \delta_{ij} = \tilde{u}_i \ . \tag{3.5.4}$$

These unknown parameters $\tilde{u}_1, \ldots, \tilde{u}_N$ can be determined by the Ritz method. This means substitution of $\tilde{u}$ into the functional which becomes a functional depending on $\tilde{u}_1, \ldots, \tilde{u}_N$ only.

Equating to zero of the derivatives of the functional with respect to each of the parameters $\tilde{u}_i$ yields a system of linear equations for the unknowns $\tilde{u}_1,\dots,\tilde{u}_N$. The restriction to a subregion e_k of the admissible approximate solution (3.5.3) is called a shape function.

By a Finite Element we shall mean the following three quantities:
a subregion e_k, the nodal points on e_k and the shape functions corresponding to these nodal points.

3.5.2 1D Poisson equation

To illustrate the described finite element procedure, we consider again the 1D Poisson equation with mixed Dirichlet-Neumann boundary conditions:

$$\left|\begin{array}{ll} -\dfrac{d^2u}{dx^2} = f & \text{on } (0,1) \\ u(0) = 0 & \\ \dfrac{du}{dx}(1) = 0 & \end{array}\right. \tag{3.5.5}$$

with associated minimization problem

$$\left|\begin{array}{l} \text{Find } u : \bar{\Omega} \to \mathbb{R} \text{ with } u(0) = 0 \text{ such that} \\ J(u) \leq J(w) \quad \text{for all} \quad w : \Omega \to \mathbb{R} \quad \text{with } w(0) = 0 \\ J(w) = \frac{1}{2} \int_0^1 \left(\frac{dw}{dx}\right)^2 dx - \int_0^1 f\, w\, dx. \end{array}\right. \tag{3.5.6}$$

According to the Ritz method we seek the approximate solution as a finite linear combination of basis functions. The basis functions will be constructed using the finite element procedure. The first step is to subdivide the interval $[0,1]$ into N intervals e_k, $k=1,\dots,N$ of arbitrary length. The second step is to choose the nodal points for which we take the extremities of the subregions e_k. We call these nodal points $\underline{x}^i$, $i=0,1,\ldots,N$ (see Fig. 3.7).

Fig. 3.7 Subregions and nodal points on $[0,1]$.

The basis functions ϕ_i, $i=0,1,\ldots,N$ are defined as follows:

(i) $\phi_i(x^j) = \delta_{ij}$

(ii) ϕ_i is linear on each subregion

(iii) ϕ_i is continuous on $[0,1]$

In Fig. 3.8 the basis functions corresponding to the points $x^0 = 0$, x^i and $x^N = 1$ are shown:

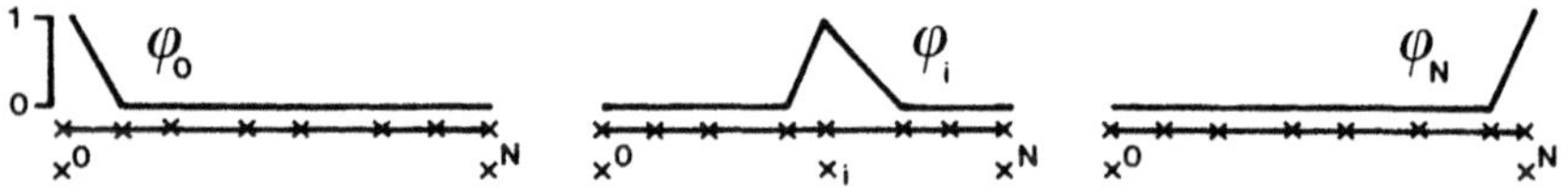

Fig. 3.8 Basis functions ϕ_0, ϕ_i and ϕ_N.

The approximate solution of the minimization problem is now assumed to take the form:

$$\tilde{u}(x) = \sum_{j=0}^{N} \tilde{u}_j \phi_j(x) \tag{3.5.7}$$

where $\tilde{u}_j$ is the unknown value of $\tilde{u}$ at $x = x^j$, and $\tilde{u}_0 = 0$ since $\tilde{u}(0) = 0$. Let us first determine the form of the shape functions. For this we consider the i-th subregion $e_i = [x^{i-1}, x^i]$. The only basis functions that differ from zero on e_i are ϕ_{i-1} and ϕ_i (see Fig. 3.9).

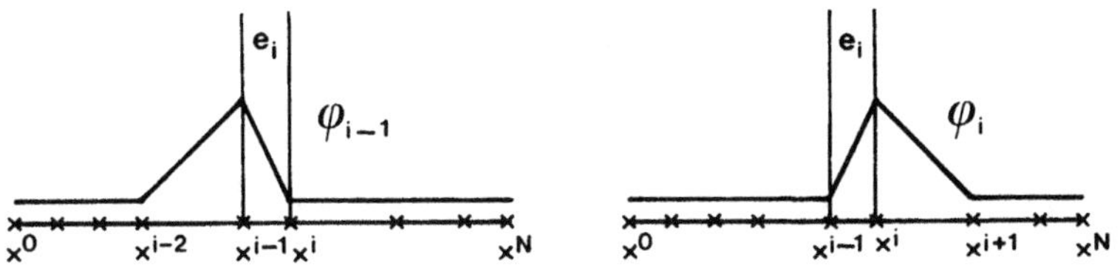

Fig. 3.9 The basis functions ϕ_{i-1} and ϕ_i.

It is easily verified that on e_i

$$\phi_{i-1}(x) = \frac{x^i - x}{x^i - x^{i-1}} \qquad\qquad \phi_i(x) = \frac{x - x^{i-1}}{x^i - x^{i-1}}$$

and that the shape function $\tilde{u}|_{e_i}$ takes the following form on e_i:

$$\begin{aligned} \tilde{u}(x) &= \sum_{j=0}^{N} \tilde{u}_j \phi_j(x) = \tilde{u}_{i-1}\phi_{i-1}(x) + \tilde{u}_i \phi_i(x) = \\ &= \tilde{u}_{i-1} + \frac{x - x^{i-1}}{x^i - x^{i-1}} (\tilde{u}_i - \tilde{u}_{i-1}), \qquad x \in [x^{i-1}, x^i]. \end{aligned} \tag{3.5.8}$$

This is a linear function on $[x^{i-1}, x^i]$ with $\tilde{u}(x^{i-1}) = \tilde{u}_{i-1}$ and $\tilde{u}(x^i) = \tilde{u}_i$. On the interval $[0,1]$ the function $\tilde{u}$ is a

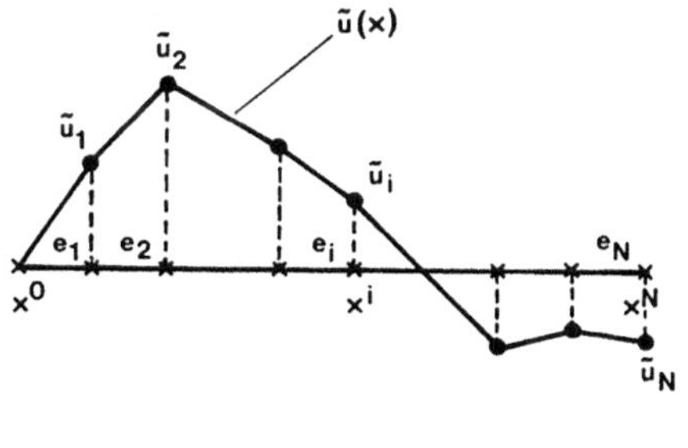

Fig. 3.10 General form of admissible approximate solution.

continuous piecewise linear function with $\tilde{u}(0) = 0$ as depicted in Fig. 3.10.

The expression for $\tilde{u}$, which depends on $\tilde{u}_1,\ldots,\tilde{u}_N$ only, is now substituted into the functional J. The minimization problem becomes

Minimize

$$J(\tilde{u}) = \tfrac{1}{2}\int_0^1 \Big(\frac{d}{dx}\sum_{j=1}^{N} \tilde{u}_j\phi_j(x)\Big)^2 dx - \int_0^1 f\Big(\sum_{j=1}^{N} \tilde{u}_j\phi_j\Big)dx$$

with respect to $\tilde{u}_1,\ldots,\tilde{u}_N$. (3.5.9)

The functional J can be written as

$$J(\tilde{u}) = \tfrac{1}{2}\int_0^1 \Big(\sum_{j=1}^{N} \tilde{u}_j \frac{d\phi_j}{dx}\Big)^2 dx - \int_0^1 f\Big(\sum_{j=1}^{N} \tilde{u}_j\phi_j\Big)dx = I(\tilde{u}_1,\ldots,\tilde{u}_N)$$

Differentiation with respect to $\tilde{u}_1,\ldots,\tilde{u}_N$ leads to the following system of equations

$$\frac{\partial I}{\partial \tilde{u}_i}(\tilde{u}_1,\ldots,\tilde{u}_N) = 0 \qquad i = 1,\ldots,N \tag{3.5.10}$$

which gives

$$\int_0^1 \Big(\sum_{j=1}^{N} \tilde{u}_j \frac{d\phi_j}{dx}\Big)\frac{d\phi_i}{dx}\,dx - \int_0^1 f\,\phi_i\,dx = 0 \qquad i = 1,\ldots,N$$

or equivalently

$$\sum_{j=1}^{N} \tilde{u}_j \int_0^1 \frac{d\phi_j}{dx}\frac{d\phi_i}{dx}\,dx = \int_0^1 f\,\phi_i\,dx \qquad i = 1,\ldots,N. \tag{3.5.11}$$

This is a system of N linear algebraic equations for N unknowns $\tilde{u}_1,\ldots,\tilde{u}_N$. In matrix notation we have

$$A\,\underline{\tilde{u}} = \underline{F}$$

with $A_{ij} = \int_0^1 \frac{d\phi_j}{dx}\frac{d\phi_i}{dx}\,dx$, $(\underline{\tilde{u}})_i = \tilde{u}_i$, $(\underline{F})_i = F_i = \int_0^1 f\,\phi_i\,dx$

We shall work out this system of equations for the special case of constant interval length $h = \frac{1}{N}$ and $f(x) = 1$. Consider the behavior of the basis function ϕ_i:

$$\phi_i(x) = \frac{x-x^{i-1}}{h} \qquad \frac{d\phi_i}{dx} = \frac{1}{h} \qquad \text{for} \quad x^{i-1} < x < x^i$$

$$\phi_i(x) = \frac{x^{i+1}-x}{h} \qquad \frac{d\phi_i}{dx} = -\frac{1}{h} \qquad \text{for} \quad x^i < x < x^{i+1}$$

$$\phi_i(x) = 0 \qquad \frac{d\phi_i}{dx} = 0 \qquad \text{for } x < x^{i-1} \quad \text{or} \quad x > x^{i+1}$$

from which we calculate, for fixed i with $2 \leqslant i \leqslant N-1$, the matrix elements A_{ij}.

$$A_{ij} = \int_0^1 \frac{d\phi_j}{dx}\frac{d\phi_i}{dx}\,dx = \int_{x^{i-1}}^{x^{i+1}} \left(\frac{d\phi_i}{dx}\right)^2 dx = 2h\,\frac{1}{h}\,\frac{1}{h} = \frac{2}{h} \quad \text{for } j=i$$

$$A_{ij} = \int_{x^i}^{x^{i+1}} \frac{d\phi_{i+1}}{dx}\,\frac{d\phi_i}{dx}\,dx = h\left(\frac{1}{h}\right)\left(-\frac{1}{h}\right) = -\frac{1}{h} \text{ for } j=i+1$$

$$A_{ij} = \int_{x^{i-1}}^{x^i} \frac{d\phi_{i-1}}{dx}\,\frac{d\phi_i}{dx}\,dx = h\left(-\frac{1}{h}\right)\left(\frac{1}{h}\right) = -\frac{1}{h} \text{ for } j=i-1$$

$$A_{ij} = 0 \qquad \text{for} \quad j < i-1 \quad \text{or} \quad j > i+1.$$

We thus have

$$A_{i\,i-1} = -\frac{1}{h}, \quad A_{ii} = \frac{2}{h}, \quad A_{i\,i+1} = -\frac{1}{h}, \quad A_{ij} = 0 \quad \text{for } |i-j|>1$$

For i=1 we find

$$A_{11} = \int_{x^0}^{x^2} \left(\frac{d\phi_1}{dx}\right)^2 dx = 2h\,\frac{1}{h^2} = \frac{2}{h}$$

$$A_{12} = \int_{x^1}^{x^2} \frac{d\phi_2}{dx}\frac{d\phi_1}{dx}\,dx = h(\frac{1}{h})(-\frac{1}{h}) = -\frac{1}{h}\,.$$

For j=N we find

$$A_{NN} = \int_{x^{N-1}}^{x^N} (\frac{d\phi_N}{dx})^2 dx = h.\frac{1}{h^2} = \frac{1}{h} \qquad \text{(see Fig. 3.8)}$$

$$A_{N\ N-1} = \int_{x^{N-1}}^{x^N} \frac{d\phi_{N-1}}{dx}\frac{d\phi_N}{dx}\,dx = h(-\frac{1}{h})(\frac{1}{h}) = -\frac{1}{h}\,.$$

Notice also that the matrix A is symmetric: $A_{ij} = A_{ji}$.
For the right hand side of (3.5.11) with f = 1 we obtain

$$F_i = \int_0^1 f\,\phi_i\,dx = \int_{x^{i-1}}^{x^i} \phi_i dx = h \qquad 1 \leqslant i \leqslant N-1$$

$$F_N = \int_0^1 f\,\phi_N\,dx = \int_{x^{N-1}}^{x^N} \phi_N\,dx = \tfrac{1}{2}h.$$

In matrix notation, the system of equations for $\tilde{u}_1,\ldots,\tilde{u}_N$ reads

$$A\,\underline{\tilde{u}} = \underline{F} \qquad (3.5.12)$$

with

$$A = \frac{1}{h}\begin{vmatrix} 2 & -1 & & & \\ -1 & 2 & -1 & & \\ & \cdot & \cdot & \cdot & \\ & & -1 & 2 & -1 \\ & & & -1 & 1 \end{vmatrix}, \quad \underline{\tilde{u}} = \begin{vmatrix} \tilde{u}_1 \\ \tilde{u}_2 \\ \cdot \\ \cdot \\ \tilde{u}_N \end{vmatrix}, \quad \underline{F} = \begin{vmatrix} h \\ h \\ \cdot \\ h \\ \tfrac{1}{2}h \end{vmatrix}.$$

The system of equations (3.5.12) is exactly the same as the one obtained by using the FDM to the same problem (see Section 2.2). The matrix A is sparse and positive definite, thus (3.5.12) can be solved in an efficient way.

So it turns out that the FEM is a method to transform a BVP for which a minimization formulation exists into a system of linear algebraic equations for the approximate solution. Due to the special choice of the basis functions, the matrix A is sparse (i.e. has many zero matrix elements) and is band structured when the nodal points are numbered appropriately. Although we have discussed a 1D example, the general procedure also applies to more dimensional problems as will be studied in Section 3.6 and Chapter 12.

The FEM has been introduced as a particular example of the Ritz method which is applicable to extremal formulations of PDEs. In the next section we shall study the Galerkin method which is a generalization of the Ritz method and also applicable to PDEs for which no extremal formulation exists (like, for instance, the convection-diffusion equation and the Navier-Stokes equations). Approximate solutions will then be studied using the same finite element procedure for the construction of the basis functions.

3.6 The Galerkin method

3.6.1 General procedure

In this section we shall study the Galerkin method, a method which is directly applicable to the BVP irrespective of the existence of an equivalent extremal formulation.
Let us consider the following linear PDE with homogeneous (essential) Dirichlet boundary conditions on Γ:

$$\left| \begin{array}{ll} \mathcal{L}u = f & \text{in } \Omega \\ u = 0 & \text{on } \Gamma \,. \end{array} \right. \tag{3.6.1}$$

The solution u is sought in a function space V_0 consisting of sufficiently smooth functions satisfying homogeneous

Dirichlet conditions on Γ. A precise formulation will be given in Chapter 11. We assume that the space V_0 has a countable basis $\phi_1, \phi_2, \ldots$, which means that any function $w \in V_0$ can be expressed as an infinite linear combination of basis functions. Formally we can write

$$w = \sum_{j=1}^{\infty} \alpha_j \phi_j$$

for any $w \in V_0$.

When an arbitrary $w \in V_0$ is substituted into (3.6.1), the equality is not satisfied. In fact,

$$\mathcal{L}w - f = R$$

where $R = R(w)$ is called the residual or error that results from taking w in stead of the solution u.

We now try to select an element $u \in V_0$ for which the residual $R(u)$ is zero. The residual is identically zero if its projection on each basis function is equal to zero. So we require that $R(u)$ satisfies:

$$\int_\Omega R(u)\, \phi_i \, d\Omega = 0 \qquad i = 1,2,\ldots$$

which implies

$$\int_\Omega (\mathcal{L}u - f)\, \phi_i \, d\Omega = 0 \qquad i = 1,2,\ldots \tag{3.6.2}$$

The order of differentiation in (3.6.2) can be lowered by performing an integration by parts (Green's formula). Substitution of the boundary conditions leads to an expression of the form

$$a(u, \phi_i) = \int_\Omega f\, \phi_i \, d\Omega \tag{3.6.3}$$

which must be satisfied for $i = 1,2,\ldots$ The form $a(.,.)$ is bilinear (see the examples below). Let us remark that by the linearity of (3.6.3) with respect to ϕ_i, (3.6.3) is equivalent to

$$\left| \begin{array}{l} \text{Find } u \in V_0 \text{ such that} \\ a(u,\phi) = \int_\Omega f\,\phi\,d\Omega \quad \text{for all } \phi \in V_0 \end{array} \right. \tag{3.6.4}$$

where the functions ϕ are called test functions.

The Galerkin method now consists in taking a finite dimensional subspace V_{ON} of V_0 spanned by N basis functions, say $\phi_1,\ldots,\phi_N$. The approximate problem can then be defined as

$$\left| \begin{array}{l} \text{Find } \tilde{u} \in V_{ON} \text{ such that} \\ a(\tilde{u},\phi) = \int_\Omega f\,\phi\,d\Omega \quad \text{for all } \phi \in V_{ON} \end{array} \right. \tag{3.6.5}$$

which is equivalent to

$$\left| \begin{array}{l} \text{Find } \tilde{u} \in V_{ON} \text{ such that} \\ a(\tilde{u},\phi_i) = \int_\Omega f\,\phi_i\,d\Omega \quad i = 1,2,\ldots,N\,. \end{array} \right. \tag{3.6.6}$$

The approximate solution $\tilde{u}$ being a function in V_{ON} has the form

$$\tilde{u} = \sum_{j=1}^{N} \alpha_j \phi_j. \tag{3.6.7}$$

Substitution of (3.6.7) into (3.6.6) leads to the following system of linear algebraic equations for $\alpha_1,\ldots,\alpha_N$:

$$\sum_{j=1}^{N} \alpha_j\, a(\phi_j,\phi_i) = \int_\Omega f\,\phi_i\,d\Omega \qquad i = 1,2,\ldots,N.$$

It may be clear that for the actual construction of the basis functions $\phi_1,\phi_2,\ldots,$ the FEM can be used most efficiently.

Problem (3.6.4) is known as the weak (or variational) formulation of problem (3.6.1), and it is obtained starting from the BVP formulation. In order to prove that both formulations, the BVP and the variational problem, are equivalent, we still have to prove that a solution of the variational problem is actually a solution of the original BVP. This point will be clarified in the examples (see also chapter 11).

When, instead of homogeneous Dirichlet conditions, the solution must satisfy inhomogeneous Dirichlet conditions $u = g_0$ on Γ, we write u as

$$u = G_0 + \sum_{j=1}^{\infty} \alpha_j \phi_j \tag{3.6.8}$$

where G_0 is such that $G_0|_\Gamma = g_0$ and $\phi_1, \phi_2, \ldots$ vanish on Γ. The (equivalent) weak formulation then will be

$$\left| \begin{array}{l} \text{Find } u \in V_{g_0} \text{ such that} \\ \\ a(u,\phi) = \int_\Omega f \, \phi \, d\Omega \qquad \text{for all } \phi \in V_0 \end{array} \right. \tag{3.6.9}$$

where V_{g_0} denotes the set of functions of the form (3.6.8):

$$V_{g_0} = G_0 + V_0$$

The approximate problem is formulated as follows:

$$\left| \begin{array}{l} \text{Find } \tilde{u} \in V_{g_0N} \text{ such that} \\ \\ a(\tilde{u},\phi) = \int_\Omega f \, \phi \, d\Omega \qquad \text{for all } \phi \in V_{0N} \end{array} \right. \tag{3.6.10}$$

where $V_{g_0N} = G_0 + V_{0N}$.

3.6.2 1D Poisson equation; Homogeneous boundary conditions

Consider the following Poisson equation in 1D

$$-\frac{d^2u}{dx^2} = f \qquad \text{on } (0,1) \tag{3.6.11}$$

with boundary conditions

$$u(0) = 0 \qquad \frac{du}{dx}(1) = 0 \ . \tag{3.6.12}$$

We consider the function space V of sufficiently smooth functions that vanish at x = 0. Let $\phi \in V$ be arbitrary. It follows from (3.6.11) that

$$-\int_0^1 \frac{d^2u}{dx^2}\,\phi\,dx = \int_0^1 f\,\phi\,dx \ .$$

Integration by parts of the left hand side gives:

$$\int_0^1 \frac{du}{dx}\frac{d\phi}{dx}\,dx - \left[\frac{du}{dx}\,\phi\right]_0^1 = \int_0^1 f\,\phi\,dx \ . \tag{3.6.13}$$

Since $\phi(0) = 0$ and $\frac{du}{dx}(1) = 0$, (3.6.13) reduces to

$$\int_0^1 \frac{du}{dx}\frac{d\phi}{dx}\,dx = \int_0^1 f\,\phi\,dx.$$

The variational formulation of problem (3.6.11), (3.6.12) reads now

$$\left|\begin{array}{l} \text{Find } u \in V \text{ such that} \\ a(u,\phi) = \int_0^1 f\,\phi\,dx \qquad \text{for all } \phi \in V \end{array}\right. \tag{3.6.14}$$

with $a(u,\phi) = \int_0^1 \frac{du}{dx}\frac{d\phi}{dx}\,dx.$

The justification for this formulation is the equivalence of (3.6.11), (3.6.12) to (3.6.14). It is evident that a solution of (3.6.11), (3.6.12) is a solution of (3.6.14). Let on the other hand u be a solution of (3.6.14), then integration by parts gives

$$\int_0^1 \left(-\frac{d^2u}{dx^2} - f\right)\phi\,dx + \left[\frac{du}{dx}\,\phi\right]_0^1 = 0$$

for all $\phi \in V$, from which we deduce, since $\phi(0) = 0$,

$$\int_0^1 \left(-\frac{d^2u}{dx^2} - f\right) \phi \, dx + \frac{du}{dx}(1)\, \phi(1) = 0 \tag{3.6.15}$$

for all $\phi \in V$. Relations (3.6.15) are certainly verified for all ϕ satisfying $\phi(0) = \phi(1) = 0$, so that

$$\int_0^1 \left(-\frac{d^2u}{dx^2} - f\right) \phi \, dx = 0 \qquad \text{for all } \phi \text{ with } \phi(0) = \phi(1) = 0$$

from which follows that

$$-\frac{d^2u}{dx^2} = f \quad \text{on } (0,1).$$

Substitution into (3.6.15) leads to

$$\frac{du}{dx}(1)\, \phi(1) = 0 \qquad \text{for all } \phi \in V$$

consequently

$$\frac{du}{dx}(1) = 0.$$

This proves the equivalence of (3.6.11), (3.6.12) to (3.6.14). Notice that the Neumann condition at $x = 1$ is a natural boundary condition since it is automatically satisfied.

In the space V we choose a finite number of basis functions $\phi_1,\ldots,\phi_N$. When the function space spanned by these basis functions is denoted by V_N, the approximate variational formulation reads:

$$\left|\begin{array}{ll} \text{Find } \tilde{u} \in V_N \text{ such that} & \\ a(\tilde{u},\phi) = \int_0^1 f\, \phi\, dx & \text{for all } \phi \in V_N. \end{array}\right. \tag{3.6.16}$$

Writing $\tilde{u}$ in the form

$$\tilde{u} = \sum_{j=1}^{N} \alpha_j \phi_j$$

then, (3.6.16) is equivalent to the following system of equations for $\alpha_1, \ldots, \alpha_N$

$$\sum_{j=1}^{N} \alpha_j \int_0^1 \frac{d\phi_j}{dx} \frac{d\phi_i}{dx} \, dx = \int_0^1 f \, \phi_i \, dx \qquad i = 1,2,\ldots,N \quad (3.6.17)$$

This is precisely the same system as obtained by using the Ritz method to the same BVP (compare (3.6.17) with (3.4.11)).

The FEM can be used to choose the basis functions. We subdivide the interval (0,1) into N subregions (subintervals), we choose the extremities of the subintervals as nodal points $0 = x^0 < x^1 < \ldots x^N = 1$. The basis functions ϕ_i, $i=1,2,\ldots,N$ are then completely determined by the following three properties

(i) $\phi_i(x^j) = \delta_{ij} \qquad i,j = 1,2,\ldots,N$

(ii) ϕ_i is linear on each subinterval

(iii) ϕ_i is continuous on $[0,1]$.

Then

$$\tilde{u}(x) = \sum_{j=1}^{N} \tilde{u}_j \phi_j(x)$$

and the system of linear equations for $\tilde{u}_1, \ldots, \tilde{u}_N$ becomes

$$\sum_{j=1}^{N} \tilde{u}_j \int_0^1 \frac{d\phi_j}{dx} \frac{d\phi_i}{dx} \, dx = \int_0^1 f \, \phi_i \, dx \qquad i = 1,2,\ldots,N \qquad (3.6.18)$$

which is exactly the same as system (3.5.11) obtained using the minimization formulation. When $f \equiv 1$ we obtain (3.5.12).

3.6.3 1D Poisson equation; non-homogeneous boundary conditions.

Consider the Poisson equation

$$-\frac{d^2u}{dx^2} = f \qquad \text{on} \quad (0,1) \tag{3.6.19}$$

with

$$u(0) = g_0 \qquad \frac{du}{dx}(1) = g_1 \tag{3.6.20}$$

We carry out the following procedure:

(i) multiply (3.6.19) by a function ϕ with $\phi(0) = 0$

(ii) integrate over (0,1)

(iii) apply Green's formula

(iv) substitute the non-homogeneous Neumann condition at x=1.

This gives subsequently

$$-\int_0^1 \frac{d^2u}{dx^2}\,\phi\,dx = \int_0^1 f\,\phi\,dx$$

$$\int_0^1 \frac{du}{dx}\frac{d\phi}{dx}\,dx - \left[\frac{du}{dx}\,\phi\right]_0^1 = \int_0^1 f\,\phi\,dx$$

$$\int_0^1 \frac{du}{dx}\frac{d\phi}{dx}\,dx = \int_0^1 f\,\phi\,dx + g_1\,\phi(1). \tag{3.6.21}$$

Let ϕ_0 be a function with $\phi_0(0) = 1$, then the variational formulation is:

Find u with $u - g_0\phi_0 \in V$ such that

$$a(u,\phi) = \int_0^1 f\,\phi\,dx + g_1\phi(1) \qquad \text{for all } \phi \in V \tag{3.6.22}$$

where: V = space of functions which vanish at $x = 0$.

$$a(u,\phi) = \int_0^1 \frac{du}{dx}\frac{d\phi}{dx}\,dx\ .$$

Using similar arguments as in Section 3.6.2 it can be shown that the BVP (3.6.19), (3.6.20) is equivalent to the variational problem (3.6.22). Notice that the Dirichlet condition is an essential boundary condition and that the Neumann condition is a natural one since it is automatically verified by the solution u.

Let us take the same finite element basis functions as in the preceding example, then

$$\tilde{u}(x) = g_0\phi_0 + \sum_{j=1}^{N} \tilde{u}_j\phi_j(x)$$

and the system of linear algebraic equations for $\tilde{u}_1,\ldots,\tilde{u}_N$ is given by

$$\sum_{j=1}^{N} \tilde{u}_j \int_0^1 \frac{d\phi_j}{dx}\frac{d\phi_i}{dx}\,dx = \int_0^1 f\,\phi_i\,dx - g_0\int_0^1 \frac{d\phi_0}{dx}\frac{d\phi_i}{dx}\,dx + g_1\phi_i(1)$$

$$i = 1,2,\ldots,N. \qquad (3.6.23)$$

This system of equations differs only slightly from the one obtained with homogeneous Dirichlet-Neumann boundary conditions. Only the terms

$$g_0\int_0^1 \frac{d\phi_0}{dx}\frac{d\phi_i}{dx}\,dx \qquad \text{and} \qquad g_1\phi_i(1) \qquad (3.6.24)$$

have been added to the right hand side.
When we take for ϕ_0 the basis function corresponding to point $x_0 = 0$ (see Fig. 3.8), then the first term of (3.6.24) is different from zero only when $i = 1$. The second term of (3.6.24) has a non-zero contribution to the right hand side of (3.6.23) when $i = N$.
Thus, compared with the homogeneous boundary conditions case only the right hand side of the first and of the N-th equation is changed. We calculate these quantities for the special case of an equidistant mesh: $h = \frac{1}{N}$ and $f \equiv 1$. Then for $i = 1$

$$\int_0^1 f\,\phi_1\,dx - g_0\int_0^1 \frac{d\phi_0}{dx}\frac{d\phi_1}{dx}\,dx + g_1\phi_1(1) = h + \frac{g_0}{h} + 0 = h + \frac{g_0}{h}$$

and for $i = N$

$$\int_0^1 f\ \phi_N dx - g_0 \int_0^1 \frac{d\phi_0}{dx}\frac{d\phi_N}{dx}\,dx + g_1\phi_N(1) = \tfrac{1}{2}h - 0 + g_1 = \tfrac{1}{2}h + g_1$$

When we compare these results to those obtained in Section 2.2 using the FDM we see that the systems of equations are identical.

3.6.4 2D problem

Consider the following two-dimensional problem in $\Omega \subset \mathbb{R}^2$ with mixed Dirichlet-Neumann-Robbins boundary conditions. The boundary Γ of Ω consists of three disjoint parts: Γ_0, Γ_1 and Γ_2, as indicated in Fig. 3.11.

$$\left|\begin{array}{ll} -\Delta u + u = f & \text{in } \Omega \\ u = 0 & \text{on } \Gamma_0 \\ \dfrac{\partial u}{\partial n} = g_1 & \text{on } \Gamma_1 \\ \dfrac{\partial u}{\partial n} + u = g_2 & \text{on } \Gamma_2 \end{array}\right. \qquad (3.6.25)$$

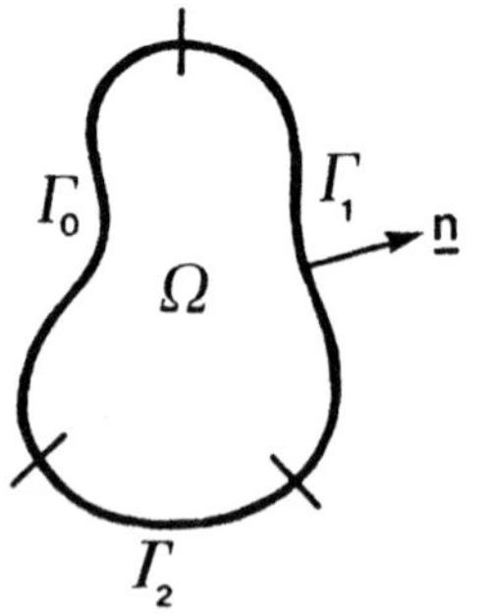

Fig. 3.11 Region Ω and boundary $\Gamma = \Gamma_0 \cup \Gamma_1 \cup \Gamma_2$.

In order to find the variational formulation corresponding to this BVP, we

(i) multiply the PDE by a function ϕ which vanishes on Γ_0

(ii) integrate over Ω

(iii) apply Green's formula

(iv) substitute the Neumann and Robbins boundary conditions.

This gives

$$\int_\Omega (-\Delta u+u)\ \phi\ d\Omega = \int_\Omega f\ \phi\ d\Omega$$

$$\int_\Omega (\text{grad } u.\text{grad}\phi + u\phi)d\Omega - \int_\Gamma \frac{\partial u}{\partial n}\ \phi\ d\Gamma = \int_\Omega f\ \phi\ d\Omega.$$

Concerning the integral over Γ we have

$$\int_{\Gamma_0} \frac{\partial u}{\partial n}\ \phi\ d\Gamma = 0 \qquad \text{since} \quad \phi = 0 \quad \text{on} \quad \Gamma_0$$

$$\int_{\Gamma_1} \frac{\partial u}{\partial n}\ \phi\ d\Gamma = \int_{\Gamma_1} g_1\ \phi\ d\Gamma$$

$$\int_{\Gamma_2} \frac{\partial u}{\partial n}\ \phi\ d\Gamma = -\int_{\Gamma_2} u\ \phi\ d\Gamma + \int_{\Gamma_2} g_2\ \phi\ d\Gamma\ .$$

When V denotes the function space consisting of functions which are equal to zero on Γ_0, then the variational formulation is as follows:

Find $u \in V$ such that

$$a(u,\phi) = \int_\Omega f\ \phi\ d\Omega + \int_{\Gamma_1} g_1\ \phi\ d\Gamma + \int_{\Gamma_2} g_2\ \phi\ d\Gamma \qquad (3.6.26)$$

for all $\phi \in V$

with $a(u,\phi) = \int_\Omega (\text{grad } u.\text{grad } \phi + u\phi)d\Omega + \int_{\Gamma_2} u\ \phi\ d\Gamma$

Using techniques similar to those employed in Section 3.2 starting at formula (3.2.9), the equivalence of (3.6.25) to (3.6.26) can be proved.

Next we restrict the formulation (3.6.26) to a finite number of basis functions $\phi_1,\phi_2,\ldots\phi_N$ in V. We assume that $\phi_i|_{\Gamma_0} = 0$ for $i = 1,2,\ldots,N$ and denote by V_N the space spanned by $\phi_1,\ldots,\phi_N$.

The approximate variational formulation is

$$\left|\begin{array}{l} \text{Find } \tilde{u} \in V_N \text{ such that} \\ a(\tilde{u},\phi) = \int_\Omega f\,\phi\,d\Omega + \int_{\Gamma_1} g_1\phi\,d\Gamma + \int_{\Gamma_2} g_2\phi\,d\Gamma \\ \text{for all } \phi \in V_N \end{array}\right. \tag{3.6.27}$$

which amounts to saying that (3.6.27) must be verified for each basis function ϕ_i in V_N, $i = 1,2,\ldots,N$.

The solution $\tilde{u}$, being a function in V_N, can be written as

$$\tilde{u}(x) = \sum_{j=1}^{N} \alpha_j\phi_j(x)$$

from which we deduce the following system of linear algebraic equations for the unknowns $\alpha_1,\alpha_2,\ldots,\alpha_N$:

$$\sum_{j=1}^{N} \alpha_j\left\{\int_\Omega \text{grad}\,\phi_j\cdot\text{grad}\,\phi_i d\Omega + \int_\Omega \phi_j\phi_i d\Omega + \int_{\Gamma_2} \phi_j\phi_i d\Gamma\right\} =$$
$$= \int_\Omega f\,\phi_i d\Omega + \int_{\Gamma_1} g_1\phi_i d\Gamma + \int_{\Gamma_2} g_2\phi_i d\Gamma \qquad i = 1,2,\ldots,N. \tag{3.6.28}$$

A finite element approximation of the problem can be defined as follows. We assume that Ω is a polygonal region which is subdivided into a finite number of triangles satisfying (3.5.1) (see Fig. 3.12).

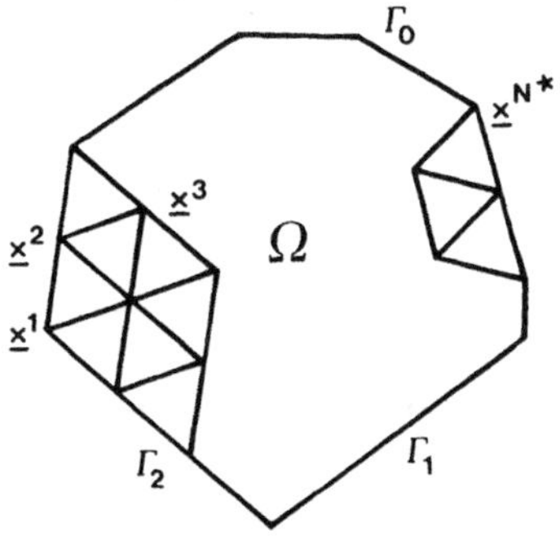

Fig. 3.12 Triangulation of Ω.

Assume that there are N^* vertices of triangles: $\underline{x}^1, \underline{x}^2, \ldots, \underline{x}^{N^*}$. These N^* points form the collection of nodal points. For each nodal point $\underline{x}^i$ we define a basis function ϕ_i by the following three properties (see Fig. 3.13):

(i) $\phi_i(\underline{x}^j) = \delta_{ij} \qquad i,j = 1,2,\ldots,N^*$

(ii) ϕ_i is linear on each triangle

(iii) ϕ_i is continuous on $\bar{\Omega}$.

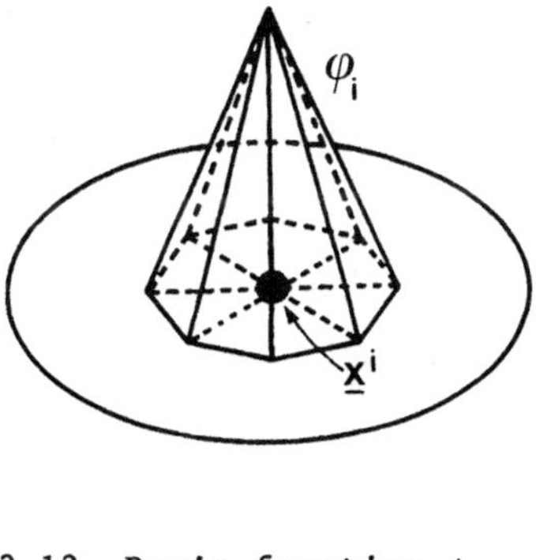

Fig. 3.13 Basis function ϕ_i.

An arbitrary linear combination of the functions ϕ_i can be written as

$$\tilde{u}(x) = \sum_{j=1}^{N^*} \tilde{u}_j \phi_j(x)$$

with $\tilde{u}_j = \tilde{u}(\underline{x}^j)$.

Since the approximate solution must satisfy the Dirichlet conditions on Γ_0, we take $\tilde{u}_j = 0$ for those values of j for which $\underline{x}^j \in \Gamma_0$. A renumbering of the nodal points now leads to the following general form of an approximate solution

$$\tilde{u}(x) = \sum_{j=1}^{N} \tilde{u}_j \phi_j(x) \qquad (3.6.29)$$

where N is the number of nodal points which do not belong to Γ_0. The final system of linear algebraic equations for $\tilde{u}_1,\ldots,\tilde{u}_N$ is, of course, identical to (3.6.28) with $\alpha_i = \tilde{u}_i$. In matrix notation we have

$$A\,\underline{\tilde{u}} = \underline{F} \qquad \tilde{u} = (\tilde{u}_1,\ldots,\tilde{u}_N)^T$$

where the ij-th matrix element of A is given by

$$A_{ij} = \int_\Omega \text{grad}\ \phi_j \cdot \text{grad}\ \phi_i d\Omega + \int_\Omega \phi_j \phi_i d\Omega + \int_{\Gamma_2} \phi_j \phi_i d\Gamma\ .$$

The line integral corresponds to the Robbins boundary condition on Γ_2.

The right hand side vector $\underline{F} = (F_1,\ldots,F_N)^T$ is given by

$$F_i = \int_\Omega f\ \phi_i d\Omega + \int_{\Gamma_1} g_1 \phi_i d\Gamma + \int_{\Gamma_2} g_2 \phi_i d\Gamma$$

where the line integral over Γ_1 is due to the Neumann condition on Γ_1 and the line integral over Γ_2 to the Robbins boundary condition on Γ_2. It is easy to see that the matrix A is positive definite. In fact: A is symmetric and

$$\tilde{\underline{u}}^T A \tilde{\underline{u}} = \sum_{i=1}^{N} \tilde{u}_i [\sum_{j=1}^{N} \tilde{u}_j \{\int_\Omega \text{grad } \phi_j \cdot \text{grad } \phi_i d\Omega + \int_\Omega \phi_j \phi_i d\Omega + \int_{\Gamma_2} \phi_j \phi_i d\Gamma\}] =$$

$$= \int_\Omega |\text{grad } \tilde{u}|^2 d\Omega + \int_\Omega \tilde{u}^2 d\Omega + \int_{\Gamma_2} \tilde{u}^2 d\Gamma > 0$$

for all $\tilde{u}$ of the form (3.6.29). Moreover the matrix A is sparse and has a band structure when the nodal points are numbered appropriately.

For this 2D example with mixed boundary conditions, we see again that a combination of the Galerkin method and the FEM leads to a system of linear equations with nice numerical properties. Notice also that we did not use the minimization formulation of the problem, but that we worked directly on the BVP (3.6.25) itself.

3.7 Comments

We can conclude that the FEM is a most attractive method to define the basis functions for the Ritz and the Galerkin method. The resulting system of equations for the unknown parameters has a sparse coefficient matrix and properties which can be proved easily; the matrix can, for instance, be symmetric, positive definite and can have a band structure.

The examples have shown also that if the Ritz method is applicable (i.e. if the BVP admits a minimization formulation), then the Ritz method combined with the FEM leads to the same system of equations as the Galerkin method combined with the FEM. The Galerkin method, however, is applicable to a larger class of problems.

For the 1D Poisson equation with right hand side $f \equiv 1$, we have seen that the FEM results into the same system of equations as the FDM. In this case the accuracy of the approximate solution is of the same order for both methods.

From the point of view of numerical efficiency there are however practical differences between the FEM and FDM. Complex geometries can be handled by the FDM using complicated interpolation formulae (especially for Neumann and Robbins boundary conditions), while for the FEM a geometrically complex region Ω gives no essential problems. An other difference between the FEM and the FDM is that the FEM is well suited for being automized, because, as we shall see in the next chapter, a repeated computation of the same type of integrals must be performed. The FDM is difficult to automize because of the interpolation formulae at the boundary of the region.

In the next chapter we shall construct various kinds of finite elements, while, in chapter 5 we shall deal with the practical aspects of the FEM, such as the most efficient way to build up the system of equations, error estimates and solution methods for systems of (non-) linear equations.

Finally let us remark that in the literature the matrix A is known as the stiffness (or large) matrix and that $\underline{F}$ is called the right hand side (or load) vector. The unknown parameters are stored in the vector $\underline{\tilde{u}}$ of which we sometimes drop the tilde.

Chapter 4

Construction of Finite Elements

Introduction

In this chapter we shall construct various finite elements in both the 1D, 2D and 3D case. In Section 4.1 the construction of quadratic and cubic 1D elements will be discussed. To facilitate the construction of these elements we shall introduce the notion of barycentric coordinates. Using barycentric coordinates in 2D we shall construct, in Section 4.2, triangular finite elements with linear, quadratic and extended quadratic basis functions. The 3D situation will be treated in Section 4.3, where two linear tetrahedral finite elements will be introduced. In Section 4.4 we introduce coordinate and element transformation and the notion of isoparametric finite element. Section 4.5 discusses two types of 2D quadrilateral finite elements, the 3D hexahedral case will be the subject of Section 4.6.

4.1 Linear, quadratic and cubic basis functions in 1D

In Chapter 3 we have introduced the 1D finite element with piecewise linear basis functions. This means: the shape functions (or admissible approximate solutions) are continuous on $\bar{\Omega}$ and linear on each subregion.

In this section we shall introduce 1D polynomial basis functions of second and third degree. It will be seen that the choice of barycentric (or natural, or area) coordinates influences the amount of computations required for the construction of basis functions. The use of these barycentric coordinates in deriving shape functions is particular advantageous because special closed-form integration formulas can be used to evaluate the integrals over subregions (see Chapter 5).

Let us consider an arbitrary interval $e = [x^1, x^2] \subset [0,1]$. We define the following functions λ_1 and λ_2 from $[x^1, x^2]$ into $\mathbb{R}^n$ by

$$\lambda_1(x) = \frac{x^2 - x}{x^2 - x^1} \qquad \lambda_2(x) = \frac{x - x^1}{x^2 - x^1} \,. \tag{4.1.1}$$

These functions have the following properties:

(i) λ_i is linear on e, $i = 1,2$

(ii) $\lambda_i(x^j) = \delta_{ij}$ $\quad i,j = 1,2$

(iii) $\lambda_1(x) + \lambda_2(x) = 1$ $\quad$ for all $x \in e$.

The relation of the functions λ_1, λ_2 to the 1D piecewise linear basis functions is evident. In fact, the basis function ϕ_i corresponding to x^i, $i = 1,2$ satisfies:

(i) ϕ_i is linear on each subinterval

(ii) ϕ_i is continuous on $[0,1]$

(iii) $\phi_i(x^j) = \delta_{ij}$ $\quad i,j = 1,2$

so that the restriction of ϕ_i to e is precisely the function λ_i (see Fig. 4.1 (a) and (b))

$$\phi_i|_e = \lambda_i.$$

For any point $x \in e$ we can calculate the values of the functions λ_1 and λ_2. The pair $\{\lambda_1(x), \lambda_2(x)\}$ is called the barycentric coordinates of the point x on e with respect to the

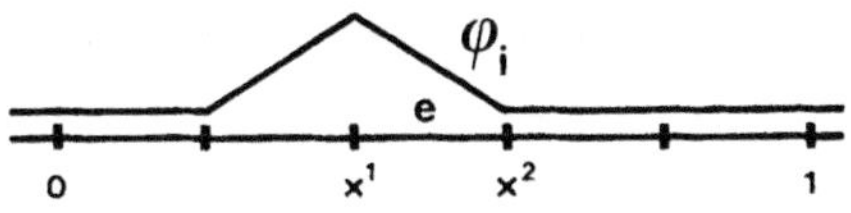

Fig. 4.1 (a) Basis function ϕ_i

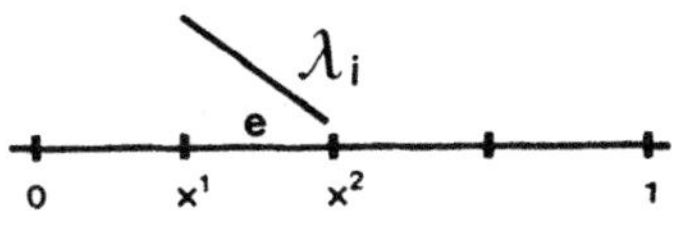

(b) The function λ_i.

points x^1 and x^2. The 1D linear finite element is now defined as: (i) a subdivision of $[0,1]$ into subintervals, (ii) on each subinterval we choose two nodal points: the end points x^1 and x^2 of the subinterval, (iii) on each subinterval we define for each nodal point its basis function: on $[x^1,x^2]$ for instance

$$\phi_1 = \lambda_1 \qquad \phi_2 = \lambda_2 \,. \tag{4.1.2}$$

The function space spanned by ϕ_1 and ϕ_2 on e is denoted by $P_1(e)$, and contains all polynomials of degree ≤ 1 in x_1. Generally we denote by $P_k(e)$, k non-negative integer, $e \subset \mathbb{R}^n$, the function space of polynomials of degree $\leq k$ in $x_1, x_2, \ldots, x_n$. In other words, $P_k(e)$ is the function space spanned by $x_1^{k_1} x_2^{k_2} \ldots x_n^{k_n}$ with $k_i \geq 0$, $i = 1,\ldots,n$, $k_1+k_2+\ldots+k_n \leq k$. It follows that the shape function $\tilde{u}$ takes the following form on e (see Fig. 4.2):

$$\begin{aligned}\tilde{u}(x) &= \tilde{u}_1 \phi_1(x) + \tilde{u}_2 \phi_2(x)\\ &= \tilde{u}_1 \lambda_1(x) + \tilde{u}_2 \lambda_2(x) \,.\end{aligned}$$

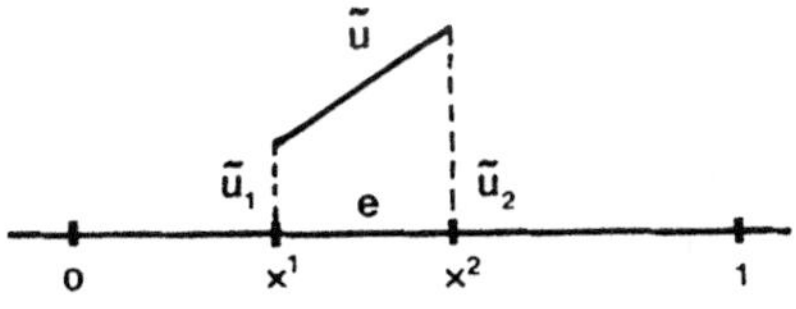

Fig. 4.2 Linear shape function $\tilde{u}$ on e.

Notice that, since the basis functions are linear on e, their derivatives are constant:

$$\frac{d\phi_1}{dx} = \frac{d\lambda_1}{dx} = -\frac{1}{x^2-x^1} \qquad \frac{d\phi_2}{dx} = \frac{d\lambda_2}{dx} = \frac{1}{x^2-x^1}\,. \tag{4.1.3}$$

The barycentric coordinates $\{\lambda_1,\lambda_2\}$ can be used to define higher order basis functions in a very efficient way.

Let us consider 1D basis functions that are quadratic on each subinterval. We assume that the interval $[0,1]$ is subdivided into subintervals. To define uniquely a quadratic function on a subinterval $e = [x^1,x^2]$ we must fix its values in three different points. For this we choose on each subinterval e the following nodal points: the two end points which we call x^1 and x^2 and the mid-point $x^{12} = \frac{1}{2}(x^1+x^2)$ (see Fig. 4.3).

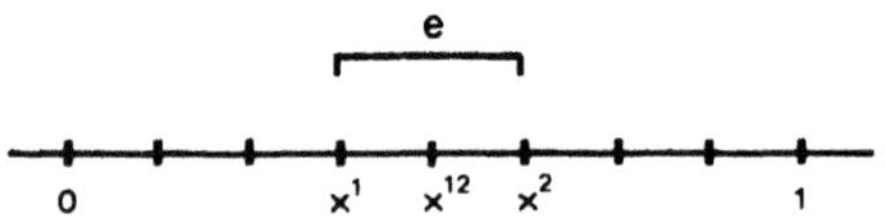

Fig. 4.3 Element e with nodal points x^1, x^2 and x^{12}.

Let $\{\lambda_1(x),\lambda_2(x)\}$ be the barycentric coordinates of a point $x \in e$ with respect to x^1, x^2. The function λ_i is linear and $\lambda_i(x^j) = \delta_{ij}$, $i,j = 1,2$. Moreover $\lambda_1(x^{12}) = \lambda_2(x^{12}) = \frac{1}{2}$. Next we notice that $\lambda_1(x^2) = 0$ and $\lambda_1(x^{12})-\frac{1}{2} = 0$. Hence the function $\lambda_1(x)(\lambda_1(x)-\frac{1}{2})$ vanishes at $x = x^2$ and $x = x^{12}$ and is quadratic since both λ_1 and $\lambda_1-\frac{1}{2}$ are linear. For $x = x^1$ the function takes the value $\lambda_1(x^1)(\lambda_1(x^1)-\frac{1}{2}) = 1\,(1-\frac{1}{2}) = \frac{1}{2}$.
From this we deduce that the function

$$\phi_1(x) = 2\lambda_1(x)(\lambda_1(x)-\tfrac{1}{2}) \tag{4.1.4}$$

is the quadratic basis function on e corresponding to point x^1.

Similarly we find that

$$\phi_2(x) = 2\lambda_2(x)(\lambda_2(x)-\tfrac{1}{2}) \tag{4.1.5}$$

is the quadratic basis function on e corresponding to point x^2. For point x^{12} we remark that $\lambda_1(x^2) = 0$, $\lambda_2(x^1) = 0$ and that $\lambda_1(x^{12})\ \lambda_2(x^{12}) = \tfrac{1}{2}\ \tfrac{1}{2} = \frac{1}{4}$ which makes that the function

$$\phi_{12}(x) = 4\lambda_1(x)\ \lambda_2(x) \tag{4.1.6}$$

is the quadratic basis function on e corresponding to point x^{12}. The three basis functions are depicted in Fig. 4.4.

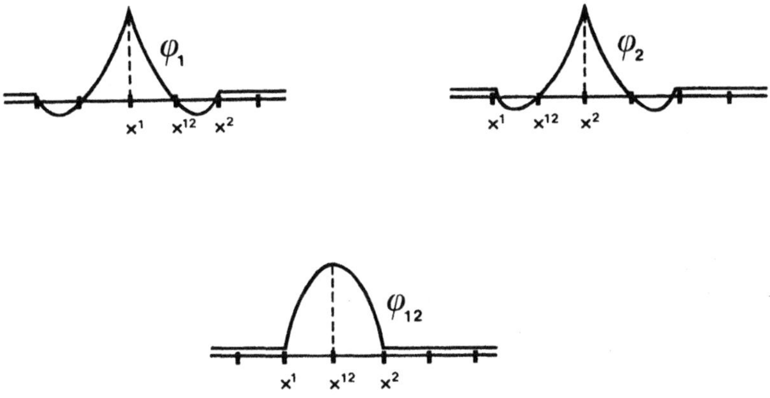

Fig. 4.4 Quadratic basis functions.

The function space spanned by ϕ_1, ϕ_2 and ϕ_{12} on e is denoted by $P_2(e)$ and consists of all polynomials of degree ≤ 2 in x_1.
The quadratic shape function takes now the following parabolic form on e (see Fig. 4.5)

$$\tilde{u}(x) = \tilde{u}_1\phi_1(x) + \tilde{u}_2\phi_2(x) + \tilde{u}_{12}\phi_{12}(x)$$

$$= \tilde{u}_1\lambda_1(x)(2\lambda_1(x)-1) + \tilde{u}_2\lambda_2(x)(2\lambda_2(x)-1) + 4\tilde{u}_{12}\lambda_1(x)\lambda_2(x).$$

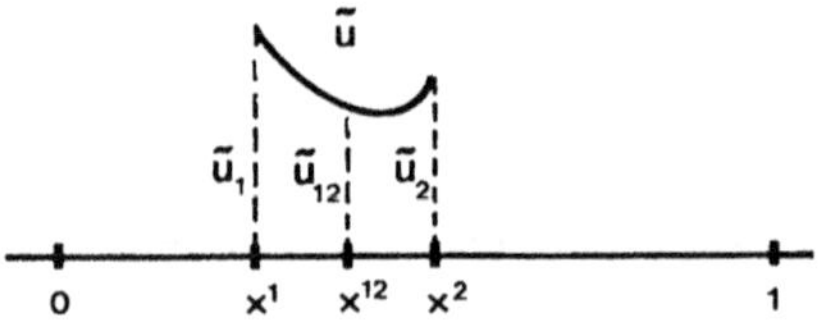

Fig. 4.5 Quadratic shape function $\tilde{u}$ on e.

One easily verifies that the 1D cubic basis functions on an element $e = [x^1, x^2]$ and nodal points x^1, x^2, $x^{12} = \frac{1}{3}(2x^1+x^2)$, $x^{21} = \frac{1}{3}(x^1+2x^2)$ (see Fig. 4.6)

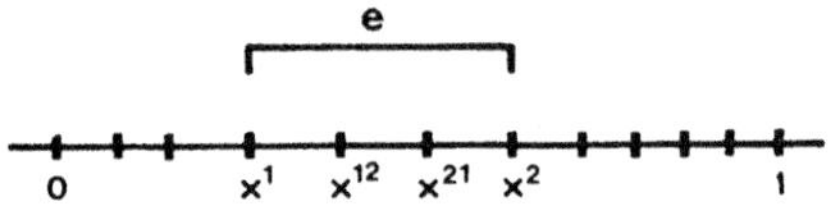

Fig. 4.6 Subinterval e with four nodal points.

are given by

$$\begin{aligned}
\phi_1 &= \tfrac{1}{2}\lambda_1(3\lambda_1-2)(3\lambda_1-1) \\
\phi_2 &= \tfrac{1}{2}\lambda_2(3\lambda_2-2)(3\lambda_2-1) \\
\phi_{12} &= \frac{9}{2}\lambda_1\lambda_2(3\lambda_1-1) \\
\phi_{21} &= \frac{9}{2}\lambda_1\lambda_2(3\lambda_2-1).
\end{aligned} \tag{4.1.7}$$

Notice that the points x^1, x^2, x^{12}, x^{21} have the following barycentric coordinates

$$\begin{aligned} x^1 &: \{1,0\} \\ x^2 &: \{0,1\} \\ x^{12} &: \{\tfrac{2}{3},\tfrac{1}{3}\} \\ x^{21} &: \{\tfrac{1}{3},\tfrac{2}{3}\}. \end{aligned} \tag{4.1.8}$$

4.2 Triangular basis functions in 2D

4.2.1 Barycentric coordinates

In this section we shall use barycentric coordinates to define linear, quadratic and extended quadratic basis functions in 2D.

Let Ω be a 2D region which is subdivided into a finite number of triangles e_k, $k = 1,2,\ldots,K$, satisfying the properties (3.5.1). A linear function on a triangle e is completely determined by its values in three non-collinear points, for example the vertices $\underline{x}^1$, $\underline{x}^2$ and $\underline{x}^3$ with coordinates $\underline{x}^1 = (x_1^1,x_2^1)$, $\underline{x}^2 = (x_1^2,x_2^2)$, $\underline{x}^3 = (x_1^3,x_2^3)$.
We shall now define three functions $\lambda_i = \lambda_i(\underline{x})$, $i = 1,2,3$, on e by the following requirements

(i) $\lambda_i(\underline{x}^j) = \delta_{ij}$ $\qquad i,j = 1,2,3$

(ii) λ_i is linear on e.

These functions λ_i can be calculated as follows: Take for instance λ_1. Since λ_1 is linear it takes the following form:

$$\lambda_1(\underline{x}) = \alpha_0 + \alpha_1 x_1 + \alpha_2 x_2 \qquad \underline{x} = (x_1,x_2)$$

Moreover we have

$$\begin{aligned}\lambda_1(\underline{x}^1) &= \alpha_0 + \alpha_1 x_1^1 + \alpha_2 x_2^1 = 1 \\ \lambda_1(\underline{x}^2) &= \alpha_0 + \alpha_1 x_1^2 + \alpha_2 x_2^2 = 0 \\ \lambda_1(\underline{x}^3) &= \alpha_0 + \alpha_1 x_1^3 + \alpha_2 x_2^3 = 0 \;.\end{aligned} \tag{4.2.1}$$

System (4.2.1) can be solved for α_0, α_1 and α_2; we find

$$\alpha_0 = \frac{x_1^2 x_2^3 - x_1^3 x_2^2}{\Delta} \qquad \alpha_1 = \frac{x_2^2 - x_2^3}{\Delta} \qquad \alpha_2 = \frac{x_1^3 - x_1^2}{\Delta}$$

with

$$\begin{aligned}\Delta &= (x_1^3 - x_1^2)(x_2^1 - x_2^2) - (x_2^2 - x_2^3)(x_1^2 - x_1^1) \\ &= \pm \text{ twice the area of } e.\end{aligned}$$

Similar expressions can be found for λ_2 and λ_3. We easily verify that the following relation holds

$$\lambda_1(\underline{x}) + \lambda_2(\underline{x}) + \lambda_3(\underline{x}) = 1 \qquad \text{for all } \underline{x} \in e. \tag{4.2.2}$$

For any point $\underline{x} \in e$ we can calculate the values of the functions λ_1, λ_2 and λ_3. Now, similar to the definition in the 1D case, we call the triple $\{\lambda_1(\underline{x}), \lambda_2(\underline{x}), \lambda_3(\underline{x})\}$ the barycentric coordinates of the point $\underline{x} \in e$ with respect to the vertices $\underline{x}^1, \underline{x}^2, \underline{x}^3$. Let us calculate the barycentric coordinates of some particular points in a triangle e (see Fig. 4.7)

$$\begin{aligned}\underline{x}^{12} &= \tfrac{1}{2}(\underline{x}^1 + \underline{x}^2) \quad \text{mid-point of segment } [\underline{x}^1, \underline{x}^2] \\ \underline{x}^{13} &= \tfrac{1}{2}(\underline{x}^1 + \underline{x}^3) \quad \text{" \quad " } [\underline{x}^1, \underline{x}^3] \\ \underline{x}^{23} &= \tfrac{1}{2}(\underline{x}^2 + \underline{x}^3) \quad \text{" \quad " } [\underline{x}^2, \underline{x}^3] \\ \underline{x}^{123} &= \tfrac{1}{3}(\underline{x}^1 + \underline{x}^2 + \underline{x}^3) \quad \text{barycentre of } e\end{aligned}$$

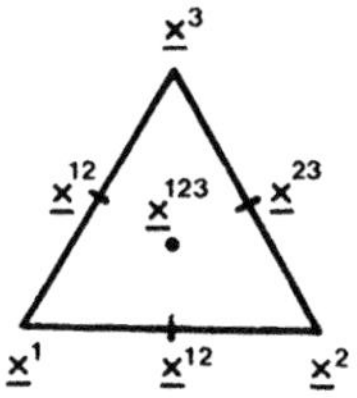

$\underline{x}^{12} = \frac{1}{2}(\underline{x}^1 + \underline{x}^2)$

$\underline{x}^{13} = \frac{1}{2}(\underline{x}^1 + \underline{x}^3)$

$\underline{x}^{23} = \frac{1}{2}(\underline{x}^2 + \underline{x}^3)$

$\underline{x}^{123} = \frac{1}{3}(\underline{x}^1 + \underline{x}^2 + \underline{x}^3)$

Fig. 4.7 Triangle e with particular points.

$$\underline{x}^1 \rightarrow \{1,0,0\}$$

$$\underline{x}^2 \rightarrow \{0,1,0\}$$

$$\underline{x}^3 \rightarrow \{0,0,1\}$$

$$\underline{x}^{12} \rightarrow \{\tfrac{1}{2},\tfrac{1}{2},0\}$$

$$\underline{x}^{13} \rightarrow \{\tfrac{1}{2},0,\tfrac{1}{2}\}$$

$$\underline{x}^{23} \rightarrow \{0,\tfrac{1}{2},\tfrac{1}{2}\}$$

$$\underline{x}^{123} \rightarrow \{\tfrac{1}{3},\tfrac{1}{3},\tfrac{1}{3}\} \ .$$

Notice also that points on the following segments satisfy a relation expressed in the barycentric coordinates:

$$\underline{x} \in [\underline{x}^1,\underline{x}^2] \iff \lambda_3(\underline{x}) = 0$$

$$\underline{x} \in [\underline{x}^1,\underline{x}^3] \iff \lambda_2(\underline{x}) = 0$$

$$\underline{x} \in [\underline{x}^2,\underline{x}^3] \iff \lambda_1(\underline{x}) = 0$$

$$\underline{x} \in [\underline{x}^{12},\underline{x}^{13}] \iff \lambda_1(\underline{x}) = \tfrac{1}{2}$$

$$\underline{x} \in [\underline{x}^{12}, \underline{x}^{23}] \iff \lambda_1(\underline{x}) = \tfrac{1}{2}$$

$$\underline{x} \in [\underline{x}^{13}, \underline{x}^{23}] \iff \lambda_3(\underline{x}) = \tfrac{1}{2}$$

The situation is sketched in Fig. 4.8.

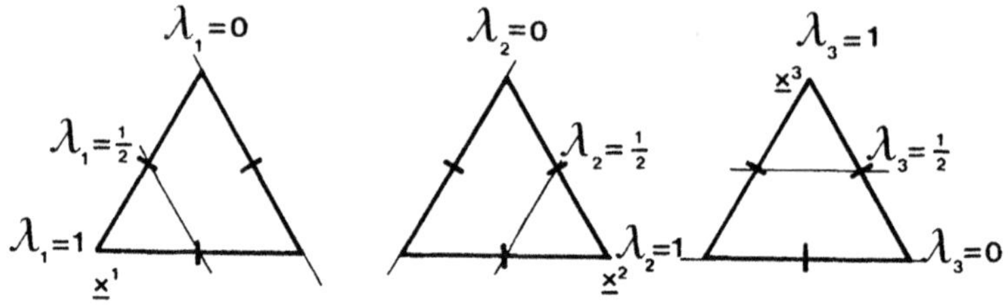

Fig. 4.8 λ-relations for some segments.

Using these barycentric coordinates we shall now construct linear, quadratic and extended quadratic basis functions.

4.2.2 Linear finite element

For the construction of piecewise linear basis functions we take all the vertices of the triangles as nodal points: $\underline{x}^1, \underline{x}^2, \ldots \underline{x}^N$. A piecewise linear basis function ϕ_i corresponding to nodal point $\underline{x}^i$ is such that

(i) $\phi_i(\underline{x}^j) = \delta_{ij}$ $\quad i,j = 1,2,\ldots,N$

(ii) ϕ_i is linear on each e_k

(iii) ϕ_i is continuous on $\bar{\Omega}$.

We plainly verify that the basis function ϕ_i is identically zero on those triangles for which $\underline{x}^i$ is no vertex. Thus, let a triangle e with vertex $\underline{x}^i$ be given. Introduce a local numbering of the vertices of e (see Fig. 4.9).

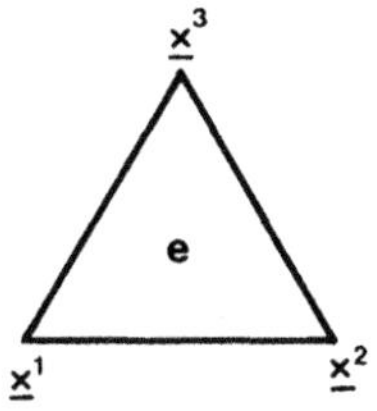

Fig. 4.9 Triangle e with nodal points $\underline{x}^1, \underline{x}^2, \underline{x}^3$.

The question now is: what is the shape on e of the basis functions ϕ_i corresponding to the points $\underline{x}^i$, $i = 1,2,3$. The function ϕ_i is linear on e and satisfies $\phi_i(\underline{x}^j) = \delta_{ij}$, $i,j = 1,2,3$. So we see immediately that

$$\phi_1 = \lambda_1 , \quad \phi_2 = \lambda_2 , \quad \phi_3 = \lambda_3 \tag{4.2.3}$$

where $\{\lambda_1(\underline{x}), \lambda_2(\underline{x}), \lambda_3(\underline{x})\}$ denotes the barycentric coordinates of $\underline{x}$ with respect to the points $\underline{x}^1$, $\underline{x}^2$, $\underline{x}^3$.

The function space spanned by ϕ_1, ϕ_2, ϕ_3 on e (the shape functions on e) is termed $P_1(e)$ and is exactly the collection of polynomials on e of degree ≤ 1 in x_1 and x_2.

4.2.3 Linear finite element (with reduced continuity)

The set of nodal points is defined as the collection of mid-points of all edges. A piecewise linear basis function ϕ_{ij} with reduced continuity corresponding to the mid-point $\underline{x}^{ij}$ is defined as follows:

(i) $\phi_{ij}(\underline{x}^{ij}) = 1$ and $\phi_{ij} = 0$ in all the other mid-points

(ii) ϕ_{ij} is linear on each triangle

(iii) ϕ_{ij} is continuous at $\underline{x}^{ij}$ in the mid-points.

The basis function ϕ_{ij} is identically zero on those triangles which do not contain the point $\underline{x}^{ij}$. Let e be a triangle which contains $\underline{x}^{ij}$ as mid-point of one of its edges. When we introduce a local numbering of the points as indicated in Figure 4.10, the basis functions are then given by

$$\begin{aligned} \phi_{12} &= 1 - 2\lambda_3 \\ \phi_{13} &= 1 - 2\lambda_2 \\ \phi_{23} &= 1 - 2\lambda_1 \end{aligned} \qquad (4.2.4)$$

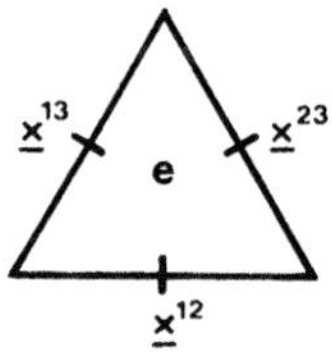

Fig. 4.10 Triangle e with three nodal points.

where $\{\lambda_1(\underline{x}),\lambda_2(\underline{x}),\lambda_3(\underline{x})\}$ denotes the barycentric coordinates of the point $\underline{x}$ with respect to the vertices $\underline{x}^1,\underline{x}^2,\underline{x}^3$ of e. The general form of an admissible approximate solution

$$\tilde{u} = \sum \tilde{u}_{ij}\,\phi_{ij} \qquad \text{(summation over all mid-points of edges)}$$

is piecewise linear, but discontinuous. This linear finite element can be used (see Sections 7.2.2, 13.3.6) to treat the incompressibility constraint of the (Navier-) Stokes equations in the 2D case.

4.2.4 Quadratic finite element

A piecewise quadratic basis function ϕ has in 2D the general form:

$$\phi(\underline{x}) = \alpha_1 x_1^2 + \alpha_2 x_1 x_2 + \alpha_3 x_2^2 + \alpha_4 x_1 + \alpha_5 x_2 + \alpha_6 \ .$$

This implies that on each triangle ϕ must be specified by six values to determine the parameters $\alpha_1, \alpha_2, \ldots, \alpha_6$. On each triangle we choose six nodal points: the three vertices and the three mid-points of sides. With local renumbering we have the situation of Fig. 4.11.

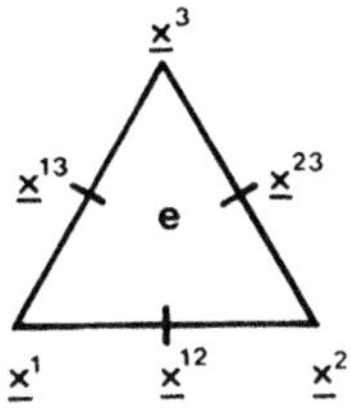

Fig. 4.11 Triangle e with six nodal points.

We shall determine the shape of the basis functions ϕ_1, ϕ_2, ϕ_3, ϕ_{12}, ϕ_{13}, ϕ_{23} corresponding to the points $\underline{x}^1$, $\underline{x}^2$, $\underline{x}^3$, $\underline{x}^{12}$, $\underline{x}^{13}$, $\underline{x}^{23}$ respectively.

Take for instance ϕ_1 which vanishes in the nodal points $\underline{x}^2$, $\underline{x}^3$, $\underline{x}^{12}$, $\underline{x}^{13}$, $\underline{x}^{23}$ and equals unity at $\underline{x}^1$. For all $\underline{x} \in [\underline{x}^2, \underline{x}^3]$ we have $\lambda_1(\underline{x}) = 0$, for all $\underline{x} \in [\underline{x}^{12}, \underline{x}^{13}]$ we have $\lambda_1(\underline{x}) = \frac{1}{2}$. Consequently the function $\lambda_1(\underline{x})(\lambda_1(\underline{x})-\frac{1}{2})$ vanishes at $\underline{x}^2$, $\underline{x}^3$, $\underline{x}^{12}$, $\underline{x}^{13}$, $\underline{x}^{23}$ and equals $\frac{1}{2}$ at $\underline{x}^1$. The basis function ϕ_1 is thus defined by

$$\phi_1 = \lambda_1(2\lambda_1 - 1) \ . \qquad (4.2.5)$$

In the same way we obtain

$$\phi_2 = \lambda_2(2\lambda_2 - 1) \qquad (4.2.6)$$

$$\phi_3 = \lambda_3(2\lambda_3 - 1) \ . \qquad (4.2.7)$$

For the basis function ϕ_{12} we notice that segment $[\underline{x}^1, \underline{x}^3]$

satisfies $\lambda_2(\underline{x}) = 0$ and that $[\underline{x}^2,\underline{x}^3]$ satisfies $\lambda_1(\underline{x}) = 0$. The function $\lambda_1(\underline{x})\lambda_2(\underline{x})$ thus vanishes at $\underline{x}^1$, $\underline{x}^2$, $\underline{x}^3$, $\underline{x}^{13}$, $\underline{x}^{23}$ and equals $\frac{1}{4}$ at $\underline{x}^{12}$. Hence ϕ_{12} is defined by

$$\phi_{12} = 4\lambda_1 \lambda_2 \; . \tag{4.2.8}$$

Similarly we have

$$\phi_{13} = 4\lambda_1 \lambda_3 \tag{4.2.9}$$

$$\phi_{23} = 4\lambda_2 \lambda_3 \; . \tag{4.2.10}$$

The function space spanned by ϕ_1, ϕ_2, ϕ_3, ϕ_{12}, ϕ_{13}, ϕ_{23} is called $P_2(e)$ and is precisely the space of polynomials of degree ≤ 2 in x_1 and x_2.

4.2.5 Extended quadratic finite element

For the definition of the so-called extended quadratic finite element, which will be used in the finite element analysis of the (Navier-) Stokes equations, we choose seven nodal points in each triangle: the three vertices, the three mid-points of sides and the barycentre (see Fig. 4.12).

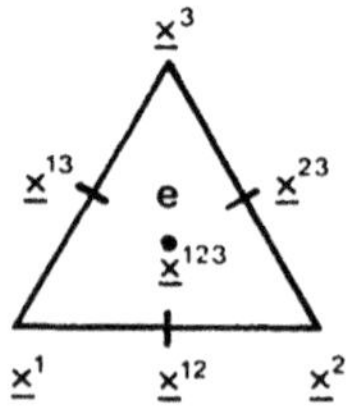

Fig. 4.12 Triangle e with seven nodal points.

Since the function $\lambda_1(\underline{x})\ \lambda_2(\underline{x})\ \lambda_3(\underline{x})$ vanishes on the sides of the triangle, it is easily verified that the basis functions corresponding to $\underline{x}^1$, $\underline{x}^2$, $\underline{x}^3$, $\underline{x}^{12}$, $\underline{x}^{13}$, $\underline{x}^{23}$, $\underline{x}^{123}$ are given by

$$\phi_i = \lambda_i(2\lambda_i - 1) + 3\lambda_1\lambda_2\lambda_3 \qquad i = 1,2,3$$

$$\phi_{ij} = 4\lambda_i\lambda_j - 12\lambda_1\lambda_2\lambda_3 \qquad \{i,j\} = \{1,2\},\ \{1,3\},\ \{2,3\}$$

$$\phi_{123} = 27\lambda_1\lambda_2\lambda_3 \ . \tag{4.2.11}$$

The function space spanned by ϕ_1, ϕ_2, ϕ_3, ϕ_{12}, ϕ_{13}, ϕ_{23}, ϕ_{123} is called $P_2^+(e)$ since it contains $P_2(e)$ and is contained in $P_3(e)$ which denotes the space of polynomials of degree ≤ 3 in x_1 and x_2. Notice that the space $P_3(e)$ needs 10 nodal points per triangle.

4.3 Triangular basis functions in 3D

4.3.1 Barycentric coordinates

Let Ω be a 3D region which is subdivided into a finite number of tetrahedrons e_k, $k = 1,2,\ldots,K$, satisfying the properties (3.5.1).

A linear function on a tetrahedron is completely determined by its values in four points which do not belong to one plane, for instance the vertices $\underline{x}^1$, $\underline{x}^2$, $\underline{x}^3$ and $\underline{x}^4$ of the tetrahedron. We define four functions $\lambda_i = \lambda_i(\underline{x})$, $i = 1,2,3,4$ on e by the following requirements (cf. sections 4.2, 4.4.1)

(i) $\lambda_i(\underline{x}^j) = \delta_{ij} \qquad i,j = 1,2,3,4$

(ii) λ_i is linear on e.

These functions can be determined as follows. Consider for instance λ_1. The general form of a linear function is:

$$\lambda_1(\underline{x}) = \alpha_0 + \alpha_1 x_1 + \alpha_2 x_2 + \alpha_3 x_3 \ , \quad \underline{x} = (x_1, x_2, x_3) \ .$$

The requirements (i) result into the following system of equations for $\alpha_0,\ldots,\alpha_3$:

$$\begin{vmatrix} 1 & x_1^1 & x_2^1 & x_3^1 \\ 1 & x_1^2 & x_2^2 & x_3^2 \\ 1 & x_1^3 & x_2^3 & x_3^3 \\ 1 & x_1^4 & x_2^4 & x_3^4 \end{vmatrix} \begin{vmatrix} \alpha_0 \\ \alpha_1 \\ \alpha_2 \\ \alpha_3 \end{vmatrix} = \begin{vmatrix} 1 \\ 0 \\ 0 \\ 0 \end{vmatrix} \tag{4.3.1}$$

This system of equations is uniquely solvable since the vertices $\underline{x}^1$, $\underline{x}^2$, $\underline{x}^3$, $\underline{x}^4$ do not belong to one plane which implies that the determinant of the coefficient matrix in (4.3.1) is different from zero.

For any point $\underline{x} \in e$ the quantities $\lambda_i(\underline{x})$ can be calculated. The quadruple $\{\lambda_1(\underline{x}), \lambda_2(\underline{x}), \lambda_3(\underline{x}), \lambda_4(\underline{x})\}$ is called the barycentric coordinates of the point $\underline{x} \in e$ with respect to the vertices $\underline{x}^1$, $\underline{x}^2$, $\underline{x}^3$, $\underline{x}^4$.

Let (see Fig. 4.13)

$$\begin{array}{lll} \underline{x}^i & & i = 1,2,3,4 \\ \underline{x}^{ij} & = \frac{1}{2}(\underline{x}^i+\underline{x}^j) & i,j = 1,2,3,4;\ i < j \\ \underline{x}^{ijk} & = \frac{1}{3}(\underline{x}^i+\underline{x}^j+\underline{x}^k) & i,j,k = 1,2,3,4;\ i < j < k\,. \end{array}$$

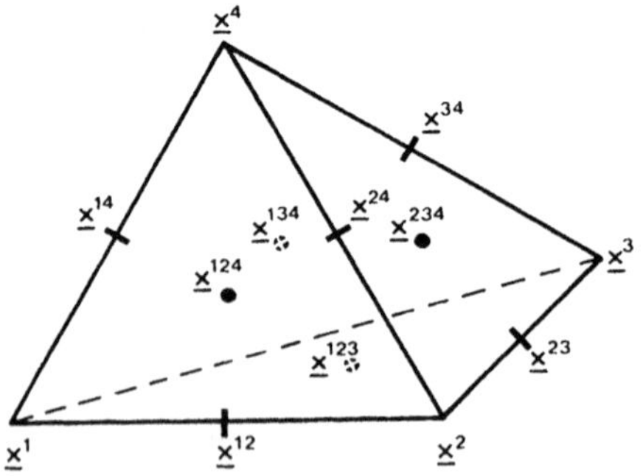

Fig. 4.13 Tetrahedron e with nodal points.

One easily verifies that the following relations hold

$$\sum_{i=1}^{4} \lambda_i(\underline{x}) = 1 \qquad \text{for all } \underline{x}$$

$$\underline{x}^1 \rightarrow \{1,0,0,0\} \;, \quad \text{etc.}$$

$$\underline{x}^{14} \rightarrow \{\tfrac{1}{2},0,0,\tfrac{1}{2}\}, \text{ etc.}$$

$$\underline{x}^{123} \rightarrow \{\tfrac{1}{3},\tfrac{1}{3},\tfrac{1}{3},0\}, \text{ etc.}$$

$$\underline{x} \in \text{plane } (\underline{x}^1,\underline{x}^2,\underline{x}^3) \rightarrow \lambda_4(\underline{x}) = 0 \quad \text{etc.}$$

$$\underline{x} \in \text{edge } [\underline{x}^1,\underline{x}^2] \rightarrow \lambda_3(\underline{x}) = \lambda_4(\underline{x}) = 0 \quad \text{etc.}$$

Using the barycentric coordinates we shall now construct two types of linear basis functions.

4.3.2 Linear finite element

Let $\underline{x}^1,\underline{x}^2,\ldots,\underline{x}^N$ be the collection of all vertices of the (finite number of) tetrahedrons subdividing the region Ω. A piecewise linear basis function ϕ_i corresponding to nodal point $\underline{x}^i$ is defined by

(i) $\phi_i(\underline{x}^j) = \delta_{ij} \qquad i,j = 1,2,\ldots,N$

(ii) ϕ_i is linear on each tetrahedron

(iii) ϕ_i is continuous on $\bar{\Omega}$.

Just as in section 4.4.2 for the 2D situation, we see immediately that on a tetrahedron e with a local numbering of the nodal points $\underline{x}^1$, $\underline{x}^2$, $\underline{x}^3$, $\underline{x}^4$, the basis functions are given by

$$\phi_i = \lambda_i \qquad i = 1,2,3,4 \; . \tag{4.3.2}$$

The function space spanned by ϕ_1, ϕ_2, ϕ_3, ϕ_4 on e is termed $P_1(e)$.

4.3.3 Linear finite element (with reduced continuity)

For the construction of piecewise linear basis functions with reduced continuity requirements, we take the collection of all barycentres of faces of the tetrahedrons. A piecewise linear basis function ϕ_{ijk} with reduced continuity corresponding to the face-barycentre $\underline{x}^{ijk}$ is defined as follows (cf. Section 4.2.3)

(i) $\phi_{ijk}(\underline{x}^{ijk}) = 1$ and $\phi_{ijk} = 0$ in all other face-barycentres

(ii) ϕ_{ijk} is linear on each tetrahedron

(iii) ϕ_{ijk} is continuous in all face-barycentres.

We easily verify that the basis function ϕ_{ijk} is identically zero on those tetrahedrons which do not contain $\underline{x}^{ijk}$.
Let a tetrahedron e be given which contains the point $\underline{x}^{ijk}$. Introduce a local numbering of the nodal points of e as indicated in Figure 4.14. The basis functions are then given by

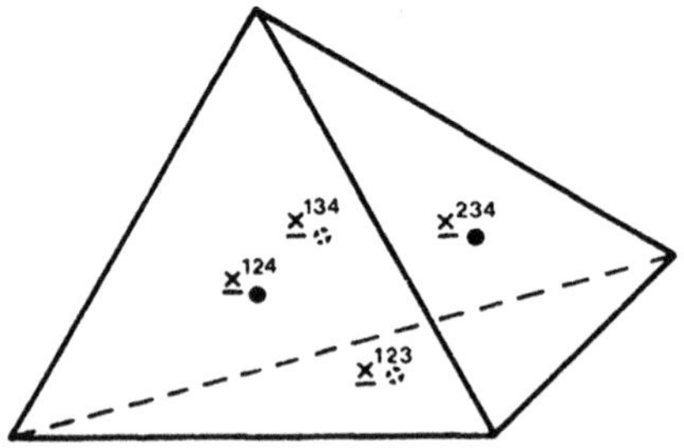

Fig. 4.14 Tetrahedron e with 4 nodal points.

$$
\begin{aligned}
\phi_{123} &= 1 - 3\lambda_4 \\
\phi_{124} &= 1 - 3\lambda_3 \\
\phi_{134} &= 1 - 3\lambda_2 \\
\phi_{234} &= 1 - 3\lambda_1
\end{aligned}
\tag{4.3.3}
$$

where $\{\lambda_1(\underline{x}),\lambda_2(\underline{x}),\lambda_3(\underline{x}),\lambda_4(\underline{x})\}$ denotes the barycentric coordinates of the point $\underline{x}$ with respect to the vertices $\underline{x}^1$, $\underline{x}^2$, $\underline{x}^3$, $\underline{x}^4$ of e.

Let us emphasize that the general form of an admissible approximate solution

$$\tilde{u} = \sum \tilde{u}_{ijk}\ \phi_{ijk} \quad \text{(summation over face-barycentres)}$$

is piecewise linear, but discontinuous. This finite element can be used to handle the incompressibility constraint of the (Navier-) Stokes equations in the 3D situation (see Section 7.2.2 for the 2D analogue).

4.4 Coordinate and element transformation

In the preceding sections we have constructed the basis functions for an element e and nodal points $\underline{x}^1$, $\underline{x}^2,\ldots,\underline{x}^N$ by working directly on the element e. A basis function ϕ_i was defined as a function of prescribed type (e.g. linear, quadratic, etc.) which equals unity in the nodal point $\underline{x}^i$ and vanishes in all the other nodal points.

A second procedure to introduce basis functions on an element e with nodal points $\underline{x}^1,\underline{x}^2,\ldots,\underline{x}^N$ is to make use of a so-called reference (or standard, or master) element $\hat{e}$. The basis functions are defined on $\hat{e}$ and then transformed into basis functions on e. This procedure emerges as extremely powerful since all computations (for all elements e) can be carried out on one and the same reference element.

In 1D the reference element is the interval $\hat{e} = [\hat{x}^1,\hat{x}^2]$ with

$\hat{x}^1 = 0$, $\hat{x}^2 = 1$, in 2D $\hat{e}$ is a triangle with vertices $\hat{\underline{x}}^1 = (0,0)$, $\hat{\underline{x}}^2 = (1,0)$, $\hat{\underline{x}}^3 = (0,1)$, in 3D $\hat{e}$ is a tetrahedron with vertices $\hat{\underline{x}}^1 = (0,0,0)$, $\hat{\underline{x}}^2 = (1,0,0)$, $\hat{\underline{x}}^3 = (0,1,0)$, $\hat{\underline{x}}^4 = (0,0,1)$ (see Fig. 4.15).

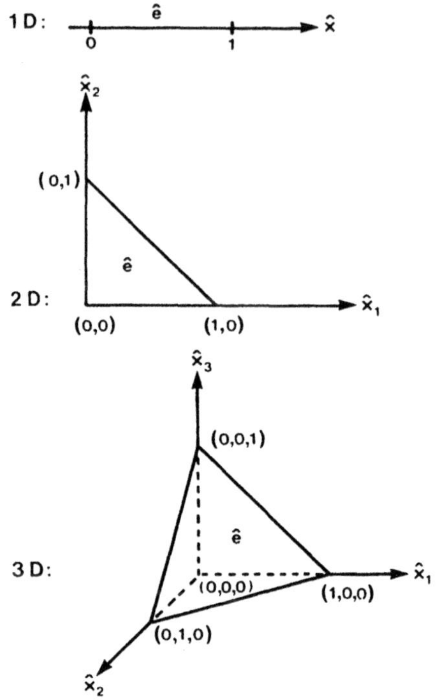

Fig. 4.15 Reference elements.

Any arbitrary straight-sided element $e \subset \Omega \subset \mathbb{R}^n$ (n=1: interval, n=2: triangle, n=3: tetrahedron) can be written as the image of the reference element $\hat{e}$ under the mapping F_e defined by

$$e = F_e(\hat{e})$$

with

$$\underline{x} = F_e(\hat{\underline{x}}) = \sum_{i=1}^{n+1} \underline{x}^i \lambda_i(\hat{\underline{x}}) \tag{4.4.1}$$

where n denotes the space dimension and where $\{\lambda_1(\hat{\underline{x}}),\ldots,\lambda_{n+1}(\hat{\underline{x}})\}$ denotes the barycentric coordinates of $\hat{\underline{x}}$ with respect to $\hat{\underline{x}}^1,\ldots,\hat{\underline{x}}^{n+1}$ (see Fig. 4.16).

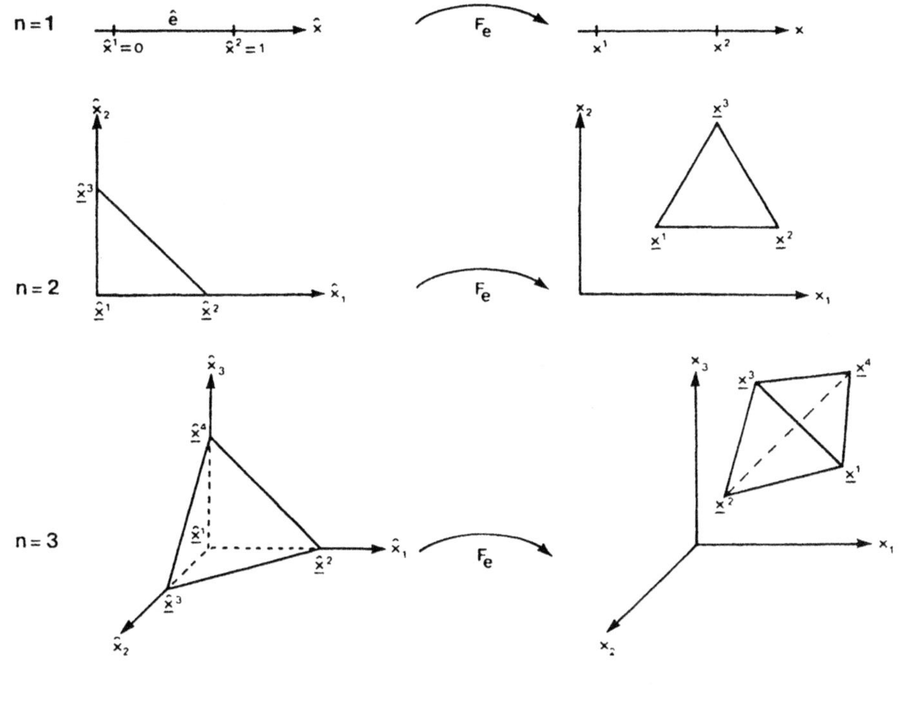

Fig. 4.16 Mapping from reference elements to arbitrary elements.

Notice that F_e is a linear (in fact an affine) mapping, which means that F_e has the form

$$F_e(\hat{\underline{x}}) = B_e\hat{\underline{x}} + \underline{b}_e \tag{4.4.2}$$

where B_e is a (constant) n×n matrix and $\underline{b}_e$ a (constant) n-vector:

$$B_e = x^2 - x^1 , \quad \underline{b}_e = x^1 \qquad \text{for } n = 1 \tag{4.4.3}$$

$$B_e = \begin{vmatrix} x_1^2 - x_1^1 & x_1^3 - x_1^1 \\ x_2^2 - x_2^1 & x_2^3 - x_2^1 \end{vmatrix} , \quad \underline{b}_e = \begin{vmatrix} x_1^1 \\ x_2^1 \end{vmatrix} \qquad \text{for } n = 2 \tag{4.4.4}$$

$$B_e = \begin{vmatrix} x_1^2 - x_1^1 & x_1^3 - x_1^1 & x_1^4 - x_1^1 \\ x_2^2 - x_2^1 & x_2^3 - x_2^1 & x_2^4 - x_2^1 \\ x_3^2 - x_3^1 & x_3^3 - x_3^1 & x_3^4 - x_3^1 \end{vmatrix} , \quad \underline{b}_e = \begin{vmatrix} x_1^1 \\ x_2^1 \\ x_3^1 \end{vmatrix} \qquad \text{for } n = 3 \tag{4.4.5}$$

The mapping F_e is chosen such that vertices are mapped onto vertices:

$$\underline{x}^i = F_e(\hat{x}^i)$$

mid-points of sides are mapped onto mid-points of sides:

$$\underline{x}^{ij} = \tfrac{1}{2}(\underline{x}^i + \underline{x}^j) = F_e(\hat{\underline{x}}^{ij}) = F_e(\tfrac{1}{2}(\hat{\underline{x}}^i + \hat{\underline{x}}^j))$$

and barycentres are mapped onto barycentres:

$$\underline{x}^{ijk} = \frac{1}{3}(\underline{x}^i + \underline{x}^j + \underline{x}^k) = F_e(\hat{\underline{x}}^{ijk}) = F_e(\frac{1}{3}(\hat{\underline{x}}^i + \hat{\underline{x}}^j + \hat{\underline{x}}^k)).$$

Let us denote by $\hat{\phi}_i$, i=1,2,...,N, the basis functions defined on the reference element $\hat{e}$ corresponding to the nodal points $\hat{\underline{x}}^i$, i=1,2,...,N (including eventually: mid-points of sides, barycentres etc.). Then the basis functions ϕ_i on the element e corresponding to the points $\underline{x}^i = F_e(\hat{x}^i)$ are given by

$$\phi_i(\underline{x}) = \hat{\phi}_i(F_e^{-1}(\underline{x})) \qquad \text{i.e. } \phi_i(\underline{x}) = \hat{\phi}_i(\hat{\underline{x}}), \ \underline{x} = F_e(\hat{\underline{x}}).$$

One easily verifies that

$$\phi_i(\underline{x}^j) = \hat{\phi}_i(\hat{\underline{x}}^j) = \delta_{ij}$$

and since $\underline{x}$ depends linearly on $\hat{\underline{x}}$, the degree of the basis functions (linear, quadratic, etc.) is preserved.

So it turns out that an arbitrary element e (an interval, triangle or tetrahedron) can be defined as the image of a reference element $\hat{e}$ under the mapping F_e. Since the mapping F_e, defined by (4.4.1), is linear, the images e are straight sides elements.
When, however, a mapping F_e is chosen that is of the same degree as the basis functions on the reference element $\hat{e}$, the image $e = F_e(\hat{e})$ is in general a curved sided element and we speak of isoparametric finite elements.
Let us give the example of the extended quadratic triangular element in 2D. A curved sided triangle with vertices $\underline{x}^1$, $\underline{x}^2$, $\underline{x}^3$ and inner point $\underline{x}^{123} \equiv \underline{x}^7$, whose sides fit the points $\underline{x}^4 \equiv \underline{x}^{23}$, $\underline{x}^5 \equiv x^{13}$, $\underline{x}^6 \equiv \underline{x}^{12}$ can be considered as the isoparametric image of the reference triangle $\hat{e}$ if

$$\underline{x} = F_e(\hat{\underline{x}}) \tag{4.4.6}$$

with

$$F_e(\hat{\underline{x}}) = \sum_{i=1}^{7} \underline{x}^i \phi_i(\hat{\underline{x}}) \tag{4.4.7}$$

where $\hat{\phi}_i$ denotes the $P_2^+(\hat{e})$ basis function corresponding to point $\hat{\underline{x}}^i$ in the reference element, $i=1,2,\ldots,7$. Notice that

$$F_e(\hat{\underline{x}}^i) = \underline{x}^i \qquad i = 1,2,\ldots,7 \tag{4.4.8}$$

with $\hat{\underline{x}}^4 = \frac{1}{2}(\hat{\underline{x}}^2+\hat{\underline{x}}^3)$, $\hat{\underline{x}}^5 = \frac{1}{2}(\hat{\underline{x}}^1+\hat{\underline{x}}^3)$, $\hat{\underline{x}}^6 = \frac{1}{2}(\hat{\underline{x}}^1+\hat{\underline{x}}^2)$, $\hat{\underline{x}}^7 = \frac{1}{3}(\hat{\underline{x}}^1+\hat{\underline{x}}^2+\hat{\underline{x}}^3)$. The situation is illustrated in Fig. 4.17.
The basis functions ϕ_i on e are defined as follows

$$\phi_i(\underline{x}) = \hat{\phi}_i(F_e^{-1}(\underline{x})) \quad \text{i.e.} \quad \phi_i(\underline{x}) = \hat{\phi}_i(\hat{\underline{x}}),\ \underline{x} = F_e(\hat{\underline{x}}). \tag{4.4.9}$$

Observe that in this case, even if the point $\underline{x}^7$ plays no role

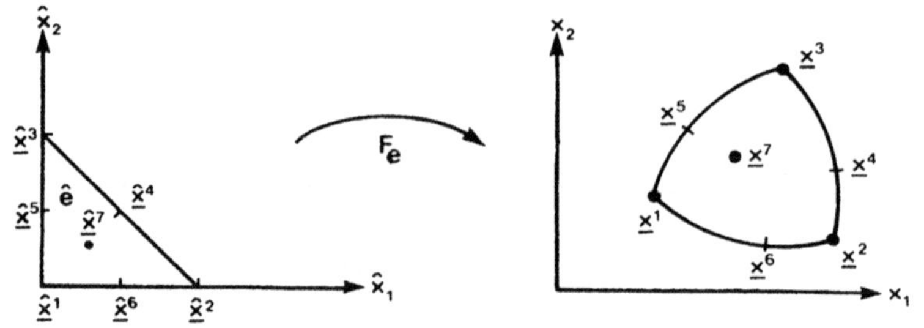

Fig. 4.17 Isoparametric mapping from $\hat{e}$ onto e.

in the definition of the boundaries of the triangle e, the basis functions ϕ_i still depend on its position. In fact, an isoparametric finite element is not directly determined by a mapping F_e but, rather, by the position of the nodal points $\underline{x}^1, \underline{x}^2, \ldots$ on e, which in turn uniquely determine the mapping F_e. When the mapping F_e is chosen of the same degree as the basis functions ϕ_i, the element is termed isoparametric. If, however, the mapping is defined by a lower-degree basis, the element is called subparametric. This would be the case, if, in the example just given, the mapping F_e of (4.4.7) were defined by

$$F_e(\hat{x}) = \sum_{i=1}^{3} \underline{x}^i \hat{\phi}_i(\hat{\underline{x}}) \qquad (4.4.10)$$

where $\hat{\phi}_i$ denotes the linear basis function corresponding to point $\hat{\underline{x}}^i$, i=1,2,3. In this case the image e is a straight-sided element with vertices $\underline{x}^i$, i=1,2,3.

To assure that the mapping F_e is invertible, which is necessary for the definition of the basis functions, we require that F_e has a non-vanishing Jacobian. For a linear mapping F_e, such as (4.4.1), this condition is satisfied if the n+1 vertices of the element e are not lying in one hyperplane (i.e. if e is not degenerated). For the general isoparametric case, consider the element e and the straight-sided element $\tilde{e}$ with the same vertices as e. The condition of non-vanishing Jacobian is satisfied if the distances between nodal points of e and corresponding nodal points of $\tilde{e}$ are small with respect to the size of the element (see Ciarlet (1978)).

Finally we mention that the technique of working on reference elements is extremely useful when studying quadrilateral (cf. Section 4.5) and hexahedral (cf. Section 4.6) finite elements. Reference elements are also used for the more theoretical aspects of the FEM, like, for instance, the study of error estimates, convergence and the theory of isoparametric finite elements (cf. Ciarlet (1978), Strang, Fix (1973), Raviart, Thomas (1983)).

4.5 Quadrilateral finite element

Concerning the 2D case, we have studied until now triangular finite elements. A second type of finite elements is known as the quadrilateral finite elements. For this we subdivide the region $\Omega \subset \mathbb{R}^2$ into a finite number of straight-sided quadrilaterals e_k, k=1,2,...,K and we assume that this subdivision satisfies condition (3.5.1). Moreover we assume that each e_k is convex and that no angle of e_k is too close to 0°.

In Fig. 4.18 an example of a subdivision of Ω into quadrilaterals is given.

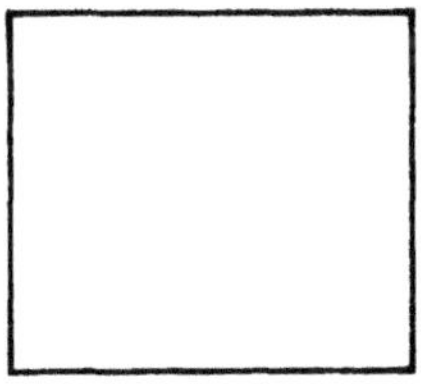

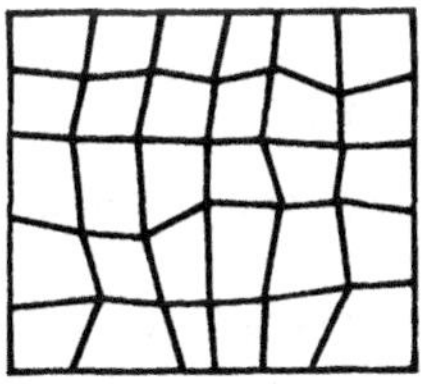

Fig. 4.18 Region Ω. Subdivision of Ω.

In each quadrilateral we choose a finite number of nodal points. See Fig. 4.9 for examples of 4 and 9 nodal points per quadrilateral e

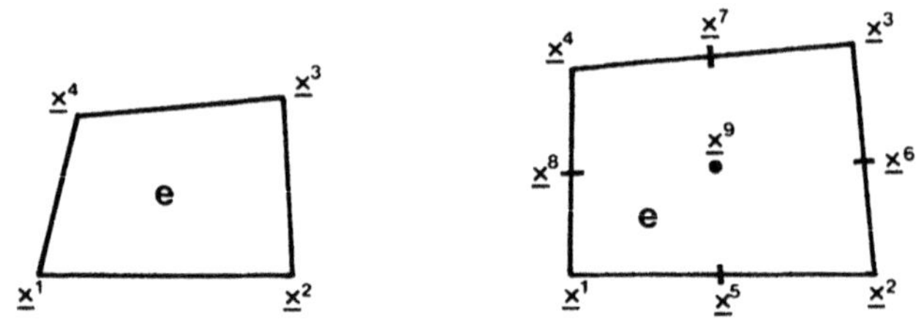

Fig. 4.19 Four nodal points. Nine nodal points.

where

$$
\begin{aligned}
&\underline{x}^1,\ \underline{x}^2,\ \underline{x}^3 \text{ and } \underline{x}^4 \text{ are vertices of e} \\
&\underline{x}^5 = \tfrac{1}{2}(\underline{x}^1+\underline{x}^2) \\
&\underline{x}^6 = \tfrac{1}{2}(\underline{x}^2+\underline{x}^3) \\
&\underline{x}^7 = \tfrac{1}{2}(\underline{x}^3+\underline{x}^4) \\
&\underline{x}^8 = \tfrac{1}{2}(\underline{x}^4+\underline{x}^1) \\
&\underline{x}^9 = \tfrac{1}{4}(\underline{x}^1+\underline{x}^2+\underline{x}^3+\underline{x}^4).
\end{aligned}
\tag{4.5.1}
$$

The next step is to define for each nodal point $\underline{x}^i$ a basis function $\phi_i : \bar{\Omega} \to \mathbb{R}$ which satisfies

(i) $\phi_i(\underline{x}^j) = \delta_{ij}$

(ii) ϕ_i has prescribed behavior on each quadrilateral

(iii) ϕ_i is continuous on $\bar{\Omega}$.

The particular behavior of the basis functions ϕ_i is defined using a reference element $\hat{e}$ which is defined as the unit-square with vertices

$\hat{\underline{x}}^1 = (0,0)$, $\hat{\underline{x}}^2 = (1,0)$, $\hat{\underline{x}}^3 = (1,1)$, $\hat{\underline{x}}^4 = (0,1)$ (see Fig.4.20).

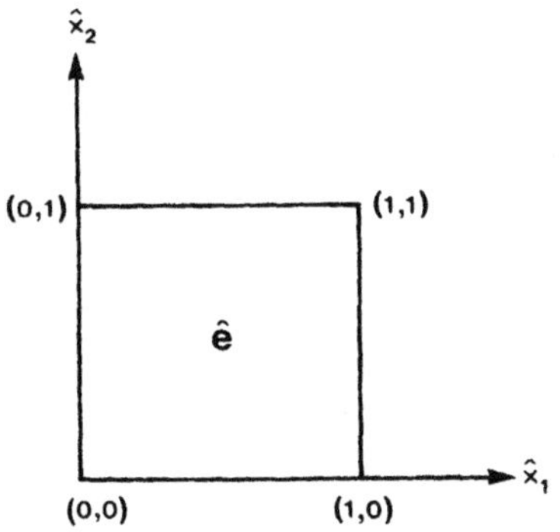

Fig. 4.20 Reference element.

4.5.1 Bilinear finite element

The first order basis functions on this reference element, i.e. basis functions with respect to four nodal points, are:

$$
\begin{aligned}
\hat{\phi}_1(\hat{\underline{x}}) &= \lambda_1(\hat{x}_1)\lambda_1(\hat{x}_2) \\
\hat{\phi}_2(\hat{\underline{x}}) &= \lambda_2(\hat{x}_1)\lambda_1(\hat{x}_2) \\
\hat{\phi}_3(\hat{\underline{x}}) &= \lambda_2(\hat{x}_1)\lambda_2(\hat{x}_2) \\
\hat{\phi}_4(\hat{\underline{x}}) &= \lambda_1(\hat{x}_1)\lambda_2(\hat{x}_2)
\end{aligned}
\tag{4.5.2}
$$

where $\lambda_1(t) = 1-t$, $\lambda_2(t) = t$.

A simple general procedure to obtain the basis function $\hat{\phi}_i$ corresponding to nodal point $\hat{\underline{x}}^i$ is to multiply the 1D basis function corresponding to nodal point $(\hat{x}_1^i,0)$ in the $\hat{x}_1$-direction by the 1D basis function corresponding to nodal point $(0,\hat{x}_2^i)$ in the $\hat{x}_2$-direction.

The function space spanned by these four basis functions is the space of all bilinear functions in $\hat{x}_1$ and $\hat{x}_2$, i.e. all polynomials in $\hat{x}_1$ and $\hat{x}_2$ of degree ≤ 1 with respect to each of the variables $\hat{x}_1$ and $\hat{x}_2$. This space is denoted by $Q_1(\hat{e})$. Generally we denote by $Q_k(\hat{e})$, k non-negative integer, $\hat{e} \in \mathbb{R}^n$, the function space of polynomials in $\hat{x}_1, \hat{x}_2, \ldots, \hat{x}_n$ of degree $\leq k$ in each of the variables $\hat{x}_1, \hat{x}_2, \ldots, \hat{x}_n$. In other words, $Q_k(\hat{e})$ is the function space spanned by

$$\hat{x}_1^{k_1} \hat{x}_2^{k_2} \ldots \hat{x}_n^{k_n}, \qquad 0 \leq k_i \leq k, \qquad i = 1,2,\ldots,n$$

In order to determine the basis functions on an arbitrary straight-sided quadrilateral e with vertices $\underline{x}^1, \underline{x}^2, \underline{x}^3, \underline{x}^4$, we define a mapping F_e from $\hat{e}$ onto e by

$$\underline{x} = F_e(\hat{\underline{x}}) = \sum_{i=1}^{4} \underline{x}^i \hat{\phi}_i(\hat{\underline{x}}) \; . \tag{4.5.3}$$

One easily verifies that (see also Fig. 4.21)

$$e = F_e(\hat{e})$$

$$F_e(\hat{\underline{x}}^i) = \underline{x}^i \qquad i = 1,2,3,4 \; .$$

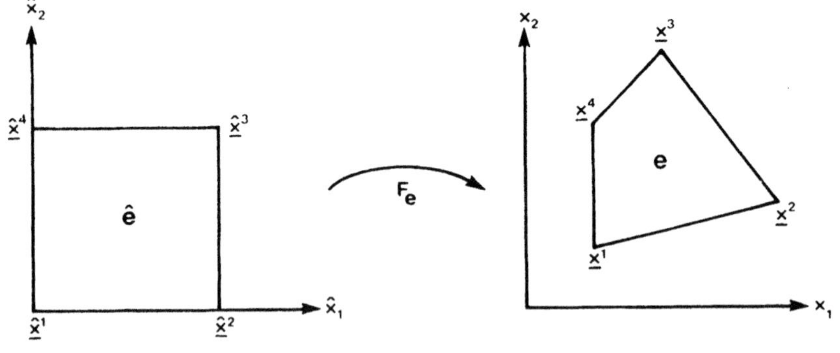

Fig. 4.21 Mapping F_e from $\hat{e}$ onto e.

In fact the element e is isoparametric since the mapping F_e is not linear and belongs to the function space spanned by $\hat{\phi}_1, \hat{\phi}_2, \hat{\phi}_3, \hat{\phi}_4$. Nevertheless the sides of e are straight.

Next we define the basis functions on e by

$$\phi_i(\underline{x}) = \hat{\phi}_i(F_e^{-1}(\underline{x})) , \quad \text{i.e. } \phi_i(\underline{x}) = \hat{\phi}_i(\hat{\underline{x}}) \quad \text{with } \underline{x} = F_e(\hat{\underline{x}}) \tag{4.5.4}$$

To assure that the mapping F_e is invertible, which amounts to saying that F_e has a non-vanishing Jacobian, the quadrilateral e must be convex.

The Q_1-quadrilateral finite element for n=2 is now defined as follows:

(i) a straight-sided quadrilateral with nodal points $\underline{x}^i$, $i=1,\ldots,4$

(ii) the basis functions are given by (4.5.4) on e.

Notice that ϕ_i is identically zero on those quadrilaterals which do not contain $\underline{x}^i$ and that ϕ_i is continuous on the whole set $\bar{\Omega}$. This last property is a direct consequence of the construction procedure using a reference element.

4.5.2 Biquadratic finite element

A second order quadrilateral finite element is defined as follows. In the reference element $\hat{e}$ with vertices $\hat{\underline{x}}^1$, $\hat{\underline{x}}^2$, $\hat{\underline{x}}^3$, $\hat{\underline{x}}^4$ we choose nine nodal points $\hat{\underline{x}}^1 = (0,0)$, $\hat{\underline{x}}^2 = (1,0)$, $\hat{\underline{x}}^3 = (1,1)$, $\hat{\underline{x}}^4 = (0,1)$, $\hat{\underline{x}}^5 = (\frac{1}{2},0)$, $\hat{\underline{x}}^6 = (1,\frac{1}{2})$, $\hat{\underline{x}}^7 = (\frac{1}{2},1)$, $\hat{\underline{x}}^8 = (0,\frac{1}{2})$, $\hat{\underline{x}}^9 = (\frac{1}{2},\frac{1}{2})$ (see Fig. 4.22).

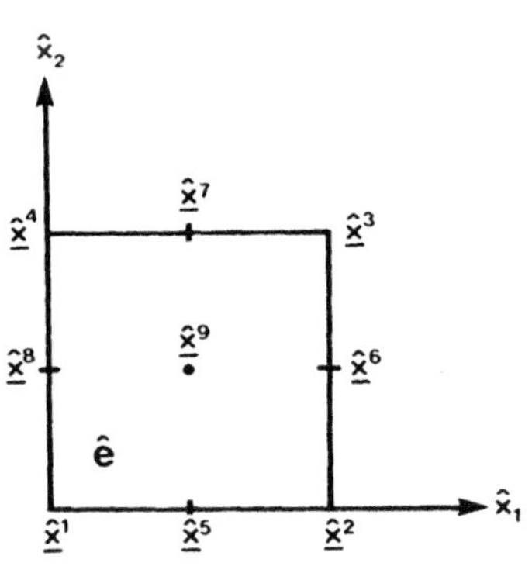

Fig. 4.22 Reference element $\hat{e}$ with nine nodal points.

and we define biquadratic basis functions on $\hat{e}$ which are polynomials in $\hat{x}_1$ and $\hat{x}_2$ of degree ≤ 2 with respect to $\hat{x}_1$ and to $\hat{x}_2$. The basis functions $\hat{\phi}_i$ corresponding to $\hat{\underline{x}}^i$, i=1,2,...,9, are

$$\begin{aligned}
\hat{\phi}_1(\hat{\underline{x}}) &= \Theta_1(\hat{x}_1)\ \Theta_1(\hat{x}_2)\\
\hat{\phi}_2(\hat{\underline{x}}) &= \Theta_2(\hat{x}_1)\ \Theta_1(\hat{x}_2)\\
\hat{\phi}_3(\hat{\underline{x}}) &= \Theta_2(\hat{x}_1)\ \Theta_2(\hat{x}_2)\\
\hat{\phi}_4(\hat{\underline{x}}) &= \Theta_1(\hat{x}_1)\ \Theta_2(\hat{x}_2)\\
\hat{\phi}_5(\hat{\underline{x}}) &= \Theta_3(\hat{x}_1)\ \Theta_1(\hat{x}_2)\\
\hat{\phi}_6(\hat{\underline{x}}) &= \Theta_2(\hat{x}_1)\ \Theta_3(\hat{x}_2)\\
\hat{\phi}_7(\hat{\underline{x}}) &= \Theta_3(\hat{x}_1)\ \Theta_2(\hat{x}_2)\\
\hat{\phi}_8(\hat{\underline{x}}) &= \Theta_1(\hat{x}_1)\ \Theta_3(\hat{x}_2)\\
\hat{\phi}_9(\hat{\underline{x}}) &= \Theta_3(\hat{x}_1)\ \Theta_3(\hat{x}_2)
\end{aligned} \tag{4.5.5}$$

with $\Theta_1(t) = (1-t)(1-2t)$, $\Theta_2(t) = t(1-2t)$, $\Theta_3(t) = 4t(1-t)$ which correspond to the 1D quadratic basis functions on the interval $[0,1]$.

The function space spanned by the basis functions $\hat{\phi}_i$, i=1,2,...,9 is the space of biquadratic functions on $\hat{e}$ and is denoted by $Q_2(e)$. The basis functions on e, with nodal points $\underline{x}^1, \underline{x}^2, \ldots, \underline{x}^9$ (given by (4.5.1)), are now defined by

$$\phi_i(\underline{x}) = \hat{\phi}_i(F_e^{-1}(\underline{x})) \qquad \underline{x} \in e \qquad i=1,2,\ldots,9 \tag{4.5.6}$$

The mapping F_e is given by (4.5.3), thus determined by the four vertices of e and the bilinear basis functions $\hat{\phi}_i$ corresponding to the points $\hat{\underline{x}}^i$, i=1,2,3,4. One easily verifies that

$$e = F_e(\hat{e})$$

$$F_e(\hat{\underline{x}}^i) = \underline{x}^i \qquad i = 1,2,\ldots,9$$

The Q_2-quadrilateral finite element for n=2 is defined by

(i) a straight-sided quadrilateral with nodal points $\underline{x}^1, \underline{x}^2, \ldots \underline{x}^9$

(ii) the basis functions given by (4.5.6) on e.

When the mapping F_e was given by

$$\underline{x} = F_e(\hat{\underline{x}}) = \sum_{i=1}^{9} \underline{x}^i \hat{\phi}_i(\hat{\underline{x}}) \ , \quad \hat{\phi}_i \text{ defined by (4.5.5)}$$

the image $e = F_e(\hat{e})$ would be a curved-sided (isoparametric) element (see Fig. 4.23).

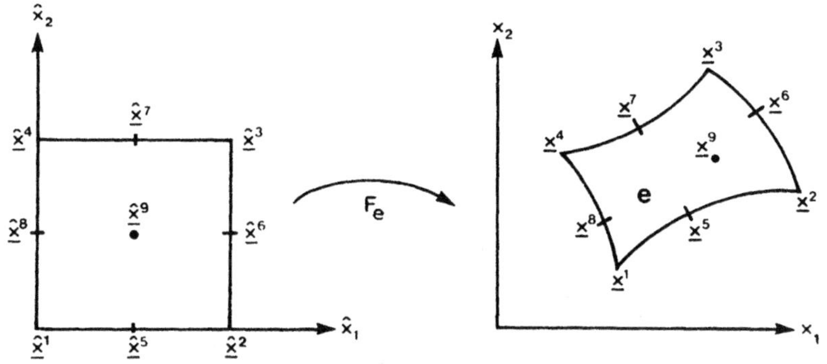

Fig. 4.23 Isoparametric mapping from $\hat{e}$ onto e.

A consequence of this construction is that all the basis functions are continuous on $\bar{\Omega}$.

4.6 Hexahedral finite elements

For the construction of hexahedral finite elements in the 3D case we proceed in the same way as for the 2D situation of Section 4.5. We assume that the 3D region $\Omega \subset \mathbb{R}^3$ is subdivided into a finite number of straight-sided hexahedra e_k, k=1,...,K, satisfying condition (3.5.1). In each hexahedron we choose a finite number of nodal points, for instance the eight vertices as depicted in Fig. 4.24

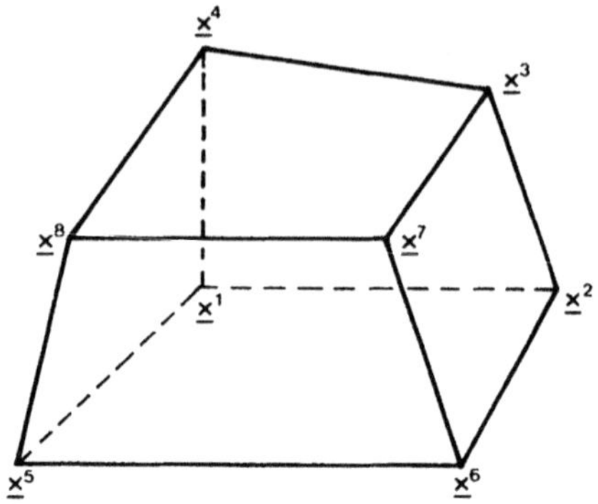

Fig. 4.24 8 nodal points.

or 27 nodal points as is shown in Fig. 4.25,

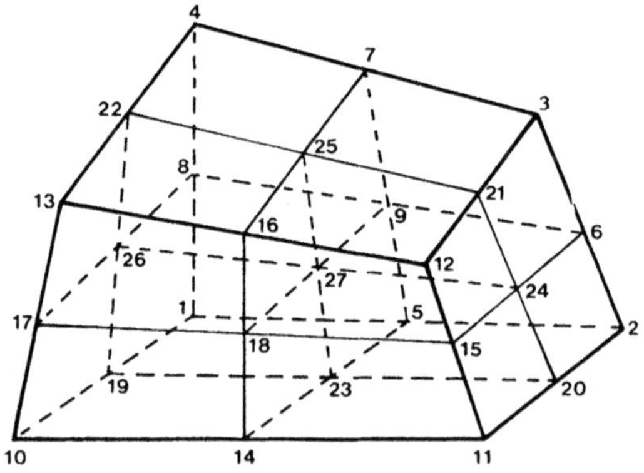

Fig. 4.25 27 Nodal points.

with

$$\left|\begin{array}{lll} \underline{x}^1, \underline{x}^2, \underline{x}^3, \underline{x}^4, \underline{x}^{10}, \underline{x}^{11}, \underline{x}^{12}, \underline{x}^{13} & \text{vertices} & \\ \underline{x}^{i+4} = \tfrac{1}{2}(\underline{x}^i+\underline{x}^{i+1}) & i = 1,2,3,10,11,12,19,20,21 & \\ & & (4.6.1) \\ \underline{x}^{i+4} = \tfrac{1}{2}(\underline{x}^i+\underline{x}^{i-3}) & i = 4,13,22 & \\ \underline{x}^{i+8} = \frac{1}{4}(\underline{x}^i+\underline{x}^{i+1}+\underline{x}^{i+2}+\underline{x}^{i+3}) & i = 1,10,19 & \\ \underline{x}^{i+18} = \tfrac{1}{2}(\underline{x}^i+\underline{x}^{i+9}) & i = 1,2,3,4 & \end{array}\right.$$

4.6.1 Trilinear finite element

In the eight nodal point situation, the basis function ϕ_i corresponding to nodal point $\underline{x}^i$ will be given using a reference element $\hat{e}$ defined as the unit-cube with vertices $\hat{\underline{x}}^1 = (0,0,0)$, $\hat{\underline{x}}^2 = (1,0,0)$, $\hat{\underline{x}}^3 = (1,1,0)$, $\hat{\underline{x}}^4 = (0,1,0)$, $\hat{\underline{x}}^5 = (0,0,1)$, $\hat{\underline{x}}^6 = (1,0,1)$, $\hat{\underline{x}}^7 = (1,1,1)$, $\hat{\underline{x}}^8 = (0,1,1)$ (see Fig. 4.26).

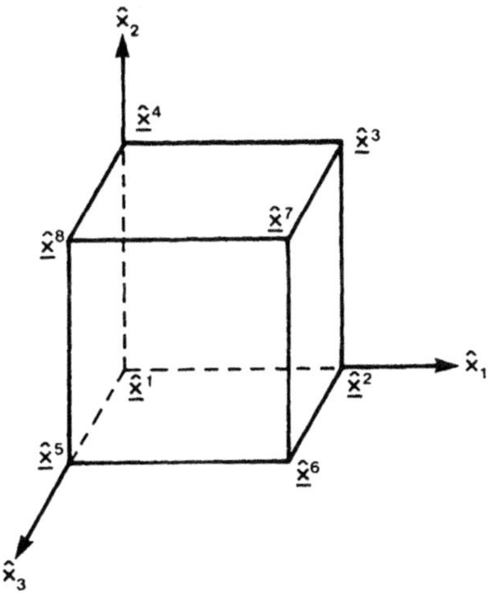

Fig. 4.26 Reference element with eight nodal points.

The basis functions take the following form:

$$\begin{aligned}
\hat{\phi}_1(\hat{\underline{x}}) &= \lambda_1(\hat{x}_1)\ \lambda_1(\hat{x}_2)\ \lambda_1(\hat{x}_3) \\
\hat{\phi}_2(\hat{\underline{x}}) &= \lambda_2(\hat{x}_1)\ \lambda_1(\hat{x}_2)\ \lambda_1(\hat{x}_3) \\
\hat{\phi}_3(\hat{\underline{x}}) &= \lambda_2(\hat{x}_1)\ \lambda_2(\hat{x}_2)\ \lambda_1(\hat{x}_3) \\
\hat{\phi}_4(\hat{\underline{x}}) &= \lambda_1(\hat{x}_1)\ \lambda_2(\hat{x}_2)\ \lambda_1(\hat{x}_3) \\
\hat{\phi}_5(\hat{\underline{x}}) &= \lambda_1(\hat{x}_1)\ \lambda_1(\hat{x}_2)\ \lambda_2(\hat{x}_3) \\
\hat{\phi}_6(\hat{\underline{x}}) &= \lambda_2(\hat{x}_1)\ \lambda_1(\hat{x}_2)\ \lambda_2(\hat{x}_3) \\
\hat{\phi}_7(\hat{\underline{x}}) &= \lambda_2(\hat{x}_1)\ \lambda_2(\hat{x}_2)\ \lambda_2(\hat{x}_3) \\
\hat{\phi}_8(\hat{\underline{x}}) &= \lambda_1(\hat{x}_1)\ \lambda_2(\hat{x}_2)\ \lambda_2(\hat{x}_3)
\end{aligned} \tag{4.6.2}$$

with $\lambda_1(t) = 1-t$, $\lambda_2(t) = t$, which correspond to the 1D linear basis functions on $[0,1]$.

The function space spanned by these eight basis functions is the space $Q_1(\hat{e})$ of all trilinear functions in $\hat{x}_1$, $\hat{x}_2$, $\hat{x}_3$, i.e. all polynomials in $\hat{x}_1$, $\hat{x}_2$, $\hat{x}_3$ of degree ≤ 1 with respect to each of the variables $\hat{x}_1$, $\hat{x}_2$, $\hat{x}_3$. The reference element $\hat{e}$ is mapped onto an arbitrary straight-sided hexahedron e satisfying (4.6.1) by the mapping F_e (see Fig. 4.27):

$$\underline{x} = F_e(\hat{\underline{x}}) = \sum_{i=1}^{8} \underline{x}^i \hat{\phi}_i(\hat{\underline{x}}) \tag{4.6.3}$$

The basis functions in the (x_1,x_2,x_3) space are now defined by

$$\phi_i(\underline{x}) = \hat{\phi}_i(F_e^{-1}(\underline{x})) \qquad i = 1,2,\ldots,8 \tag{4.6.4}$$

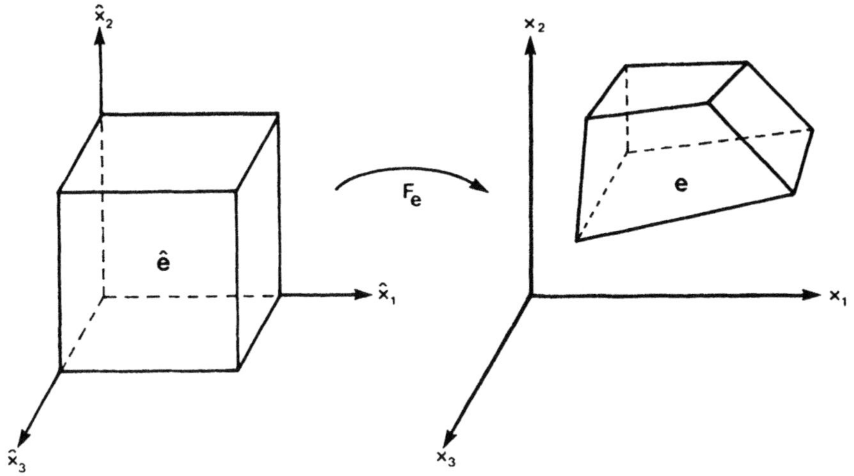

Fig. 4.27 Mapping F_e from $\hat{e}$ onto e.

4.6.2 Triquadratic finite element

A higher order finite element can be constructed using 27 nodal points (see Fig. 4.25). In the reference unit cube we choose the following 27 points:

$$\hat{\underline{x}}^1 = (0,0,0),\ \hat{\underline{x}}^2 = (1,0,0),\ \hat{\underline{x}}^3 = (1,1,0),\ \hat{\underline{x}}^4 = (0,1,0)$$

$$\hat{\underline{x}}^5 = (\tfrac12,0,0),\ \hat{\underline{x}}^6 = (1,\tfrac12,0),\ \hat{\underline{x}}^7 = (\tfrac12,1,0),\ \hat{\underline{x}}^8 = (0,\tfrac12,0)$$

$$\hat{\underline{x}}^9 = (\tfrac12,\tfrac12,0)$$

$$\hat{\underline{x}}^{10} = (0,0,1),\ \hat{\underline{x}}^{11} = (1,0,1),\ \hat{\underline{x}}^{12} = (1,1,1),\ \hat{\underline{x}}^{13} = (0,1,1)$$

$$\hat{\underline{x}}^{14} = (\tfrac12,0,1),\ \hat{\underline{x}}^{15} = (1,\tfrac12,1),\ \hat{\underline{x}}^{16} = (\tfrac12,1,1),\ \hat{\underline{x}}^{17} = (0,\tfrac12,1)$$

$$\hat{\underline{x}}^{18} = (\tfrac12,\tfrac12,1) \tag{4.6.5}$$

$$\hat{\underline{x}}^{19} = (0,0,\tfrac12),\ \hat{\underline{x}}^{20} = (1,0,\tfrac12),\ \hat{\underline{x}}^{21} = (1,1,\tfrac12),\ \hat{\underline{x}}^{22} = (0,1,\tfrac12)$$

$$\hat{\underline{x}}^{23} = (\tfrac12,0,\tfrac12),\ \hat{\underline{x}}^{24} = (1,\tfrac12,\tfrac12),\ \hat{\underline{x}}^{25} = (\tfrac12,1,\tfrac12),\ \hat{\underline{x}}^{26} = (0,\tfrac12,\tfrac12)$$

$$\hat{\underline{x}}^{27} = (\tfrac12,\tfrac12,\tfrac12).$$

Next we define triquadratic basis functions on $\hat{e}$ as the product of 1D quadratic basis functions in the three coordinate directions. The basis functions take the following form (cf. (4.5.5)):

$$\begin{aligned}
\hat{\phi}_{1+9i}(\hat{\underline{x}}) &= \Theta_1(\hat{x}_1)\ \Theta_1(\hat{x}_2)\ \Theta_{i+1}(\hat{x}_3)\\
\hat{\phi}_{2+9i}(\hat{\underline{x}}) &= \Theta_2(\hat{x}_1)\ \Theta_1(\hat{x}_2)\ \Theta_{i+1}(\hat{x}_3)\\
\hat{\phi}_{3+9i}(\hat{\underline{x}}) &= \Theta_2(\hat{x}_1)\ \Theta_2(\hat{x}_2)\ \Theta_{i+1}(\hat{x}_3)\\
\hat{\phi}_{4+9i}(\hat{\underline{x}}) &= \Theta_1(\hat{x}_1)\ \Theta_2(\hat{x}_2)\ \Theta_{i+1}(\hat{x}_3)\\
\hat{\phi}_{5+9i}(\hat{\underline{x}}) &= \Theta_3(\hat{x}_1)\ \Theta_1(\hat{x}_2)\ \Theta_{i+1}(\hat{x}_3)\\
\hat{\phi}_{6+9i}(\hat{\underline{x}}) &= \Theta_2(\hat{x}_1)\ \Theta_3(\hat{x}_2)\ \Theta_{i+1}(\hat{x}_3)\\
\hat{\phi}_{7+9i}(\hat{\underline{x}}) &= \Theta_3(\hat{x}_1)\ \Theta_2(\hat{x}_2)\ \Theta_{i+1}(\hat{x}_3)\\
\hat{\phi}_{8+9i}(\hat{\underline{x}}) &= \Theta_1(\hat{x}_1)\ \Theta_3(\hat{x}_2)\ \Theta_{i+1}(\hat{x}_3)\\
\hat{\phi}_{9+9i}(\hat{\underline{x}}) &= \Theta_3(\hat{x}_1)\ \Theta_3(\hat{x}_2)\ \Theta_{i+1}(\hat{x}_3)
\end{aligned} \qquad (4.6.6)$$

with $\Theta_1(t) = (1-t)(1-2t)$, $\Theta_2(t) = t(1-2t)$, $\Theta_3(t) = 4t(1-t)$ and $i = 0,1,2$.

The function space spanned by the functions (4.6.6) is denoted by $Q_2(\hat{e})$ and consists of all triquadratic functions in $\hat{x}_1$, $\hat{x}_2$, $\hat{x}_3$. On an arbitrary straight-sided hexahedron satisfying (4.6.1) the basis functions are obtained by transformation of the reference basis functions (4.6.6) according to the formula

$$\phi_i(\underline{x}) = \hat{\phi}_i(F_e^{-1}(\underline{x})) \qquad i = 1,2,\ldots,27 \qquad (4.6.7)$$

with $\underline{F}_e$ given by

$$F_e(\hat{\underline{x}}) = \sum_{i \in I} \underline{x}^i \hat{\phi}_i(\hat{\underline{x}})$$

where I is the index set $I = \{1,2,3,4,10,11,12,13\}$ and where $\hat{\phi}_i$

denotes the trilinear basis function corresponding to point $\hat{\underline{x}}^i$, $i \in I$.

Chapter 5

Practical aspects of the Finite Element Method

In this Chapter we shall be concerned with some practical aspects of the FEM. First of all we shall explain in Section 5.1 the so-called "finite element assembly algorithm", which is an algorithm that permits an efficient construction of the final system of equations on an element basis. This algorithm leads to the introduction of the notions of element (stiffness) matrix and element vector. Examples will be given of the 1D Poisson equation with linear, quadratic and cubic basis functions (Section 5.2). The 2D Poisson equation will be studied using linear and quadratic elements (Section 5.3). In order to reduce the computational work without reducing the accuracy of the approximate solution, a number of integration rules of Newton-Cotes and Gauss type will be introduced in Section 5.4. Concerning the estimates of the error in the approximate solution, we shall give some qualitative results in Section 5.5, references to precise results will be given in Chapter 12. Finally we shall pay attention to solution methods for systems of (linear) equations. This point will be the subject of Section 5.6.

5.1 Finite element assembly algorithm

Let us consider the following problem

$$\left| \begin{array}{ll} -\dfrac{d^2u}{dx^2} = f & \text{on } (0,1) \\[2ex] u(0) = 0 & \dfrac{du}{dx}(1) = 0 \end{array} \right. \tag{5.1.1}$$

to which correspond the following Galerkin equations (see (3.6.18)):

$$\sum_{j=0}^{N} \tilde{u}_j \{ \int_0^1 \frac{d\phi_j}{dx} \frac{d\phi_i}{dx} dx \} = \int_0^1 f \; \phi_i \; dx \qquad i = 1,2,\dots,N \quad (5.1.2)$$

where $\tilde{u}_i$ is the value of $\tilde{u}$ in the nodal point x^i. The approximate solution $\tilde{u}$ has the form

$$\tilde{u}(x) = \sum_{j=0}^{N} \tilde{u}_j \phi_j(x)$$

where ϕ_j, $j=0,1,\dots,N$, denote the piecewise linear basis functions (see Fig. 3.8). Since $u(0) = 0$ we impose $\tilde{u}_0=\tilde{u}(0)=0$.

The interval (0,1) is subdivided into N subintervals e_k, $k=1,2,\dots,N$ (see Fig. 5.1).

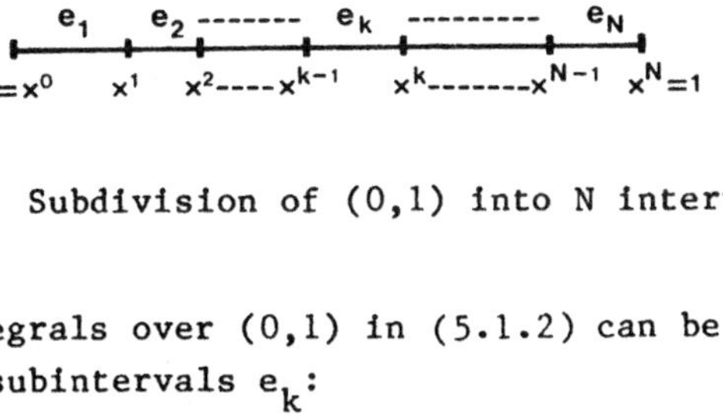

Fig. 5.1 Subdivision of (0,1) into N intervals e_k.

Next, the integrals over (0,1) in (5.1.2) can be written as the sum over the subintervals e_k:

$$\sum_{j=0}^{N} \tilde{u}_j \{ \sum_{k=1}^{N} \int_{e_k} \frac{d\phi_j}{dx} \frac{d\phi_i}{dx} dx \} = \sum_{k=1}^{N} \int_{e_k} f \; \phi_i dx \qquad (5.1.3)$$

$$i = 1,2,\dots,N$$

or equivalently

$$\sum_{k=1}^{N} \sum_{j=0}^{N} \tilde{u}_j \{ \int_{e_k} \frac{d\phi_j}{dx} \frac{d\phi_i}{dx} dx \} = \sum_{k=1}^{N} \int_{e_k} f \; \phi_i dx \qquad (5.1.4)$$

$$i = 1,2,\dots,N$$

We see that for the construction of the stiffness matrix $\underline{A}$, integrals of the type

$$\int_{e_k} \frac{d\phi_j}{dx} \frac{d\phi_i}{dx} \, dx \qquad j = 0,1,\ldots,N \qquad i = 1,2,\ldots,N \qquad (5.1.5)$$

must be calculated.
On a subinterval e_k only two basis functions are different from zero, viz. ϕ_{k-1} and ϕ_k (see Fig. 3.9). This means that there are four combinations $\{i,j\}$ that contribute to the integral (5.1.5): $\{i,j\} = \{k-1,k-1\},\{k-1,k\},\{k,k-1\},\{k,k\}$. We put these four contributions in a matrix

$$A^{e_k} = \begin{vmatrix} \int_{e_k} \frac{d\phi_{k-1}}{dx} \frac{d\phi_{k-1}}{dx} dx & \int_{e_k} \frac{d\phi_k}{dx} \frac{d\phi_{k-1}}{dx} dx \\ \int_{e_k} \frac{d\phi_{k-1}}{dx} \frac{d\phi_k}{dx} dx & \int_{e_k} \frac{d\phi_k}{dx} \frac{d\phi_k}{dx} dx \end{vmatrix} \qquad (5.1.6)$$

This matrix A^e is called the element- or elementstiffness matrix. The element matrices must be calculated for each subinterval, after which the matrix elements are put on the right place in the stiffness matrix A. The use of element matrices is well suited for automization of the assembly process of the matrix A. This technique, which is nothing else than an efficient programming trick, is called the finite element assembly algorithm.
In the same way we can assemble the right hand side vector $\underline{F}$. As we see from (5.1.3), integrals of the type

$$\int_{e_k} f \, \phi_i \, dx$$

must be calculated. On the subinterval e_k only the basis functions ϕ_{k-1} and ϕ_k have non-zero contribution. These two quantities are put into the vector $\underline{F}^{e_k}$, called element vector:

$$\underline{F}^{e_k} = \begin{vmatrix} \int_{e_k} f\ \phi_{k-1}\ dx \\ \\ \int_{e_k} f\ \phi_k\ dx \end{vmatrix} . \tag{5.1.7}$$

We calculate the element vectors $\underline{F}^e$ for all subintervals and put their components on the right place in the right hand side vector $\underline{F}$.
We shall elucidate this assembly process in the following example.

Consider problem (5.1.2) for the special case of three elements e_k, k=1,2,3 of equal length $h = \frac{1}{3}$ (see Fig. 5.2) and $f = 1$. For the basis functions which are different from zero on e_k we have:

$$\phi_{k-1}(x) = \frac{h-x}{h} \qquad \frac{d\phi_{k-1}}{dx}(x) = -\frac{1}{h} \qquad \text{for } x^{k-1} < x < x^k$$

$$\phi_k(x) = \frac{x}{h} \qquad \frac{d\phi_k}{dx}(x) = \frac{1}{h} \qquad \text{for } x^{k-1} < x < x^k$$

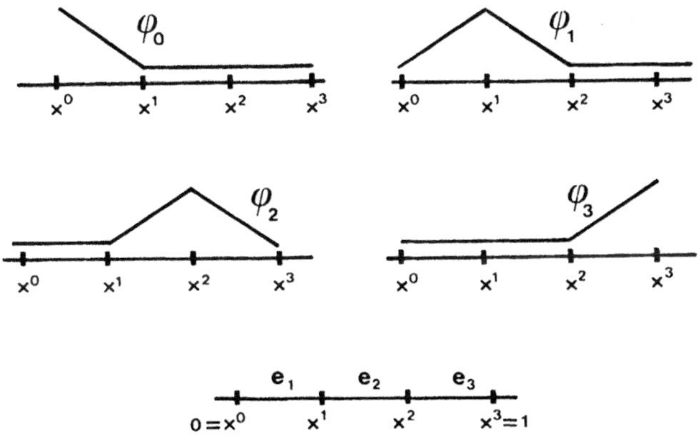

Fig. 5.2 Subdivision of (0,1) into three elements and their basis functions $\phi_0,\cdots,\phi_3$.

We find

$$A^{e_1} = \begin{vmatrix} \int_{e_1} \frac{d\phi_0}{dx}\frac{d\phi_0}{dx}\,dx & \int_{e_1} \frac{d\phi_1}{dx}\frac{d\phi_0}{dx}\,dx \\ \int_{e_1} \frac{d\phi_0}{dx}\frac{d\phi_1}{dx}\,dx & \int_{e_1} \frac{d\phi_1}{dx}\frac{d\phi_1}{dx}\,dx \end{vmatrix} = \frac{1}{h} \begin{array}{c} \begin{array}{cc} j=0 & j=1 \\ \downarrow & \downarrow \end{array} \\ \begin{vmatrix} 1 & -1 \\ -1 & 1 \end{vmatrix} \end{array} \begin{array}{l} \\ \leftarrow i=0 \\ \leftarrow i=1 \end{array}$$

$$A^{e_2} = \begin{vmatrix} \int_{e_2} \frac{d\phi_1}{dx}\frac{d\phi_1}{dx}\,dx & \int_{e_2} \frac{d\phi_2}{dx}\frac{d\phi_1}{dx}\,dx \\ \int_{e_2} \frac{d\phi_1}{dx}\frac{d\phi_2}{dx}\,dx & \int_{e_2} \frac{d\phi_2}{dx}\frac{d\phi_2}{dx}\,dx \end{vmatrix} = \frac{1}{h} \begin{array}{c} \begin{array}{cc} j=1 & j=2 \\ \downarrow & \downarrow \end{array} \\ \begin{vmatrix} 1 & -1 \\ -1 & 1 \end{vmatrix} \end{array} \begin{array}{l} \\ \leftarrow i=1 \\ \leftarrow i=2 \end{array}$$

$$A^{e_3} = \begin{vmatrix} \int_{e_3} \frac{d\phi_2}{dx}\frac{d\phi_2}{dx}\,dx & \int_{e_3} \frac{d\phi_3}{dx}\frac{d\phi_2}{dx}\,dx \\ \int_{e_3} \frac{d\phi_2}{dx}\frac{a\phi_3}{dx}\,dx & \int_{e_3} \frac{d\phi_3}{dx}\frac{d\phi_3}{dx}\,dx \end{vmatrix} = \frac{1}{h} \begin{array}{c} \begin{array}{cc} j=2 & j=3 \\ \downarrow & \downarrow \end{array} \\ \begin{vmatrix} 1 & -1 \\ -1 & 1 \end{vmatrix} \end{array} \begin{array}{l} \\ \leftarrow i=2 \\ \leftarrow i=3 \end{array}$$

We now assemble the three element matrices to form the matrix A^*:

$$A^* = A^{e_1} + A^{e_2} + A^{e_3} =$$

$$= \frac{1}{h} \begin{array}{c} \begin{array}{cccc} j=0 & j=1 & j=2 & j=3 \\ \downarrow & \downarrow & \downarrow & \downarrow \end{array} \\ \begin{vmatrix} 1 & -1 & 0 & 0 \\ -1 & 1 & 0 & 0 \\ 0 & 0 & 0 & 0 \\ 0 & 0 & 0 & 0 \end{vmatrix} \end{array} + \frac{1}{h} \begin{array}{c} \begin{array}{cccc} j=0 & j=1 & j=2 & j=3 \\ \downarrow & \downarrow & \downarrow & \downarrow \end{array} \\ \begin{vmatrix} 0 & 0 & 0 & 0 \\ 0 & 1 & -1 & 0 \\ 0 & -1 & 1 & 0 \\ 0 & 0 & 0 & 0 \end{vmatrix} \end{array}$$

$$+ \frac{1}{h} \begin{array}{c} \begin{array}{cccc} j=0 & j=1 & j=2 & j=3 \\ \downarrow & \downarrow & \downarrow & \downarrow \end{array} \\ \begin{vmatrix} 0 & 0 & 0 & 0 \\ 0 & 0 & 0 & 0 \\ 0 & 0 & 1 & -1 \\ 0 & 0 & -1 & 1 \end{vmatrix} \end{array} \begin{array}{l} \\ \leftarrow i=0 \\ \leftarrow i=1 \\ \leftarrow i=2 \\ \leftarrow i=3 \end{array}$$

$$= \frac{1}{h} \begin{vmatrix} 1 & -1 & 0 & 0 \\ -1 & 2 & -1 & 0 \\ 0 & -1 & 2 & -1 \\ 0 & 0 & -1 & 1 \end{vmatrix} .$$

Since in (5.1.4) i takes values from 1 to N, we can cancel the first row in A^* which becomes

$$A^{**} = \frac{1}{h} \begin{vmatrix} -1 & 2 & -1 & 0 \\ 0 & -1 & 2 & -1 \\ 0 & 0 & -1 & 1 \end{vmatrix} .$$

The same procedure is followed for the construction of the right hand side vector:

$$\underline{F}^{e_1} = \begin{vmatrix} \int_{e_1} \phi_0 \, dx \\ \int_{e_1} \phi_1 \, dx \end{vmatrix} = \frac{h}{2} \begin{vmatrix} 1 \\ 1 \end{vmatrix} \begin{matrix} \leftarrow i=0 \\ \leftarrow i=1 \end{matrix}$$

$$\underline{F}^{e_2} = \begin{vmatrix} \int_{e_2} \phi_1 \, dx \\ \int_{e_2} \phi_2 \, dx \end{vmatrix} = \frac{h}{2} \begin{vmatrix} 1 \\ 1 \end{vmatrix} \begin{matrix} \leftarrow i=1 \\ \leftarrow i=2 \end{matrix}$$

$$\underline{F}^{e_3} = \begin{vmatrix} \int_{e_3} \phi_2 \, dx \\ \int_{e_3} \phi_3 \, dx \end{vmatrix} = \frac{h}{2} \begin{vmatrix} 1 \\ 1 \end{vmatrix} \begin{matrix} \leftarrow i=2 \\ \leftarrow i=3 \end{matrix}$$

Summation of these three vectors gives:

$$\underline{F}^* = \frac{h}{2} \begin{vmatrix} 1 \\ 1 \\ 0 \\ 0 \end{vmatrix} + \frac{h}{2} \begin{vmatrix} 0 \\ 1 \\ 1 \\ 0 \end{vmatrix} + \frac{h}{2} \begin{vmatrix} 0 \\ 0 \\ 1 \\ 1 \end{vmatrix} \begin{matrix} \leftarrow i=0 \\ \leftarrow i=1 \\ \leftarrow i=2 \\ \leftarrow i=3 \end{matrix}$$

$$= \frac{h}{2} \begin{vmatrix} 1 \\ 2 \\ 2 \\ 1 \end{vmatrix} .$$

We cancel the row corresponding to i=0 and obtain

$$\underline{F}^{**} = \frac{h}{2} \begin{vmatrix} 2 \\ 2 \\ 1 \end{vmatrix} .$$

The system of equations for $\tilde{u}_0$, $\tilde{u}_1$, $\tilde{u}_2$ and $\tilde{u}_3$ is now

$$A^{**} \begin{vmatrix} \tilde{u}_0 \\ \tilde{u}_1 \\ \tilde{u}_2 \\ \tilde{u}_3 \end{vmatrix} = \underline{F}^{**} . \qquad (5.1.8)$$

Since $\tilde{u}_0 = 0$ we obtain finally the system

$$A \, \underline{\tilde{u}} = \underline{F}$$

or explicitly

$$\frac{1}{h} \begin{vmatrix} 2 & -1 & 0 \\ -1 & 2 & -1 \\ 0 & -1 & 1 \end{vmatrix} \begin{vmatrix} \tilde{u}_1 \\ \tilde{u}_2 \\ \tilde{u}_3 \end{vmatrix} = h \begin{vmatrix} 1 \\ 1 \\ \frac{1}{2} \end{vmatrix}$$

from which we deduce

$$\tilde{u}_1 = \frac{5}{2} h^2 \; , \quad \tilde{u}_2 = 4h^2 \; , \quad \tilde{u}_3 = \frac{9}{2} h^2 \qquad h = \frac{1}{3} .$$

It may be clear that this assembly algorithm is very well suitable for computer implementation. The efficiency of this procedure is even more striking for problems in more than 1D on non-rectangular regions with a non-regular distribution of the subregions.

As we have already seen in Section 3.6 the treatment of different kinds of boundary conditions does not give any difficulty at all. This remains true for the assembly algorithm for the stiffness matrix and the right hand side vector.

In summary we can say:

(i) Dirichlet boundary conditions:

The matrix A^* and vector $\underline{F}^*$ are assembled taking into account the basis functions corresponding to all nodal points. Then we cancel the rows corresponding to those nodal points on the boundary for which Dirichlet conditions are given. In terms of the above treated example we obtain (5.1.8). Finally the Dirichlet conditions are substituted into system (5.1.8). For a homogeneous Dirichlet condition at nodal point x^i, this amounts to saying that the column corresponding to x^i is cancelled. Non-homogeneous Dirichlet conditions give an extra contribution to the right hand side vector. When, instead of $u(0) = 0$ in (5.1.1), the condition was $u(0) = g_0$, we still deduce system (5.1.8). The first row reads

$$A^{**}_{10}\tilde{u}_0 + A^{**}_{11}\tilde{u}_1 + A^{**}_{12}\tilde{u}_2 + A^{**}_{13}\tilde{u}_3 = F^{**}_1 .$$

Since $\tilde{u}_0 = g_0$ is known, we put it on the right hand side:

$$A^{**}_{11}\tilde{u}_1 + A^{**}_{12}\tilde{u}_2 + A^{**}_{13}\tilde{u}_3 = F^{**}_1 - A^{**}_{10}g_0 \qquad (\equiv F_1).$$

Writing F_1 explicitly we get

$$F_1 = \int_{e_1} f\,\phi_1\,dx - g_0 \int_{e_1} \frac{d\phi_0}{dx}\frac{d\phi_1}{dx}\,dx$$

which is exactly the same as obtained in (3.6.23) with $i=1$.

(ii) Neumann boundary conditions:

Homogeneous Neumann conditions are satisfied automatically and do not give any problem at all. When, for instance, a non-homogeneous Neumann condition

$$\frac{du}{dx}(1) = g_1$$

is imposed at x=1, then, according to (3.6.23) the element vector for subinterval e_k, k = 1,2,...,N-1, is given by (5.1.7). However element vector $\underline{F}^{e_N}$ becomes

$$\underline{F}^{e_N} = \begin{vmatrix} \int_{e_N} f \, \phi_{N-1} dx \\ \\ \int_{e_N} f \, \phi_N dx + g_1 \end{vmatrix}$$

Similar procedures are followed for the treatment of boundary conditions in 2D situations, see Section 5.3.

5.2 1D Poisson equation; quadratic finite elements

We consider the one-dimensional Poisson equation

$$\left| \begin{array}{ll} -\dfrac{d^2u}{dx^2} = f & \text{on } (0,1) \\ \\ u(0) = g_0 & \dfrac{du}{dx}(1) = g_1 \end{array} \right. \tag{5.2.1}$$

and we shall define an approximate solution using quadratic finite elements.

The weak formulation of this problem is given by (3.6.22):

$$\left| \begin{array}{l} \text{Find } u \quad \text{with } u - g_0\phi_0 \in V \quad \text{such that} \\ \\ \int_0^1 \dfrac{du}{dx}\dfrac{d\phi}{dx}\, dx = \int_0^1 f \, \phi \, dx + g_1 \phi(1) \qquad \text{for all } \phi \in V \end{array} \right. \tag{5.2.2}$$

where V denotes the space of functions which vanish at x=0, and where ϕ_0 is a function with $\phi_0(0) = 1$. In order to define a quadratic finite element approximation of the space V, we divide the region $[0,1]$ into subregions $0 = x^0 < x^1 < \ldots < x^N = 1$ and on each subinterval $[x^{k-1}, x^k]$ we choose three nodal points: x^{k-1}, x^k and $x^{k-1\,k} = \frac{1}{2}(x^{k-1}+x^k)$ (see Fig. 5.3).

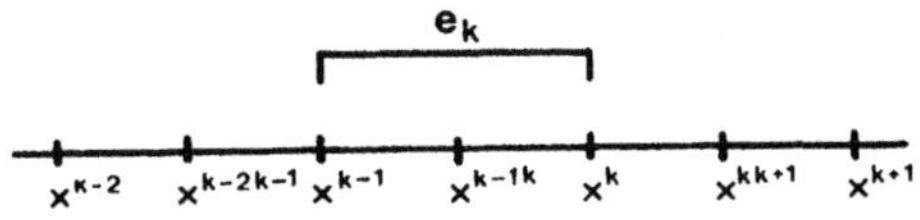

Fig. 5.3 Subinterval e_k with three nodal points.

An approximate solution $\tilde{u}$ can now be written as a linear combination of basis functions associated with the nodal points $x^0, x^{01}, x^1, x^{12}, x^2, \ldots, x^{N-1\,N}, x^N$:

$$\tilde{u}(x) = \sum_{j\in I} \tilde{u}_j \phi_j(x)$$

$$= \tilde{u}_0\phi_0(x) + \tilde{u}_{01}\phi_{01}(x) + \tilde{u}_1\phi_1(x) + \tilde{u}_{12}\phi_{12}(x) \qquad (5.2.3)$$

$$+ \ldots + \tilde{u}_{N-1\,N}\phi_{N-1N}(x) + \tilde{u}_N\phi_N(x)$$

where I denotes the index set $\{0,01,1,12,2,\ldots,N-1\,N,N\}$ and where the basis functions ϕ_j are depicted in Fig. 4.4. Since $u(0) = g_0$ we take $\tilde{u}_0 = g_0$.

We now replace in (5.2.2) the space V by the finite dimensional function space V_N spanned by the basis functions $\phi_{01}, \phi_1, \phi_{12}, \ldots, \phi_{N-1\,N}, \phi_N$, which leads to the following Galerkin equations:

Find $\tilde{u}$ of the form (5.2.3) i.e. $\tilde{u} \in V_N$, such that (5.2.4)

$$\int_0^1 \frac{d\tilde{u}}{dx}\frac{d\phi_i}{dx}\,dx = \int_0^1 f\,\phi_i dx + g_1\phi_i(1) \qquad \text{for } i \in I.$$

Substitution of (5.2.3) into (5.2.4) and splitting the integral over $[0,1]$ into a sum of integrals over the subintervals $e_k = [x^{k-1}, x^k]$, $k = 1,2,\ldots,N$, we obtain

$$\sum_{k=1}^{N} \sum_{j\in I} \tilde{u}_j \{\int_{e_k} \frac{d\phi_j}{dx}\frac{d\phi_i}{dx}\,dx\} = \sum_{k=1}^{N} \int_{e_k} f\,\phi_i dx + g_1\phi_i(1) \text{ for } i \in I. \tag{5.2.5}$$

On the interval e_k only three basis functions are different from zero, viz. ϕ_{k-1}, $\phi_{k-1\,k}$ and ϕ_k. This means that for each k, $k=1,2,\ldots,N$, there are nine combinations $\{i,j\}$, $i \in I$, $j \in I$, that contribute to the integral over e_k in (5.2.5).

These nine quantities can be arranged in the element stiffness matrix

$$A^{e_k} = \begin{vmatrix} \int_{e_k} \frac{d\phi_{k-1}}{dx}\frac{d\phi_{k-1}}{dx}\,dx & \int_{e_k} \frac{d\phi_{k-1\,k}}{dx}\frac{d\phi_{k-1}}{dx}\,dx & \int_{e_k} \frac{d\phi_{k}}{dx}\frac{d\phi_{k-1}}{dx}\,dx \\ \int_{e_k} \frac{d\phi_{k-1}}{dx}\frac{d\phi_{k-1\,k}}{dx}\,dx & \int_{e_k} \frac{d\phi_{k-1\,k}}{dx}\frac{d\phi_{k-1\,k}}{dx}\,dx & \int_{e_k} \frac{d\phi_{k}}{dx}\frac{d\phi_{k-1\,k}}{dx}\,dx \\ \int_{e_k} \frac{d\phi_{k-1}}{dx}\frac{d\phi_{k}}{dx}\,dx & \int_{e_k} \frac{d\phi_{k-1\,k}}{dx}\frac{d\phi_{k}}{dx}\,dx & \int_{e_k} \frac{d\phi_{k}}{dx}\frac{d\phi_{k}}{dx}\,dx \end{vmatrix} \tag{5.2.6}$$

All these integrals can be evaluated easily using the barycentric coordinate expressions for the basis functions. Consider for instance subinterval e_k with nodal points x^{k-1},

$x^{k-1} \rightarrow x^1$

$x^{k-1\,k} \rightarrow x^{12}$

$x^k \rightarrow x^2$

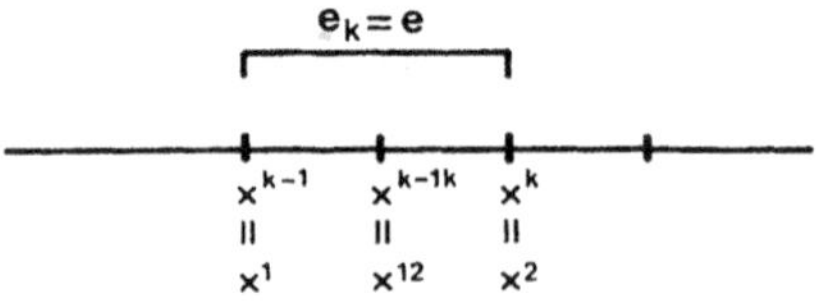

Fig. 5.4 Global and local numbering.

x^{k-1} ^{k}, x^k. Let us introduce a local numbering of these nodal points as clarified by Fig. 5.4.
In the new numbering the basis functions are given by (4.1.4), (4.1.5) and (4.1.6) where $\{\lambda_1(x),\lambda_2(x)\}$ denote the barycentric coordinates of the point $x \in e$ with respect to x^1 and x^2. The derivatives of these basis functions with respect to x can be expressed in the barycentric coordinates too. One easily finds:

$$\frac{d\phi_1}{dx} = \frac{d\lambda_1}{dx}(2\lambda_1-1) + \lambda_1(2\frac{d\lambda_1}{dx}) = -\frac{4\lambda_1-1}{x^2-x^1}$$

$$\frac{d\phi_2}{dx} = \frac{d\lambda_2}{dx}(2\lambda_2-1) + \lambda_2(2\frac{d\lambda_2}{dx}) = \frac{4\lambda_2-1}{x^2-x^1} \qquad (5.2.7)$$

$$\frac{d\phi_{12}}{dx} = 4\frac{d\lambda_1}{dx}\lambda_2 + 4\lambda_1\frac{d\lambda_2}{dx} = 4\frac{\lambda_1-\lambda_2}{x^2-x^1} .$$

Using the following integration formula

$$\int_e \lambda_1^{m_1}\lambda_2^{m_2}\,dx = \frac{m_1!m_2!}{(m_1+m_2+1)!}h \qquad h = |x^2-x^1| \qquad (5.2.8)$$

m_1, m_2 non-negative integers

integrals of the type

$$\int_e \frac{d\phi_{..}}{dx}\frac{d\phi_{..}}{dx}\,dx$$

can be computed easily. Let us calculate for instance $(A^{e_k})_{21}$:

$$\int_{e_k}\frac{d\phi_{k-1}}{dx}\frac{d\phi_{k-1\,k}}{dx}\,dx \rightarrow \int_e \frac{d\phi_1}{dx}\frac{d\phi_{12}}{dx}\,dx = \int_e \left(-\frac{4\lambda_1-1}{x^2-x^1}\right)\left(4\frac{\lambda_1-\lambda_2}{x^2-x^1}\right)dx =$$

$$= \frac{4}{h^2}\int_e(-4\lambda_1+1)(\lambda_1-\lambda_2)dx = \frac{4}{h^2}\int_e(-4\lambda_1^2+4\lambda_1\lambda_2+\lambda_1-\lambda_2)dx =$$

$$= \frac{4}{h^2}\left[-4\frac{2}{3!}h + 4\frac{1}{3!}h + \frac{1}{2!}h - \frac{1}{2!}h\right] = -\frac{8}{3h}$$

with $h = |x^2-x^1|$.

For the element matrix A^{e_k} we find

$$A^{e_k} = \frac{1}{3h} \begin{vmatrix} 7 & -8 & 1 \\ -8 & 16 & -8 \\ 1 & -8 & 7 \end{vmatrix} \begin{matrix} \leftarrow i=k-1 \\ \leftarrow i=k-1\ k \\ \leftarrow i=k \end{matrix} \tag{5.2.9}$$

(columns: $j=k-1$, $j=k-1\ k$, $j=k$)

with $h = |e_k|$ = length of element e_k.

The procedure is now to calculate the element matrices for all subintervals e_k, $k=1,2,\ldots,N$, and to assemble them to form the large stiffness matrix A. The essential boundary condition $u_0 = g_0$ is taken into account in exactly the same way as described in Section 5.1. The right hand side vector is evaluated as follows. On e_k only three basis functions differ from zero. They give rise to the following element vector

$$\underline{F}^{e_k} = \begin{vmatrix} \int_{e_k} f\ \phi_{k-1} dx + g_1 \phi_{k-1}(1) \\ \int_{e_k} f\ \phi_{k-1\ k} dx + g_1 \phi_{k-1\ k}(1) \\ \int_{e} f\ \phi_k dx + g_1 \phi_k(1) \end{vmatrix} . \tag{5.2.10}$$

Renumbering the points x^{k-1}, $x^{k-1\ k}$, x^k of e_k into x^1, x^{12}, x^2 and taking into account that $\phi_k(1) \neq 0$ only when $k = N$ (or after renumbering: for $k = 2$ and $e = e_N$), we find

$$\underline{F}^{e} = \begin{vmatrix} \int_{e} f\ \phi_1\ dx \\ \int_{e} f\ \phi_{12}\ dx \\ \int_{e} f\ \phi_2\ dx \end{vmatrix} \qquad \text{for } e = e_k,\ k = 1,2,\ldots,N-1$$

and

$$\underline{F}^{e} = \left| \begin{array}{l} \int_e f\ \phi_1 dx \\ \\ \int_e f\ \phi_{12} dx \\ \\ \int_e f\ \phi_2 dx + g_1 \end{array} \right| \qquad \text{for } e = e_N .$$

If we take for example f = 1, we obtain, using (5.2.8) for the computation of the integrals

$$\underline{F}^{e_k} = \left| \begin{array}{c} \frac{h}{6} \\ \\ \frac{4h}{6} \\ \\ \frac{h}{6} \end{array} \right| , \ h = 1,2,\ldots,N-1; \quad \underline{F}^{e_N} = \left| \begin{array}{l} \frac{h}{6} \\ \\ \frac{4h}{6} \\ \\ \frac{h}{6} + g_1 \end{array} \right| . \qquad (5.2.11)$$

By assembling the element vectors, the right hand side vector $\underline{F}$ can be constructed. Finally the system of linear algebraic equations

$$A\ \underline{\tilde{u}} = \underline{F}\ , \qquad \tilde{u} = (\tilde{u}_{01}, \tilde{u}_1, \tilde{u}_{12}, \ldots, \tilde{u}_{N-1\ N}, \tilde{u}_N)^T$$

can be solved.

Just as in Section 3.6 we can prove that A is positive definite.

5.3 2D Poisson equation; linear and quadratic triangular elements

Let us consider the following problem on a 2D region Ω with boundary Γ

$$\left| \begin{array}{ll} -\ \Delta u = f & \text{in } \Omega \\ \\ u = 0 & \text{on } \Gamma_0 \\ \\ \frac{\partial u}{\partial n} = g_1 & \text{on } \Gamma_1 \\ \\ \frac{\partial u}{\partial n} + u = g_2 & \text{on } \Gamma_2 \end{array} \right. \qquad (5.3.1)$$

where Γ_0, Γ_1 and Γ_2 are disjoint and satisfy $\Gamma = \Gamma_0 \cup \Gamma_1 \cup \Gamma_2$. The weak or variational formulation of (5.3.1) is

Find $u : \bar{\Omega} \to \mathbb{R}$ such that $u|_{\Gamma_0} = 0$ and

$$\int_\Omega \text{grad } u . \text{grad } v \, d\Omega + \int_{\Gamma_2} uv \, d\Gamma = \tag{5.3.2}$$

$$= \int_\Omega fv \, d\Omega + \int_{\Gamma_1} g_1 v \, d\Gamma + \int_{\Gamma_2} g_2 v \, d\Gamma$$

for all v with $v|_{\Gamma_0} = 0$

On Ω we choose a triangulation e_k, $k=1,2,...,K$. Let Γ_1 consist of P edges ℓ_{1p}, $p=1,2,...,P$, and let Γ_2 consist of Q edges ℓ_{1q}, $q=1,2,...,Q$, as sketched in fig. 5.5.

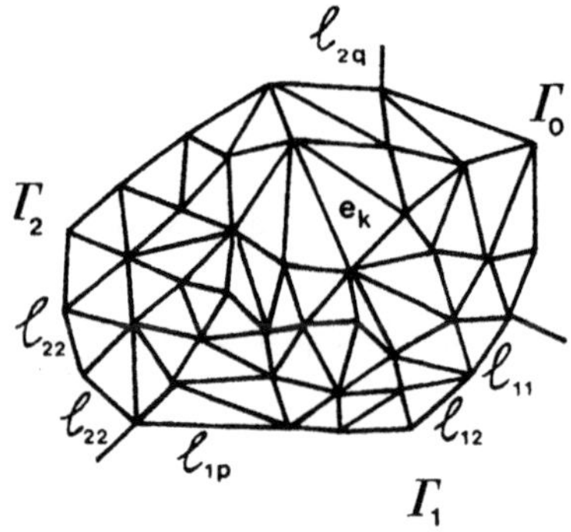

Fig. 5.5 Triangulation of Ω.

5.3.1 Linear finite element

For the construction of a linear finite element approximation we choose as nodal points all the vertices of the triangles. Then, if ϕ_i denotes the piecewise linear basis function corresponding to nodal point $\underline{x}^i$, the Galerkin equations for the approximate solution are

$$\sum_{j\in I} \tilde{u}_j \int_\Omega \text{grad } \phi_j \cdot \text{grad } \phi_i \, d\Omega + \sum_{j\in I} \tilde{u}_j \int_{\Gamma_2} \phi_j \phi_i \, d\Gamma =$$
$$= \int_\Omega f\phi_i \, d\Omega + \int_{\Gamma_1} g_1\phi_i \, d\Gamma + \int_{\Gamma_2} g_2\phi_i \, d\Gamma \tag{5.3.3}$$

for all $i \in I$, where I denotes the set of indices corresponding to those nodal points which do not belong to Γ_0. The approximate solution $\tilde{u}$ has the following form:

$$\tilde{u}(\underline{x}) = \sum_{j\in I} \tilde{u}_j \phi_j(\underline{x})$$

The integrals appearing in the system of equations (5.3.3) can be split up into a sum of integrals over subregions to give:

$$\sum_{k=1}^{K} \sum_{j\in I} \tilde{u}_j \Big\{ \int_{e_k} \text{grad } \phi_j \cdot \text{grad } \phi_i \, d\Omega \Big\} + \sum_{q=1}^{Q} \sum_{j\in I} \tilde{u}_j \Big\{ \int_{\ell_{2q}} \phi_j \phi_i \, d\Gamma \Big\} =$$
$$= \sum_{k=1}^{K} \int_{e_k} f\phi_i \, d\Omega + \sum_{p=1}^{P} \int_{\ell_{1p}} g_1\phi_i \, d\Gamma + \sum_{q=1}^{Q} \int_{\ell_{2q}} g_2\phi_i \, d\Gamma. \tag{5.3.4}$$

The subregions e_k are called triangular elements, the subregions ℓ_{1p} and ℓ_{2q} are termed line elements.
Consider triangle e_k and renumber the nodal points on e_k into $\underline{x}^1$, $\underline{x}^2$, $\underline{x}^3$ (see Fig. 5.6).

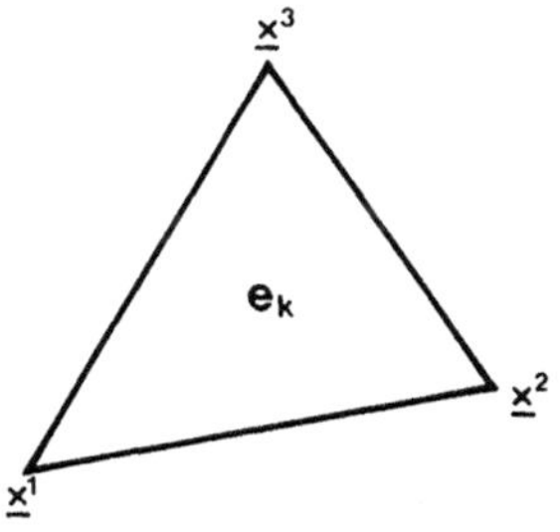

Fig. 5.6 Triangle e_k with nodal points $\underline{x}^1$, $\underline{x}^2$, $\underline{x}^3$.

On e_k only the basis functions ϕ_1, ϕ_2, ϕ_3 differ from zero. This implies that for each k = 1,2,...,K,the integral in the first term of (5.3.4) gives rise to the following 3×3 element matrix A^{e_k}:

$$(A^{e_k})_{ij} = \int_{e_k} \text{grad } \phi_j \cdot \text{grad } \phi_i \, d\Omega \quad i,j = 1,2,3. \tag{5.3.5}$$

Since on e_k the basis functions ϕ_1, ϕ_2, ϕ_3 satisfy (4.2.3), it is, using the following relations

$$\frac{\partial\lambda_1}{\partial x_1} = \frac{x_2^2-x_2^3}{\Delta} \qquad \frac{\partial\lambda_2}{\partial x_1} = \frac{x_2^3-x_2^1}{\Delta} \qquad \frac{\partial\lambda_3}{\partial x_1} = \frac{x_2^1-x_2^2}{\Delta} \tag{5.3.6}$$

$$\frac{\partial\lambda_1}{\partial x_2} = \frac{x_1^3-x_1^2}{\Delta} \qquad \frac{\partial\lambda_2}{\partial x_2} = \frac{x_1^1-x_1^3}{\Delta} \qquad \frac{\partial\lambda_3}{\partial x_2} = \frac{x_1^2-x_1^1}{\Delta}, \tag{5.3.7}$$

an easy exercise to compute the element matrix:

$$A^{e_k} = \frac{1}{2|\Delta|}\begin{vmatrix} e^{22}e^{22}+e^{21}e^{21} & & \text{symmetric} \\ e^{22}e^{32}+e^{21}e^{31} & e^{32}e^{32}+e^{31}e^{31} & \\ e^{22}e^{12}+e^{21}e^{11} & e^{32}e^{12}+e^{31}e^{11} & e^{12}e^{12}+e^{11}e^{11} \end{vmatrix} \tag{5.3.8}$$

with

$$e^{ij} = x_j^{i+1}-x_j^i \qquad i = 1,2,3 \text{ (cyclic)}$$
$$j = 1,2$$

$$\Delta = e^{11}e^{22}-e^{12}e^{21}.$$

Next we calculate the contribution to the stiffness matrix of the second term in (5.3.4). Consider a line-element ℓ_{2q} which is for instance an edge of triangle e_k (see Fig. 5.7).

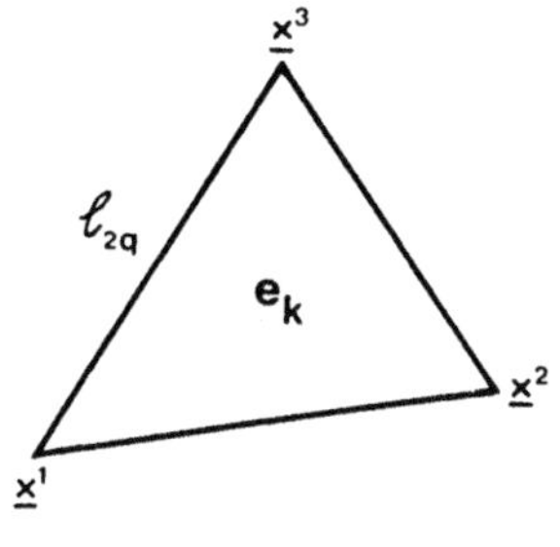

Fig. 5.7 Triangle e_k with edge ℓ_{2q}.

The only basis functions which do not vanish on ℓ_{2q} are ϕ_1 and ϕ_2 corresponding to the nodal points $\underline{x}^1$ and $\underline{x}^2$ respectively. This implies that the line-element ℓ_{2q} gives rise to a 2×2 (line-) element matrix $A^{\ell_{2q}}$ the components of which are

$$(A^{\ell_{2q}})_{ij} = \int_{\ell_{2q}} \phi_j \phi_i \, d\Gamma \ . \qquad (5.3.9)$$

Since $\phi_1 = \lambda_1$ and $\phi_2 = \lambda_2$ on ℓ_{2q}, we can use relation (5.2.8) to compute the matrix $A^{\ell_{2q}}$. We obtain

$$A^{\ell_{2q}} = \frac{h}{6} \begin{vmatrix} 2 & 1 \\ 1 & 2 \end{vmatrix} \quad \text{with } h = \text{length of } \ell_{2q}. \qquad (5.3.10)$$

Assembling now all the element matrices A^{e_k}, k=1,2,...,K,and $A^{\ell_{2q}}$, q=1,2,...,Q, we obtain finally the stiffness matrix A.

Concerning the right hand side vector, we notice that the integral over e_k in the first term of the right hand side of (5.3.4) is different from zero when ϕ_i corresponds to one of the three nodal points (vertices) of e_k. For e_k, with nodal points $\underline{x}^1$, $\underline{x}^2$, $\underline{x}^3$, this leads to a 3-component element vector $\underline{F}^{e_k}$:

$$\underline{F}^{e_k} = \begin{vmatrix} \int_{e_k} f\ \phi_1\ d\Omega \\ \int_{e_k} f\ \phi_2\ d\Omega \\ \int_{e_k} f\ \phi_3\ d\Omega \end{vmatrix} \qquad k = 1,2,\ldots,K \qquad (5.3.11)$$

The second and third term on the right hand side of (5.3.4) give rise to the (line-) element vectors

$$\underline{F}^{\ell_{1p}} = \begin{vmatrix} \int_{\ell_{1p}} g_1\ \phi_1\ d\Gamma \\ \int_{\ell_{1p}} g_1\ \phi_2\ d\Gamma \end{vmatrix} \qquad p = 1,2,\ldots,P \qquad (5.3.12)$$

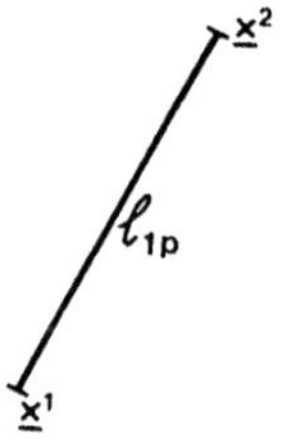

Fig. 5.8 Line element ℓ_{1p}.

and

$$\underline{F}^{\ell_{2q}} = \begin{vmatrix} \int_{\ell_{2q}} g_2\ \phi_1\ d\Gamma \\ \int_{\ell_{2q}} g_2\ \phi_2\ d\Gamma \end{vmatrix} \qquad q = 1,2,\ldots,Q \qquad (5.3.13)$$

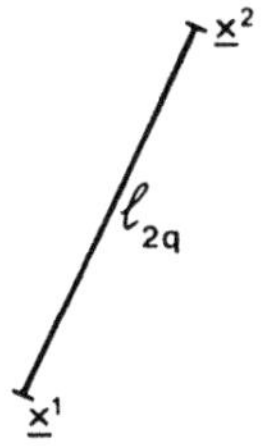

Fig. 5.9 Line element ℓ_{2q}.

Assemblage of all $\underline{F}^{e_k}$, k=1,2,...,K, $\underline{F}^{\ell_{1p}}$, p=1,2,...,P, and $\underline{F}^{\ell_{2q}}$, q=1,2,...,Q, results in the right hand side vector $\underline{F}$.

Finally we have to solve the following system of linear algebraic equations

$$A\,\underline{\tilde{u}} = \underline{F} \tag{5.3.14}$$

where $\underline{\tilde{u}}$ denotes the unknown parameters $\underline{\tilde{u}} = (\ldots,\tilde{u}_i,\ldots)_{i\in I}$. Let us emphasize that (5.3.14) is nothing else than system (5.3.3). The introduction of the notions of element matrix and element vector is only a programming trick for an efficient assemblage of the matrix A and the vector $\underline{F}$. The stiffness matrix A is positive definite as can be seen from the following. We multiply the left hand side of relation (5.3.3) by $\tilde{u}_i$ and add the equations for all $i \in I$, this gives

$$\underline{\tilde{u}}^T A\,\underline{\tilde{u}} = \sum_{i\in I}\tilde{u}_i\Big[\sum_{j\in I}\tilde{u}_j\int_\Omega \text{grad}\,\phi_j\cdot\text{grad}\,\phi_i\,d\Omega + \sum_{j\in I}\tilde{u}_j\int_{\Gamma_2}\phi_j\phi_i\,d\Gamma\Big] =$$

$$= \int_\Omega \text{grad}\,\tilde{u}\cdot\text{grad}\,\tilde{u}\,d\Omega + \int_{\Gamma_2}(\tilde{u})^2 d\Gamma > 0 \tag{5.3.15}$$

for all $\tilde{u}$ which do not vanish identically. Moreover the matrix A is symmetric since

$$A_{ji} = \int_\Omega \text{grad}\,\phi_i\cdot\text{grad}\,\phi_j\,d\Omega + \int_{\Gamma_2}\phi_i\phi_j\,d\Gamma = A_{ij}$$

The matrix A is sparse since $A_{ij} = 0$ except when the nodal points $\underline{x}^i$ and $\underline{x}^j$ are vertices of a same triangle. When the nodal points $\underline{x}^i$, $i \in I$, are numbered appropriately the matrix A has a band structure, with band width b_w

$$b_w = 2b_{hw} + 1$$

where the half band width b_{hw} is defined as the maximum over all triangles of the absolute value of the difference of the numbers of two nodal points which belong to the same triangle:

$$b_{hw} = \max_k \max_{\substack{\underline{x}^i \in e_k \\ \underline{x}^j \in e_k}} |i-j|$$

5.3.2 Quadratic finite element

For the definition of a quadratic finite element approximation we choose in each triangle e_k six nodal points: three vertices and three mid-points of sides. The Galerkin equations for the approximate solution are given by (5.3.4), where I denotes now the set of indices corresponding to the nodal points (i.e. vertices and mid-points of sides) which do not belong to Γ_0. For the construction of the stiffness matrix and the right hand side vector, we follow the same procedure as for the linear finite element approximation. The (triangular-) element matrix A^{e_k} for triangle e_k with nodal points $\underline{x}^1, \underline{x}^2, \underline{x}^3, \underline{x}^4 = \underline{x}^{23}$, $\underline{x}^5 = \underline{x}^{13}$, $\underline{x}^6 = \underline{x}^{12}$ (see Fig. 5.10)

is a 6×6 matrix since there are 6 basis functions different from zero on e_k. The matrix-elements of A^{e_k} are given by

$$(A^{e_k})_{ij} = \int_{e_k} \operatorname{grad} \phi_j \cdot \operatorname{grad} \phi_i \, d\Omega \qquad i,j = 1,2,\ldots,6$$

where ϕ_i is the piecewise quadratic basis function corresponding to nodal point $\underline{x}^i$, $i=1,2,\ldots,6$.

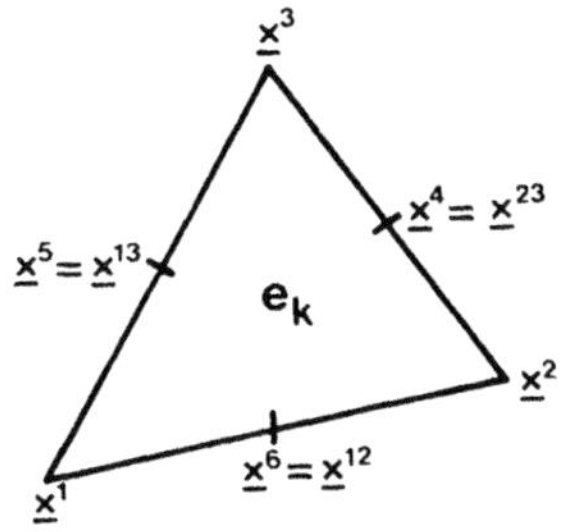

Fig. 5.10 Triangle e_k with nodal points $\underline{x}^1,\underline{x}^2,\ldots,\underline{x}^6$.

The shape of the functions $\phi_1,\phi_2,\phi_3,\phi_4 = \phi_{23}$, $\phi_5 = \phi_{13}$, $\phi_6 = \phi_{12}$ is given by (4.2.5),...,(4.2.10) in terms of the barycentric coordinates.

The (line-) element matrix $A^{\ell_{2q}}$ for edge ℓ_{2q} with nodal points $\underline{x}^1,\underline{x}^2,\underline{x}^3 = \underline{x}^{12}$ (see Fig. 5.11)

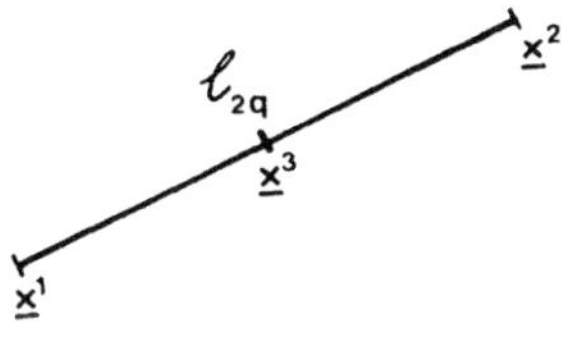

Fig. 5.11 Edge ℓ_{2q} with nodal points $\underline{x}^1,\underline{x}^2,\underline{x}^3$.

is a 3×3 matrix given by

$$(A^{\ell_{2q}})_{ij} = \int_{\ell_{2q}} \phi_j\phi_i d\Gamma \qquad i,j = 1,2,3 \ .$$

Using (4.1.4),...,(4.1.6) and (5.2.8) we find

$$A^{\ell_{2q}} = \frac{h}{30} \begin{vmatrix} 4 & -1 & 2 \\ -1 & 4 & 2 \\ 2 & 2 & 16 \end{vmatrix} \qquad h = \text{length of } \ell_{2q}.$$

The (triangular-) element vector $\underline{F}^{e_k}$ and the (line-) element vectors $\underline{F}^{\ell_{1p}}$ and $\underline{F}^{\ell_{2q}}$ are defined by

$$(\underline{F}^{e_k})_i = \int_{e_k} f\ \phi_i\ d\Omega \qquad i = 1,2,\ldots,6$$

$$(\underline{F}^{\ell_{1p}})_i = \int_{\ell_{1p}} g_1 \phi_i d\Gamma \qquad i = 1,2,3$$

$$(\underline{F}^{\ell_{2q}})_i = \int_{\ell_{2q}} g_2 \phi_i d\Gamma \qquad i = 1,2,3$$

with e_k, ℓ_{1p} and ℓ_{2q} depicted in Fig. 5.12.

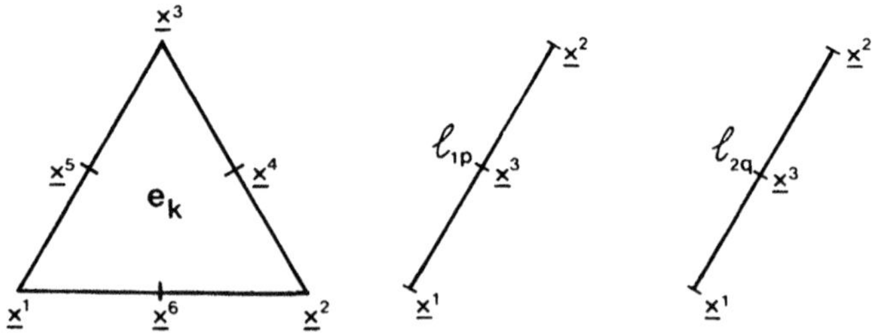

Fig. 5.12 Triangular element e_k, line elements ℓ_{1p} and ℓ_{2q}.

Once the matrix A and the right hand side vector $\underline{F}$ have been assembled, we can solve the system $A\ \underline{\tilde{u}} = \underline{F}$ for the unknown parameters $\tilde{u}_i$, $i \in I$.

Analogous to the linear finite element case we verify that A is symmetric, positive definite, sparse and has a band structure when the nodal points are numbered appropriately.

Finally we remark that for the computation of integrals of the following form

$$\int_e \frac{\partial\phi_{..}}{\partial x_k} \frac{\partial\phi_{..}}{\partial x_\ell} \, d\Omega \qquad\qquad k,\ell = 1,2$$

$$\int_e \phi_{..} \phi_{..} \, d\Omega$$

we can use the expressions of $\phi_{..}$ in terms of the barycentric coordinates, together with the relations (5.3.6), (5.3.7) and the integration formula

$$\int_e \lambda_1^{m_1} \lambda_2^{m_2} \lambda_3^{m_3} d\Omega = 2 \frac{m_1!m_2!m_3!}{(m_1+m_2+m_3+2)!} |e| \qquad (5.3.16)$$

where m_1, m_2 and m_3 are non-negative integers and $|e|$ denotes the area of e (cf. (5.2.8)).
The integration rules (5.2.8) for n=1 and (5.3.16) for n=2 can be generalized to the n-dimensional case (see Section 12.3.1). For the 3D case of a tetrahedron e one has

$$\int_e \lambda_1^{m_1} \lambda_2^{m_2} \lambda_3^{m_3} \lambda_4^{m_4} d\Omega = 3! \frac{m_1!m_2!m_3!m_4!}{(m_1+m_2+m_3+m_4+3)!} |e| \qquad (5.3.17)$$

with m_1, m_2, m_3, m_4 non-negative integers and $|e|$ the volume of e.

5.4 Numerical integration formulas

As we have seen in the preceding section, the computation of element vectors involves the computation of integrals over subregions of the following form:

$$\int_e f \, \phi_i \, d\Omega \qquad \text{(cf. (5.1.7), (5.2.10), (5.3.11))} \qquad (5.4.1)$$

$$\int_e g \, \phi_i \, d\Gamma \qquad \text{(cf. (5.3.12), (5.3.13))} \qquad (5.4.2)$$

When f and g are arbitrary functions, it could be very complicated or even impossible to evaluate these integrals exactly. Numerical integration formulas must be used. A similar situation occurs for the computation of element matrices. Consider for instance a Poisson type PDE with non-constant coefficients (for instance the heat conduction equation with non-constant thermal conductivity, see Section 1.2).

$$-\operatorname{div}(k \operatorname{grad} u) = f \quad \text{in } \Omega$$

with $k = k(\underline{x})$. This equation leads to the following finite element Galerkin equations:

$$\sum_{j \in I} \tilde{u}_j \int_\Omega k \operatorname{grad} \phi_j \cdot \operatorname{grad} \phi_i \, d\Omega = \int_\Omega f \, \phi_i \, d\Omega$$

The element matrix for subregion e contains integrals of the type:

$$\int_e k \operatorname{grad} \phi_j \cdot \operatorname{grad} \phi_i \, d\Omega \tag{5.4.3}$$

which, in general, can only be approximated using numerical integration formulas.
This means that we replace an integral of the form

$$\int_e \psi(\underline{x}) \, d\Omega$$

by a finite sum of the form

$$\sum_{r=1}^{R} \omega_r^e \, \psi(\underline{b}_r^e)$$

The real numbers $\omega_r^e > 0$ are called the weights and the points $\underline{b}_r^e \in e$ are called the integration nodes. R is an integer independent of e.
A theory and tables of formulas can be found in Stroud (1971).

5.4.1 Numerical integration on intervals, triangles and tetrahedra

In 1D the subregion e is an interval $[x^1,x^2]$; in 2D e represents a triangle with vertices $\underline{x}^1,\underline{x}^2,\underline{x}^3$, in 3D a tetrahedra with vertices $\underline{x}^1,\underline{x}^2,\underline{x}^3,\underline{x}^4$. A class of numerical integration formulas is known as the Newton-Cotes formulas which are derived by integrating exactly a Lagrangian interpolation polynomial of degree R-1 that is fitted to R values of ψ. We shall use the integration nodes indicated in Fig. 5.13.

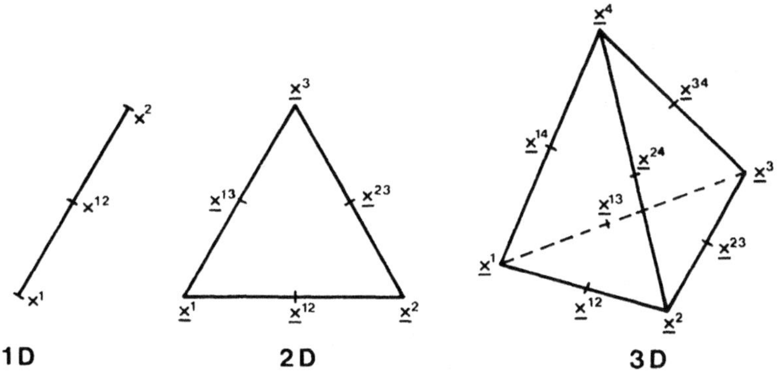

Fig. 5.13 Subregions e with integration points.

A few examples of Newton-Cotes formulas in 1D, 2D and 3D will be given. By $|e|$ we represent the length, the area or the volume of e for the 1D, 2D, 3D case respectively. By ψ_α we denote the value of ψ at the point $\underline{x}^\alpha$: $\psi_\alpha = \psi(\underline{x}^\alpha)$.

1D; e = $[x^1,x^2]$ (see Fig. 5.13)

$$\int_e \psi(x)\, dx \approx |e|\, \psi_{12} \tag{5.4.4}$$

This formula is exact for all $\psi \in P_1(e)$ and is called the mid-point rule.

$$\int_e \psi(x)\, dx \approx \frac{|e|}{2} \{\psi_1+\psi_2\} \tag{5.4.5}$$

This formula is known as the trapezoidal rule and is exact for all $\psi \in P_1(e)$

$$\int_e \psi(x)\, dx \approx \frac{|e|}{6} \{\psi_1+4\psi_{12}+\psi_2\} \tag{5.4.6}$$

Simpson's rule; exact for $\psi \in P_3(e)$

2D; e is triangle (see fig. 5.13)

$$\int_e \psi(\underline{x})\, d\Omega \approx \frac{|e|}{3} \{\psi_1+\psi_2+\psi_3\}, \text{ exact for } \psi \in P_1(e) \tag{5.4.7}$$

$$\int_e \psi(\underline{x})\, d\Omega \approx |e|\, \psi_{123}, \text{ exact for } \psi \in P_1(e) \tag{5.4.8}$$

$$\int_e \psi(\underline{x})\, d\Omega \approx \frac{|e|}{3} \{\psi_{12}+\psi_{13}+\psi_{23}\}, \text{ exact for } \psi \in P_2(e) \tag{5.4.9}$$

$$\int_e \psi(\underline{x})\, d\Omega \approx \frac{|e|}{60} [3\{\psi_1+\psi_2+\psi_3\}+8\{\psi_{12}+\psi_{13}+\psi_{23}\}+27\psi_{123}], \text{ exact for } \psi \in P_3(e) \tag{5.4.10}$$

3D; e is tetrahedron (see fig. 5.13)

$$\int_e \psi(\underline{x})d\Omega \approx \frac{|e|}{4} \{\psi_1+\psi_2+\psi_3+\psi_4\}, \text{ exact for } \psi \in P_1(e) \tag{5.4.11}$$

$$\int_e \psi(\underline{x})d\Omega \approx |e|\, \psi_{1234},\ \underline{x}^{1234} = \tfrac{1}{4}(\underline{x}^1+\underline{x}^2+\underline{x}^3+\underline{x}^4) \text{ is barycentre of e, exact for } \psi \in P_1(e) \tag{5.4.12}$$

$$\int_e \psi(\underline{x})d\Omega \approx \frac{|e|}{20} \{16\psi_{1234}+(\psi_1+\psi_2+\psi_3+\psi_4)\} \text{ exact for } \psi \in P_2(e) \tag{5.4.13}$$

Which integration rule should be used for a specific problem depends on the shape of the basis functions (linear, quadratic, ...). We will come back to this point in Section 5.5.

Using these formulas, element vectors can be written as follows:

1D, linear basis functions, formula (5.4.5):

$$\underline{F}^e = \begin{vmatrix} \int_e f\phi_1 dx \\ \int_e f\phi_2 dx \end{vmatrix} \simeq \frac{|e|}{2} \begin{vmatrix} f(x^1) \\ f(x^2) \end{vmatrix} \quad \text{(cf. (5.1.7)).}$$

1D, quadratic basis functions, formula (5.4.6):

$$\underline{F}^e = \begin{vmatrix} \int_e f\phi_1 dx \\ \int_e f\phi_{12} dx \\ \int_e f\phi_2 dx \end{vmatrix} \simeq \frac{|e|}{6} \begin{vmatrix} f(x^1) \\ 4f(x^{12}) \\ f(x^2) \end{vmatrix} \quad \text{(cf. (5.2.10, (5.2.11))}$$

2D, linear basis functions, formula (5.4.7):

$$\underline{F}^e = \begin{vmatrix} \int_e f\phi_1 d\Omega \\ \int_e f\phi_2 d\Omega \\ \int_e f\phi_3 d\Omega \end{vmatrix} \simeq \frac{|e|}{3} \begin{vmatrix} f(\underline{x}^1) \\ f(\underline{x}^2) \\ f(\underline{x}^3) \end{vmatrix} \quad \text{(cf. (5.3.11))}$$

Using formula (5.4.5) we find

$$\underline{F}^\ell = \begin{vmatrix} \int_\ell g_1\phi_1 d\Gamma \\ \int_\ell g_1\phi_2 d\Gamma \end{vmatrix} \simeq \frac{|\ell|}{2} \begin{vmatrix} g(\underline{x}^1) \\ g(\underline{x}^2) \end{vmatrix} \quad \begin{matrix}\text{(cf. (5.3.12)} \\ \text{(5.3.13)}\end{matrix}$$

2D, extended quadratic basis functions, formula (5.4.10):

$$\underline{F}^e = \begin{vmatrix} \int_e f\phi_1 d\Omega \\ \int_e f\phi_2 d\Omega \\ \int_e f\phi_3 d\Omega \\ \int_e f\phi_{23} d\Omega \\ \int_e f\phi_{13} d\Omega \\ \int_e f\phi_{12} d\Omega \\ \int_e f\phi_{123} d\Omega \end{vmatrix} \approx \frac{|e|}{60} \begin{vmatrix} 3f(\underline{x}^1) \\ 3f(\underline{x}^2) \\ 3f(\underline{x}^3) \\ 8f(\underline{x}^{23}) \\ 8f(\underline{x}^{13}) \\ 8f(\underline{x}^{12}) \\ 27f(\underline{x}^{123}) \end{vmatrix} .$$

3D, linear basis functions, formula (5.4.11):

$$\underline{F}^e = \begin{vmatrix} \int_e f\phi_1 d\Omega \\ \int_e f\phi_2 d\Omega \\ \int_e f\phi_3 d\Omega \\ \int_e f\phi_4 d\Omega \end{vmatrix} \simeq \frac{|e|}{4} \begin{vmatrix} f(\underline{x}^1) \\ f(\underline{x}^2) \\ f(\underline{x}^3) \\ f(\underline{x}^4) \end{vmatrix} .$$

Let us calculate finally the element matrix component (5.4.3) for linear basis functions in 2D using formula (5.4.7). Since $\phi_i = \lambda_i$ and grad λ_i is constant, we find:

$$\int_e k \text{ grad } \phi_j \cdot \text{grad } \phi_i \, d\Omega = \int_e k \text{ grad } \lambda_j \cdot \text{grad } \lambda_i \, d\Omega \simeq$$

$$\simeq \text{grad } \lambda_j \cdot \text{grad } \lambda_i \, \{\frac{|e|}{3} (k(\underline{x}^1) + k(\underline{x}^2) + k(\underline{x}^3))\}$$

which can be evaluated further using the expressions (5.3.6), (5.3.7).

Another approach to introduce numerical integration is to transform an integral over a subregion e into an integral over

the reference element $\hat{e}$ (see Section 4.4) and then to apply a numerical integration formula over the reference element.
Let an element e be the image of the reference element $\hat{e}$ under the mapping $\underline{F}_e$, then an integral over e of the type

$$\int_e \psi(\underline{x})\, d\Omega \qquad (n = 1,2,3) \tag{5.4.14}$$

can be transformed into an integral over $\hat{e}$ by the following equality

$$\int_e \psi(\underline{x})\, d\Omega = \int_{\hat{e}} \psi(F_e(\hat{\underline{x}}))J_{F_e}(\hat{\underline{x}})\, d\hat{\Omega} \equiv \int_{\hat{e}} \hat{\psi}(\hat{\underline{x}})\, d\hat{\Omega} \tag{5.4.15}$$

with

$$\hat{\psi}(\hat{\underline{x}}) = \psi(F_e(\hat{\underline{x}}))J_{F_e}(\hat{\underline{x}})$$

where $J_{F_e}(\hat{\underline{x}})$ denotes the Jacobian of the mapping F_e at the point $\hat{\underline{x}}$. When F_e is a linear mapping, it can be written as (4.4.2) and (5.4.15) simplifies to

$$\begin{aligned}\int_e \psi(\underline{x})\, d\Omega &= \int_{\hat{e}} \psi(F_e(\hat{\underline{x}}))\, \det(B_e)\, d\hat{\Omega} = \\ &= \det(B_e) \int_{\hat{e}} \psi(F_e(\hat{\underline{x}}))\, d\hat{\Omega}\end{aligned} \tag{5.4.16}$$

Next, the integrals over $\hat{e}$ in (5.4.15) and in (5.4.16) can be approximated using one of the formulas (5.4.4),...,(5.4.13).
For a linear mapping F_e of the type (4.4.2), integrals of the form

$$\int_e k \frac{\partial \phi_j}{\partial x_\ell} \frac{\partial \phi_i}{\partial x_m} \qquad k = k(\underline{x})$$

are transformed into integrals over $\hat{e}$ by

$$\int_e k(\underline{x}) \frac{\partial \phi_j}{\partial x_\ell}(\underline{x}) \frac{\partial \phi_i}{\partial x_m}(\underline{x})\, d\Omega =$$

$$= \sum_{r,s=1}^{n} \int_{\hat{e}} \det(B_e)\, k(F_e(\hat{\underline{x}}))\, \frac{\partial\hat{\phi}_j}{\partial\hat{x}_r} \frac{\partial\hat{\phi}_i}{\partial\hat{x}_s} \frac{\partial\hat{x}_r}{\partial x_j} \frac{\partial\hat{x}_s}{\partial x_i}\, d\hat{\Omega} = \tag{5.4.17}$$

$$= \det(B_e) \sum_{r,s=1}^{n} \frac{\partial\hat{x}_r}{\partial x_j} \frac{\partial\hat{x}_s}{\partial x_i} \int_{\hat{e}} k(F_e(\hat{\underline{x}}))\, \frac{\partial\hat{\phi}_j}{\partial\hat{x}_r} \frac{\partial\hat{\phi}_i}{\partial\hat{x}_s}\, d\hat{\Omega}.$$

Integration rules based on specific points in the reference element $\hat{e}$ are called Gauss integration formulas, of which we give the following examples:

1D; $\hat{e}$ = $[0,1]$ (see fig. 4.15)

$$\int_{\hat{e}} \hat{\psi}(\hat{x})\, d\hat{x} \approx \tfrac{1}{2}\{\hat{\psi}(\eta_1) + \hat{\psi}(\eta_2)\}, \text{ exact for } \hat{\psi} \in P_3(\hat{e}) \tag{5.4.18}$$

with $\eta_1 = \frac{1}{2} - \frac{1}{6}\sqrt{3}$, $\eta_2 = \frac{1}{2} + \frac{1}{6}\sqrt{3}$

$$\int_{\hat{e}} \hat{\psi}(\hat{x})d\hat{x} \approx \frac{1}{18}\{5\hat{\psi}(\xi_1) + 8\hat{\psi}(\xi_2) + 5\hat{\psi}(\xi_3)\}, \tag{5.4.19}$$

exact for $\hat{\psi} \in P_5(\hat{e})$, with

$\xi_1 = \frac{1}{2} - \frac{1}{10}\sqrt{15}$, $\xi_2 = \frac{1}{2}$, $\xi_3 = \frac{1}{2} + \frac{1}{10}\sqrt{15}$

2D; $\hat{e}$ is triangle (see fig. 4.15)

$$\int_{\hat{e}} \hat{\psi}(\hat{x})\, d\hat{\Omega} \approx \frac{1}{6}\{\hat{\psi}(\tfrac{1}{6},\tfrac{1}{6}) + \hat{\psi}(\tfrac{2}{3},\tfrac{1}{6}) + \hat{\psi}(\tfrac{1}{6},\tfrac{2}{3})\}, \tag{5.4.20}$$

exact for $\hat{\psi} \in P_2(\hat{e})$.

$$\int_{\hat{e}} \hat{\psi}(\hat{x})\, d\hat{\Omega} \approx \frac{1}{60}\, [\hat{\psi}(\tfrac{1}{2},\tfrac{1}{2}) + \hat{\psi}(0,\tfrac{1}{2}) + \hat{\psi}(\tfrac{1}{2},0) + 9\{\hat{\psi}(\tfrac{1}{6},\tfrac{1}{6}) + \hat{\psi}(\tfrac{1}{6},\tfrac{2}{3}) + \hat{\psi}(\tfrac{2}{3},\tfrac{1}{6})\}], \tag{5.4.21}$$

exact for $\hat{\psi} \in P_3(\hat{e})$.

3D; $\hat{e}$ is tetrahedron (see fig. 4.15)

$$\int_{\hat{e}} \hat{\psi}(\hat{x})d\hat{\Omega} \approx \frac{1}{24} \{\hat{\psi}(\gamma_1,\gamma_1,\gamma_1)+\hat{\psi}(\gamma_1,\gamma_1,\gamma_2)+\hat{\psi}(\gamma_1,\gamma_2,\gamma_1)+\hat{\psi}(\gamma_2,\gamma_1,\gamma_1)\} \tag{5.4.22}$$

with $\gamma_1 = \frac{1}{20}(5-\sqrt{5})$, $\gamma_2 = \frac{1}{20}(5+3\sqrt{5})$,

exact for $\hat{\psi} \in P_2(\hat{e})$

$$\int_{\hat{e}} \hat{\psi}(\hat{x})d\hat{\Omega} \approx \frac{1}{240} [\hat{\psi}(0,0,0)+\hat{\psi}(1,0,0)+\hat{\psi}(0,1,0)+\hat{\psi}(0,0,1)+ \tag{5.4.23}$$

$$+ 9\{\hat{\psi}(\tfrac{1}{3},\tfrac{1}{3},0)+\hat{\psi}(\tfrac{1}{3},0,\tfrac{1}{3})+\hat{\psi}(0,\tfrac{1}{3},\tfrac{1}{3})+\hat{\psi}(\tfrac{1}{3},\tfrac{1}{3},\tfrac{1}{3})\}],$$

exact for $\hat{\psi} \in P_3(\hat{e})$.

5.4.2 Numerical integration on quadrilaterals and hexahedra

Since the basis functions for the quadrilateral and hexahedral case are defined on reference elements (see Sections 4.5, 4.6), it is quite natural to introduce the numerical integration on these reference elements too. The Newton-Cotes and the Gauss integration rules are easily obtained by a repeated application of the 1D Newton Cotes and Gauss rules (i.e. formulas (5.4.5), (5.4.6), (5.4.18), (5.4.19)) in the two or three coordinate directions.

For the Newton-Cotes formulas in two dimensions we shall use the integration points $\hat{\underline{x}}^1, \hat{\underline{x}}^2, \ldots, \hat{\underline{x}}^9$ indicated in Fig. 4.22. For the 3D case, the integration points $\hat{\underline{x}}^1, \hat{\underline{x}}^2, \ldots, \hat{\underline{x}}^{27}$ are given by (4.6.5). For the Gauss quadrature rules we use the following quantities (cf. (5.4.18), (5.4.19)):

$$\eta_1 = \frac{1}{2} - \frac{1}{6}\sqrt{3} \qquad \eta_2 = \frac{1}{2} + \frac{1}{6}\sqrt{3}$$

$$\xi_1 = \frac{1}{2} - \frac{1}{10}\sqrt{15} \qquad \xi_2 = \frac{1}{2} \qquad \xi_3 = \frac{1}{2} + \frac{1}{10}\sqrt{15}$$

2D; $\hat{e}$ is unit square (see fig. 4.22)

$$\int_{\hat{e}} \hat{\psi}(\hat{\underline{x}})\, d\hat{\Omega} \approx \frac{1}{4} \sum_{i=1}^{4} \hat{\psi}_i \quad , \quad \text{exact for } \hat{\psi} \in Q_1(\hat{e}) \tag{5.4.24}$$

$$\int_{\hat{e}} \hat{\psi}(\hat{\underline{x}})\, d\hat{\Omega} \approx \frac{1}{36} \{ \sum_{i=1}^{4} \hat{\psi}_i + 4 \sum_{i=5}^{8} \hat{\psi}_i + 16\hat{\psi}_9 \}, \tag{5.4.25}$$

exact for $\hat{\psi} \in Q_3(\hat{e})$

$$\int_{\hat{e}} \hat{\psi}(\hat{\underline{x}}) d\hat{\Omega} \approx \frac{1}{4} \sum_{i=1}^{2} \sum_{j=1}^{2} \hat{\psi}(\eta_i, \eta_j), \text{ exact for } \hat{\psi} \in Q_3(\hat{e}) \tag{5.4.26}$$

$$\int_{\hat{e}} \hat{\psi}(\hat{\underline{x}})\, d\hat{\Omega} \approx \sum_{i=1}^{3} \sum_{j=1}^{3} \omega_i \omega_j \hat{\psi}(\xi_i, \xi_j) \tag{5.4.27}$$

with $\omega_1 = \omega_3 = \frac{5}{18}$, $\omega_2 = \frac{8}{18}$, exact for $\hat{\psi} \in Q_5(\hat{e})$

3D; $\hat{e}$ is unit cube (see Figs. 4.26, 4.27)

$$\int_{\hat{e}} \hat{\psi}(\hat{\underline{x}}) d\hat{\Omega} \approx \frac{1}{8} \sum_{i=1}^{8} \hat{\psi}_i \quad , \quad \text{exact for } \hat{\psi} \in Q_1(\hat{e}) \tag{5.4.28}$$

$$\begin{aligned}\int_{\hat{e}} \hat{\psi}(\hat{\underline{x}}) d\hat{\Omega} \approx \frac{1}{216} [\{ & \hat{\psi}_1 + \hat{\psi}_2 + \hat{\psi}_3 + \hat{\psi}_4 + 4(\hat{\psi}_5 + \hat{\psi}_6 + \hat{\psi}_7 + \hat{\psi}_8) + 16\hat{\psi}_9 \} + \\ & + 4\{\hat{\psi}_{19} + \hat{\psi}_{20} + \hat{\psi}_{21} + \hat{\psi}_{22} + 4(\hat{\psi}_{23} + \hat{\psi}_{24} + \hat{\psi}_{25} + \hat{\psi}_{26}) + 16\hat{\psi}_{27}\} + \\ & + \{\hat{\psi}_{10} + \hat{\psi}_{11} + \hat{\psi}_{12} + \hat{\psi}_{13} + 4(\hat{\psi}_{14} + \hat{\psi}_{15} + \hat{\psi}_{16} + \hat{\psi}_{17}) + 16\hat{\psi}_{18}\}],\end{aligned} \tag{5.4.29}$$

exact for $\hat{\psi} \in Q_3(\hat{e})$

$$\int_{\hat{e}} \hat{\psi}(\hat{\underline{x}}) d\hat{\Omega} \approx \frac{1}{8} \sum_{i=1}^{2} \sum_{j=1}^{2} \sum_{k=1}^{2} \hat{\psi}(\eta_i, \eta_j, \eta_k), \text{ exact for } \hat{\psi} \in Q_3(\hat{e}) \tag{5.4.30}$$

$$\int_{\hat{e}} \hat{\psi}(\hat{\underline{x}}) d\hat{\Omega} \approx \sum_{i=1}^{3} \sum_{j=1}^{3} \sum_{k=1}^{3} \omega_i \omega_j \omega_k \hat{\psi}(\xi_i, \xi_j, \xi_k), \tag{5.4.31}$$

with $\omega_1 = \omega_2 = \frac{5}{18}$, $\omega_2 = \frac{8}{18}$, exact for $\hat{\psi} \in Q_5(\hat{e})$.

5.5 Accuracy aspects of the FEM.

The purpose of this section is to give some insight into the accuracy aspects of the FEM, or in other words: What is the difference between the approximate solution $\tilde{u}$ of the finite element Galerkin equations and the (exact) solution of the original (continuous) problem.
In this section we content ourselves to give some practical results. We refer to chapter 12 for results on convergence and error estimates of a more mathematical nature.
Let us consider an elliptic PDE of order 2m, m > 0, then the corresponding Galerkin equations, which are obtained after m repeated partial integrations, contain derivatives of order less than or equal to m. A finite element approximation is called conform if the following two conditions are satisfied:
(i) the basis functions ϕ_i are m times continuously differentiable on each subregion
(ii) the basis functions ϕ_i are m-1 times continuously differentiable in Ω.
We shall restrict ourselves to second order PDEs, which means m=1. In the study of the (Navier-) Stokes problem, an equation for the pressure will appear with m=0. Consequently a finite element approximation for the pressure will then be conform if the basis functions for the pressure are continuous on each subregion without satisfying any inter-subregional continuity requirement.

In the 1D, the triangular and the tetrahedral case, a FEM is said to be of order k, if on each element e the basis functions belong to the space $P_k(e)$. This is equivalent to saying that on the reference element $\hat{e}$ the basis functions $\hat{\phi}$ belong to $P_k(\hat{e})$.
In the quadrilateral and the hexahedral case we call the FEM of order k if on the reference element $\hat{e}$ the basis functions $\hat{\phi}$ belong to $Q_k(\hat{e})$.

Let us first remark that the "optimal" accuracy of the FEM is equal to the accuracy of the interpolation on each subregion. In fact, on each subregion, the exact solution u is approximated by a (linear, quadratic,...) shape function. When h_e denotes the diameter of e and

$$h = \max_e h_e \qquad (5.5.1)$$

we deduce from Lagrangian interpolation theory (see Ciarlet, Raviart (1972), Ciarlet (1978), Raviart, Thomas (1983)) that the optimal (pointwise) finite element accuracy is $O(h^{k+1})$ where k denotes the order of the FEM.

Generally we can say that the accuracy of the FEM depends on the following conditions:

(i) Let us denote by E_h the collection of all subregions (intervals, triangles, tetrahedra, quadrilaterals or hexahedra) with diameter less than or equal to h. The collection E_h must satisfy condition (3.5.1). Moreover the subdivision into subregions must be regular which implies that there exists a constant $\gamma \geq 1$ such that

$$\frac{h_e}{h'_e} \leq \gamma \qquad \text{for all } e \in E_h \text{ , for all } h > 0 \qquad (5.5.2)$$

where h'_e denotes the supremum of the diameters of all spheres contained in e.
For the 2D situation, the quantities h_e and h'_e are geometrically given in Fig. 5.14.
For the 2D case of a triangle, condition (5.5.2) is equivalent to the condition that all the angles of all the triangles are bounded away from 0°. Jamet (1976) and Babuska, Aziz (1976) relaxed this condition and proved that the angles need to be kept away from 180° only. Thus triangles may degenerate into flat triangles with no angles of 180^0. For the 2D case of a quadrilateral, condition (5.5.2) is satisfied if no angle is too close to 0°.

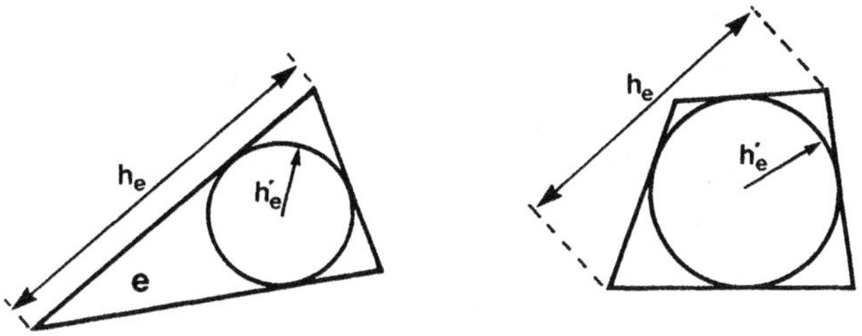

Fig. 5.14 Triangle and quadrilateral with h_e and h'_e.

(ii) The exact solution u is required to be smooth enough. In particular the accuracy of the FEM deteriorates when the exact solution contains singularities due to, for instance, discontinuous data or corners in the region Ω (for $n \geq 1$). Refinement of the element distribution in the neighbourhood of a singularity may be necessary.

(iii) All the integrals (in element matrices and vectors) are computed exactly.

When the above mentioned conditions are satisfied we have the following error estimates for k=1,2 and for n=1,2,3:

$$\{ \int_\Omega |u-\tilde{u}|^2 d\Omega \}^{\frac{1}{2}} \leq c\ h^k \tag{5.5.3}$$

$$\{ \int_\Omega |\text{grad}\ u - \text{grad}\ \tilde{u}|^2 d\Omega \}^{\frac{1}{2}} \leq c\ h^k \tag{5.5.4}$$

In some particular cases, for example when the following sufficient conditions are satisfied: smooth data, Ω is convex, Dirichlet or Neumann boundary conditions on Γ, then estimate (5.5.3) can be improved (see Ciarlet (1978), Raviart, Thomas (1983)):

$$\{ \int_\Omega |u-\tilde{u}|^2 d\Omega \}^{\frac{1}{2}} \leq c\ h^{k+1} \tag{5.5.5}$$

The constants c depend for instance on the way in which the domain Ω is subdivided into subregions. They depend also on the geometry of Ω, the smoothness of the exact solution u and other data of the problem like the right hand side f and the coefficients of the operator. Concerning the dependence of c on γ (the constant in (5.5.2)), we remark that c becomes smaller according as γ can be chosen smaller. This means that the most accurate results can be expected for equilateral triangles and squares in 2D and for equihedral tetrahedra and cubes in the 3D case.

When integration formulas are used to evaluate the integrals occurring in the element matrices and vectors, an extra error is introduced. Concerning the relative importance of this numerical integration error to the error estimates (5.5.3),...,(5.5.5), the following results can be proved (see Ciarlet (1978)):

1D, triangular and tetrahedral finite element of order k

The error estimates (5.5.3),...,(5.5.5) remain valid when an integration formula is used which is exact for polynomials in $P_{2k-2}(e)$. (5.5.6)

For the linear finite element (k=1), (5.5.6) implies the use of integration formulas which are exact for polynomials in $P_0(e)$, i.e. for constant functions. Consequently, any of the Newton-Cotes formulas (5.4.4),...,(5.4.13) or the Gauss formulas (5.4.18),...,(5.4.23) can be used.
For a second order finite element (k=2), the integration rule must be exact for functions in $P_2(e)$. The Newton-Cotes formulas (5.4.6), (5.4.9), (5.4.10), (5.4.13) and the Gauss formulas (5.4.18),...,(5.4.23) satisfy this condition.

For the case of the extended quadratic finite element of section 4.4.5, we may expect the error estimates (5.5.3),...,(5.5.5) to hold for k=2. When an integration rule

is used it must be exact for polynomials of degree 3 (for instance: (5.4.10)).
These results have been proved by Ciarlet (1978) for particular cases.

Quadrilateral and hexahedral finite element of order k

> The error estimates (5.5.3),...,(5.5.5) remain valid when an integration formula is used which is exact for functions in $Q_{2k-1}(\hat{e})$, moreover the number of integration points may not be smaller than the dimension of the function space $Q_k(\hat{e}) \cap P_{nk-1}(\hat{e})$. (5.5.7)

The last condition of (5.5.7) is satisfied if the integration rule is based on at least $(k+1)^n-1$ integration points.
From the statement (5.5.7) we deduce that for the $Q_1(\hat{e})$ finite element all the integration rules (5.4.24),...,(5.4.31) can be used. For the second order $Q_2(\hat{e})$ case the integration rules (5.4.25), (5.4.27), (5.4.29), (5.4.31) can be used.

Let us emphasize that the conditions of the statements are sufficient ones and that it is not quite clear whether they can be weakened for particular problems, like BVP with constant coefficients.

5.6 Solution methods for systems of (non-)linear equations

The aim of this section will be the treatment of methods for the solution of large sparse systems of equations which are the result of finite element discretizations. First attention will be paid to the solution of systems of linear equations, after which some linearization techniques will be considered, which transform systems of non-linear equations into a series of linear problems.
For the solution of systems of linear equations, two classes of methods are available: direct methods, which are all variations

of Gaussian elimination, and iterative methods.
A selection of both types of methods will be treated. For the solution of systems of non-linear equations, there exists a large body of literature. However, if we restrict ourselves to methods that are useful in finite element theory, Picard methods (successive substitution), Newton methods, and quasi-Newton methods appear to be the most important.

5.6.1 Direct methods to solve systems of linear equations

All direct methods to solve systems of linear algebraic equations, which arise from the discretization by FEMs or FDMs, are variations of Gaussian elimination. See for example Strang (1976). The variations arise, because use is made of the special structure of the equations. For the practical execution of the Gauss elimination process, the following theorem is used:

Theorem 5.1.
Each non-singular matrix A can be written as:

$$A = L\,U\,, \tag{5.6.1}$$

where L is a lower triangular matrix with a unit main diagonal and U is an upper triangular matrix. For a proof see Strang (1976).

The expression (5.6.1) is exploited in the following manner:
Let $A\,\underline{x} = \underline{f}$ be the system of equations to be solved.

Define $U\,\underline{x} = \underline{y}$, (5.6.2)

then $L\,\underline{y} = \underline{f}$. (5.6.3)

From (5.6.3) $\underline{y}$ can be solved directly since L is a lower triangular matrix, and subsequently $\underline{x}$ can be solved from (5.6.2) starting from the bottom. This latter process is called back-substitution.
It is remarked that when the matrix A is symmetric the LU-factorization can be written as:

$$A = L\,D\,L^T,$$

where D is a diagonal matrix, hence only half the matrix and its factorization has to be stored. When the matrix A is symmetric and positive definite then A can be written as:

$$A = G\,G^T,$$

the so-called Cholesky decomposition, where G is a lower triangular matrix. See Strang (1976).

A typical aspect of the systems of equations that arise from FEMs, is that the corresponding matrices are sparse, that is most of the entries are identically zero. Unfortunately Gaussian elimination cannot fully exploit the non-zero structure, since during elimination, zero entries are transformed into non-zero numbers. However, for some special classes of sparse matrices a part of the sparseness can be used to reduce the computing time and memory during the Gaussian elimination process. The simplest example is the band matrix. When an adequate numbering is used, then the large matrix has a band structure. All elements outside the band are zero and remain zero during the elimination process. The matrices L and U are also band matrices. See fig. 5.15. In fact the discretization by FEMs, when a "rectangular" regular mesh is used, turns out to be exactly a band matrix. For irregular meshes, however, it is useful to exploit more than the band

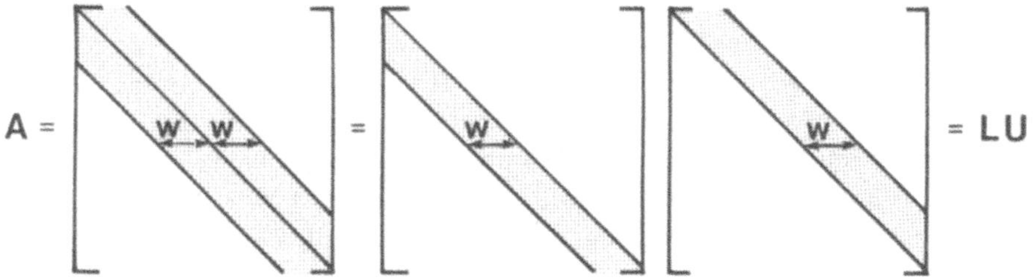

Fig. 5.15 A band matrix and its factors, with half bandwidth w.

alone. For these meshes, profile methods (or skyline methods) may be used.

Profile methods

Consider a matrix A, resulting from the discretization of a partial differential equation, either by a FEM, or a FDM. Consider nodal point i in the mesh with neighbours j,k,ℓ,...,m (See fig. 5.16)

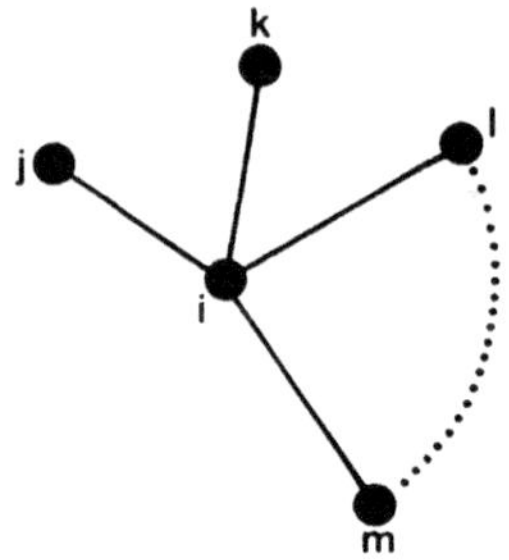

Fig. 5.16 Nodal point i "connected" with its neighbours.

Nodal point i is "connected" with its neighbours, which means that in general the entries $a_{i\ell}$, a_{ik}, a_{ii}, a_{ij} are unequal to zero. When nodal point n is not connected with i then $a_{ni} = a_{in} = 0$. These entries are called "essential zeros". A profile of a matrix is defined by the following:

Consider the ith row.
All elements of the ith row in the lower triangular matrix and all elements in the ith column of the upper triangular matrix, situated between the exterior "non-zeros", including these non-zeros itself, are said to belong to the profile. Non-zeros are those elements that are not "essential zeros"; they may be zero as the result of a computation, structurally they are unequal to zero. The profile thus defined is always symmetric. An example of the profile of a typical finite element matrix is sketched in fig. 5.17.

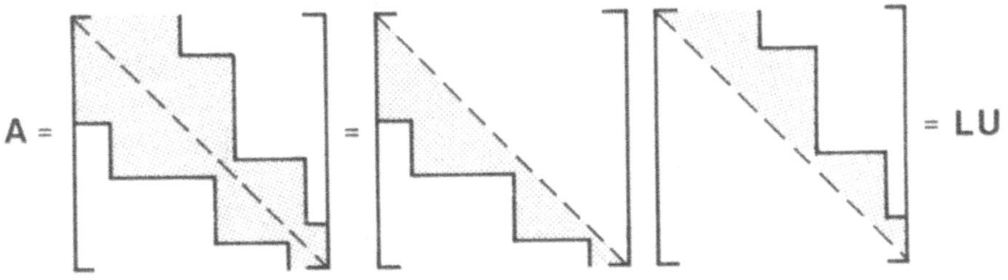

Fig. 5.17 Example of a profile matrix and its factors
The structure of L is equal to the lower triangular matrix of A, the structure of U is equal to the upper triangular matrix of A.

One easily verifies that entries outside the profile of the matrix, remain zero in the Gaussian elimination process. Profile methods take advantage of this property and use only the profile of the matrix for the factorization (see fig. 5.17). Although some extra book-keeping is necessary, practical computations have shown that even for a strict band matrix, profile methods are competitive with band methods. For a profile matrix as the one in fig. 5.17, profile methods are always cheaper than band methods. An efficient way to store and factor a profile matrix is the following:

(i) Store the matrix in a one-dimensional array.
(ii) First the diagonal element of the first row is stored, then the second row of the lower triangular matrix, followed by the second column of the upper triangular matrix from below. In the same way, alternatively rows of the lower triangular matrix and columns of the upper triangular matrix must be stored. See fig. 5.18. The advantage of this storage is that only entries in the profile must be stored. In order to find entries in the matrix two extra arrays of length N are needed (where N is the order of the matrix), one indicating the position of the diagonal elements in the one-dimensional array and one giving the number of the left most non-zero column of each row.

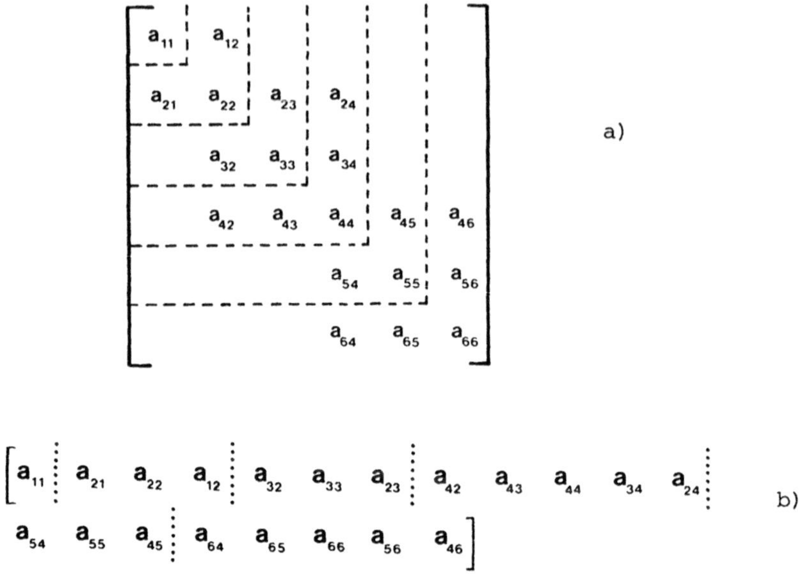

Fig. 5.18 Example of a profile matrix: a) two-dimensional storage, b) one-dimensional storage. The cuts in the matrix have been indicated.

The L and U matrices can also be computed in the alternating sequence row-column, however, columns must be computed from the top to the main diagonal. Hence for the example in fig. 5.18, the L and U matrices may be computed in the sequence:

$$u_{11} \vdots \ell_{21}\ u_{12}\ u_{22} \vdots \ell_{32}\ u_{23}\ u_{33} \vdots \ell_{42}\ \ell_{43}\ u_{24}\ u_{34}\ u_{44} \vdots \ell_{54}\ u_{45}\ u_{55} \vdots \ell_{64}\ \ell_{65}\ u_{46}\ u_{56}\ u_{66}.$$

The efficiency of both profile and band methods is strongly dependent on the numbering of the nodal points. In the literature several algorithms have been developed for the automatic construction of an "optimal" numbering. See for example Cuthill and McKee (1969).

Among the algorithms for renumbering the nodal points, we mention the nested dissection schemes and the ordering schemes to "minimize" profile or band width. Nested dissection schemes appear to be competitive only for very large numbers of unknowns, George (1976), and will not be considered here. Profile or bandwidth minimizing schemes, of which the Cuthill-McKee algorithm, Cuthill-McKee (1969), is the one most commonly used, are all based on the development of so-called levels. The idea is the following:

(i) Start with a suitable nodal point (an algorithm to find such a point is given by Gibbs, Poole, and Stockmeyer (1976). This nodal point is called the first level.

(ii) The ith level consists of all neighbours of the $i-1^{th}$ level that are not yet numbered ($i=2,3,4,....$).

(iii) The nodal points in the ith level are numbered. The main difference in the algorithms is how the nodal points are numbered in the levels, and how new levels are formed. For example a new level may be constructed once one nodal point has been numbered or when all nodal points in the level have been numbered.

In fig. 5.19 the levels of a mesh have been indicated.

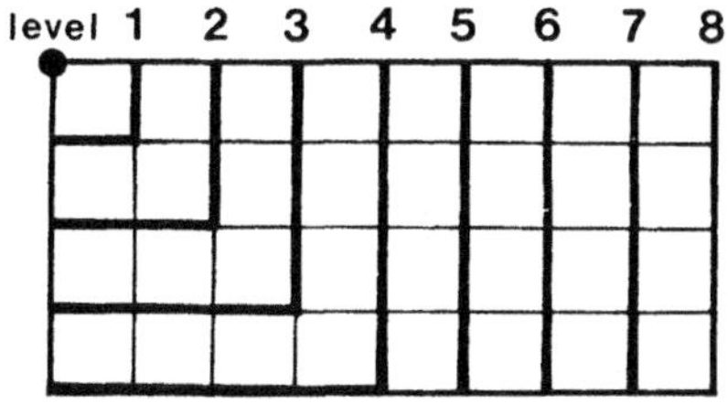

Fig. 5.19 Example of levels for a rectangular mesh.

The wave-front method

In the FEM literature the wave-front method, or frontal solution method, is a very popular technique for the solution of systems of linear equations. The frontal solution technique was first devised by Irons (1970) for symmetric positive

definite matrices and later extended by Hood (1976) for non-symmetric matrices. It is characterized by the fact that the assembly of the large matrix and the Gaussian elimination process are combined in one algorithm.
The idea is very simple. Consider a finite element mesh where the elements are numbered such that the difference between the element numbers of neighbouring elements is small. The wave-front method requires an adequate element numbering instead of nodal point numbering. However, from the description of the renumbering routines it will be clear that there is no essential difference between those two.
The wave-front method starts assembling each of the element matrices in turn. Once a row is fully summed it may be eliminated by Gaussian elimination. Thus elimination and element assembly are carried out at the same time. In this way only a number of entries of the matrix are active, that is are being assembled or must be eliminated. These elements form the front. See fig. 5.20. It is clear that all nodal points behind

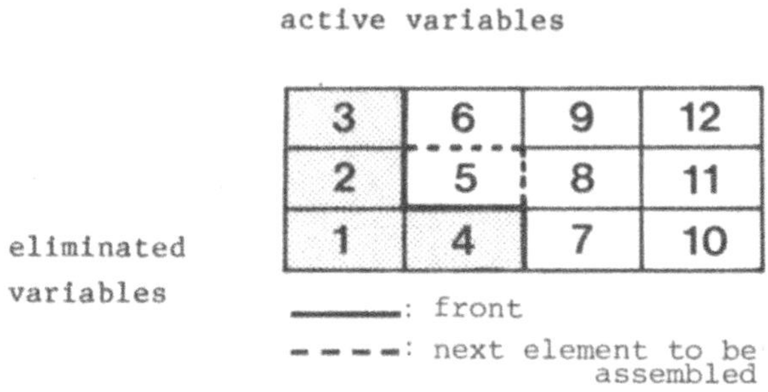

Fig. 5.20 Progress of front through a finite element mesh.

the front are fully eliminated and hence can be written to backing storage. It is only necessary to keep the nodal points in the front in-core and hence a considerable reduction of computer memory is possible. Hood (1976) describes a wave-front method for non-symmetric matrices combined with pivoting in the front. From these observations it is clear that a wave-front method is a nice algorithm when the large matrix cannot be kept in core. However, Hasbani and Engelman (1979) devised a method to solve systems of linear equations out-of-core that is based on a profile structure of the matrix. They fully exploit the levels in the renumbering routines. Their method is an attractive alternative for the wave-front method.

5.6.2 Iterative methods for the solution of systems of linear equations

An alternative for direct methods are iterative methods. These methods are characterized by the fact that the sparsity structure of the matrix is fully exploited, and there is no fill-in during the solution process.
Iterative methods can roughly be subdivided into three groups: relaxation methods, conjugate gradient type methods and defect correction methods (multigrid methods).

Classical iterative methods as Gauss-Seidel and overrelaxation, a review of which is given by Young (1971), are extensively treated in Varga (1962). These methods are relatively simple to implement, but their convergence is very slow for large problems. In fact these methods can not compete with direct methods and are therefore used only on very small computers.
In recent years conjugate gradient and defect correction methods have come to be known as powerful and rather general iterative methods. The reader is referred to the paper of Sonneveld et al (1983) where these methods are considered as convergence acceleration techniques. The implementation of multigrid methods in general finite element programs, is a heavy task. Therefore we shall limit ourselves to the conjugate gradient methods. For multigrid methods the papers of Sonneveld et al (1983) and Brandt (1977, 1982) can be used as reference.

The classical conjugate gradient method (CG) can only be used for positive definite matrices. The algorithm reads (See Ginsburg 1970):

Let A be a n×n positive matrix and $\underline{f}$ a right-hand-side vector. Let $\underline{u}$ be the solution of:

$$A\,\underline{u} = \underline{f} \,. \tag{5.6.4}$$

Then the following algorithm can be used to approximate $\underline{u}$.

(i) Let $\underline{u}_0$ be some initial guess, (5.6.5)

$$\underline{r}_0 = A\,\underline{u}_0 - \underline{f}\,, \qquad \underline{s}_0 = \underline{r}_0 \,.$$

(ii) $$\alpha_i = -\frac{(\underline{r}_i,\underline{r}_i)}{(\underline{r}_i,A\underline{s}_i)}\,,$$

$$\underline{u}_{i+1} = \underline{u}_i + \alpha_i\underline{s}_i\,,$$

$$\underline{r}_{i+1} = \underline{r}_i + \alpha_i A\,\underline{s}_i\,,$$

$$\beta_i = \frac{(\underline{r}_{i+1},\underline{r}_{i+1})}{(\underline{r}_i,\underline{r}_i)}\,,$$

$$\underline{s}_{i+1} = \underline{r}_{i+1} + \beta_i\underline{s}_i\,, \qquad i = 1,2,\ldots\,,$$

with (.,.) the usual inner product of two vectors.

The rate of convergence of this algorithm is roughly the same as the rate of convergence of overrelaxation with optimal choice of the overrelaxation parameter ω. For practical computations, however, it is very difficult to estimate the optimal value of ω. Conjugate gradients on the other hand have no parameters that have to be estimated. Hence in practice conjugate gradients tend to be faster than overrelaxation.

A non-symmetric variant of conjugate gradients, due to Sonneveld (1985), is the following algorithm called CGS (conjugate gradients squared):

(i) $$\underline{r}_0 = \underline{f} - A\,\underline{u}_0\,, \qquad \underline{g}_{-1} = \underline{h}_0 = \underline{0}\,, \tag{5.6.6}$$

(ii) $$\underline{y} = \underline{r}_i + \beta_i\underline{h}_i\,,$$

$$\underline{g}_i = \underline{v} + \beta_i(\beta_i \underline{g}_{i-1} + \underline{h}_i) ,$$

$$\underline{h}_{i+1} = \underline{v} - \alpha_i \, A \, \underline{g}_i ,$$

$$\underline{u}_{i+1} = \underline{u}_i + \alpha_i(\underline{v} + \underline{h}_{i+1}) ,$$

$$\underline{r}_{i+1} = \underline{r}_i - \alpha_i A(\underline{v} + \underline{h}_{i+1}) ,$$

with $\alpha_i = \rho_i/\sigma_i , \quad \beta_0 = 0 , \quad \beta_i = \rho_i/\rho_{i-1} ,$

$$\rho_i = \underline{r}_0^T \underline{r}_i , \quad \sigma_i = \underline{r}_0^T A \underline{g}_i ,$$

$$i = 1,2,...$$

The computing time required for this algorithm is, roughly speaking, about the same as for the CG method when the matrix is positive definite. For convergence it is necessary that the matrix A is approximately diagonally dominant.

It is clear that both CG and CGS require only matrix-vector multiplications as matrix operations. Therefore the sparsity pattern of the matrix can be fully exploited. The main reason to use CG type methods is that these methods can be easily accelerated by preconditioning of the large matrix.

The idea of preconditioning is a simple one. It is well known that the rate of convergence of most iteration processes strongly depends on the condition number of the matrix (see Varga 1962). Hence if one succeeds in decreasing this number, then faster convergence may be expected. The standard way to decrease the condition number of the matrix is by premultiplying the system of equations by a suitable matrix H, i.e.

$$H \, A \, \underline{u} = H \, \underline{f} \, . \tag{5.6.7}$$

When H is a good approximation of A^{-1}, the condition number of the matrix HA in (5.6.7) is small and fast convergence can be expected. Of all preconditionings that have been introduced in the literature we mention incomplete LU factorization, Meijerink and v.d. Vorst (1977), and incomplete line LU factorization (ILLU), Underwood (1976), Concus, Golub and Meurant (1982), and Meijerink, see Kettler (1982).

A simple ILU factorization is the following one:
Let P be a set of pairs (i,j) representing a matrix sparsity pattern, i.e.
When $(i,j) \notin P$ then $a_{ij} = 0$ (essential zero).

Define L and U as lower and upper triangular matrices satisfying:

$$\ell_{ii} = 1 \,, \tag{5.6.8}$$

$$\ell_{ij} = 0 \,, \quad (i,j) \notin P \,,$$

$$u_{ij} = 0 \,, \quad (i,j) \notin P \,,$$

$$(LU)_{ij} = A_{ij} \,, \quad (i,j) \in P \,.$$

Then the matrix H is defined as $(LU)^{-1}$.
Since only matrix-vector multiplications are necessary for the CG type methods it is sufficient that the product

$$(LU)^{-1} A \underline{u} \tag{5.6.9}$$

can be computed cheaply, which is the case for triangular L and U. In many cases the matrices L and U can be computed simply by means of (incomplete) Crout formulae, where only the elements within the sparsity pattern are used.

Incomplete line LU factorizations are more difficult to derive. In order that an ILLU factorization can be computed, it is necessary that the matrix A can be written as a block-tridiagonal matrix of the following shape:

$$A = \begin{bmatrix} B_1 & U_1 & & & \\ L_2 & B_2 & U_2 & & \\ & L_3 & B_3 & U_3 & \\ & & \cdot & \cdot & \\ & & & L_n & B_n \end{bmatrix} \tag{5.6.10}$$

with L_i, B_i and U_i matrices. Such a structure is very natural for FDMs on a rectangular grid, where the size of the matrices is m×m with m the number of points in horizontal or vertical direction. In FEMs such a structure is available when a renumbering algorithm as given in section 5.6.1, is used, where the lines are replaced by levels. In that case the matrices L_k and U_k are rectangular but not square.

The idea of ILLU is to find a matrix D such that:

$$A = (L+D)D^{-1}(D+U) \ , \tag{5.6.11}$$

where

$$L = \begin{bmatrix} 0 & & & & \\ L_2 & & & & \\ & L_3 & & & \\ & & \ddots & & \\ & & & L_n & 0 \end{bmatrix}, \quad U = \begin{bmatrix} 0 & U_1 & & & \\ & 0 & U_2 & & \\ & & \ddots & \ddots & \\ & & & \cdot & U_{n-1} \\ & & & & 0 \end{bmatrix}$$

$$D = \begin{bmatrix} D_1 & & \\ & \ddots & \\ & & D_n \end{bmatrix} \tag{5.6.12}$$

From (5.6.11) it follows that:

$$A = L + D + U + LD^{-1}U \ , \tag{5.6.13}$$

and one finds that $LD^{-1}U$ is the following block-diagonal matrix:

$$LD^{-1}U = \begin{bmatrix} 0 & & & \\ & L_2D_1^{-1}U_1 & & \\ & & \cdot & \\ & & & \cdot \\ & & & & L_nD_{n-1}^{-1}U_{n-1} \end{bmatrix} \tag{5.6.14}$$

From (5.6.13) and (5.6.14) we deduce the following algorithm for the computation of D:

$$D_1 = B_1 , \tag{5.6.15}$$

$$D_i = B_i - L_iD_{i-1}^{-1}U_{i-1} , \qquad i=2,3,\ldots,n .$$

The matrix D_i^{-1} is full, which makes a line LU decomposition as expensive as standard LU factorization. An incomplete line LU decomposition is obtained if we replace $L_iD_{i-1}^{-1}U_{i-1}$ by that part of $L_iD_{i-1}^{-1}U_{i-1}$ that has the same sparsity structure as the matrix B_i. Thus algorithm (5.6.15) is replaced by

$$\tilde{D}_1 = B_1 , \quad \tilde{D}_i = B_i - \text{Spars}(L_i\tilde{D}_{i-1}^{-1}U_{i-1}), \qquad i = 2,3,\ldots,n. \tag{5.6.16}$$

The problem of computing $\tilde{D}_i$ and $\tilde{D}_i^{-1}$ has been solved for tridiagonal matrices B_i, see Sonneveld et al (1983).
For general sparsity structures, no solution is known yet. Practical computations have shown that preconditioning may accelerate the convergence considerably, depending on the type of problem. At least for finite difference problems ILLU decompositions have proven to be more robust than ILU decompositions; moreover the convergence is often better. For a review of ILU and ILLU preconditionings on a number of test problems the reader is referred to Sonneveld et al (1983).

5.6.3 Linearization techniques for systems of non-linear equations

In this section we consider some methods to linearize systems of non-linear equations. In this way the solution of the non-linear equations is reduced to the solution of a sequence of systems of linear equations. The system of non-linear equations we consider will be written formally as:

$$K(\underline{u})\underline{u} = \underline{f} \ , \tag{5.6.17}$$

since problems in finite elements or finite differences are frequently formulated in this way. Of all methods that can be used to linearize (5.6.17) we shall treat the three most popular ones, i.e. Picard iteration (or successive substitution), Newton iteration and quasi-Newton methods.

5.6.3.1 Picard iteration

Picard iteration (also called successive substitution), is characterized by the fact that one of the vectors $\underline{u}$ in (5.6.17) is taken on a preceding iteration level, whereas the other one is taken at the new level. For example a Picard iteration could be

(i) Start: choose $\underline{u}^0$.

(ii) For i = 1,2,... solve:

$$K(\underline{u}^{i-1})\underline{u}^{i} = \underline{f} \ , \tag{5.6.18}$$

or alternatively

(ii) For i = 1,2,... solve:

$$K(\underline{u}^{i})\underline{u}^{i-1} = \underline{f} \ . \tag{5.6.19}$$

Picard methods are always simple to program and have in general a large region of convergence. However, Picard methods converge linearly and for most problems their rate of convergence is too slow. The most important application of

Picard iteration, is that it can be used as a good start for the faster Newton and quasi-Newton methods.
Furthermore, it can be used sometimes in those cases for which Newton-type methods do not converge at all. One remark concerning the stopping criterion should be made. Suppose that $\underline{u}$ is the solution of (5.6.17). Then the error $\underline{u}-\underline{u}^i$ in the linear Picard process is related to $\underline{u}-\underline{u}^{i-1}$ by:

$$\underline{u} - \underline{u}^{i-1} \simeq \lambda(\underline{u}-\underline{u}^i) , \tag{5.6.20}$$

with λ the rate of convergence.
When an accuracy of ε is required the process may be terminated when:

$$\max_i |\underline{u}^{i-1}-\underline{u}^i| < \frac{1-\lambda}{\lambda} \varepsilon , \tag{5.6.21}$$

which is found after some rearrangement of eq. (5.6.20).
Especially for $\lambda \approx 1$ this means that the difference between two succeeding iterations must be very small.
λ can be computed during the process by:

$$\lambda \simeq \frac{||\underline{u}^{i+1}-\underline{u}^i||}{||\underline{u}^i-\underline{u}^{i-1}||} , \tag{5.6.22}$$

with $||.||$ the euclidian norm, provided there is a dominating eigenvalue. (Not true in successive overrelaxation for example, see Blum (1972)).

5.6.3.2 Newton's method

Newton's method is characterized by the fact that it is a quadratically converging process. Therefore, once it converges, it requires only a few iterations. A typical disadvantage of Newton's method is that usually a good initial estimate is required. For the application of Newton's method, (5.6.17) is written as:

$$R(\underline{u}) = K(\underline{u})\underline{u} - \underline{f} \ . \tag{5.6.23}$$

Then Newton's method yields:

(i) Choose $\underline{u}^0$.

(ii) For $i = 0,1,2,\ldots$ solve:

$$J(\underline{u}^i)(\underline{u}^{i+1}-\underline{u}^i) = R(\underline{u}^i) \ , \tag{5.6.24}$$

where $J(\underline{u}^i)$ is the Jacobian matrix:

$$(J(\underline{u}^i))_{jk} = \frac{\partial R_j(\underline{u}^i)}{\partial(\underline{u}^i)_k} \ . \tag{5.6.25}$$

Newton's method requires the construction and factorization of a new Jacobian at each iteration step. With a good initial guess a small number of iterations is in general sufficient for convergence. Engelman et al (1981) report that for Navier-Stokes problems, the result of one Picard iteration is usually a good starting value for the Newton process.

5.6.3.3 The quasi-Newton method

A major drawback of both Picard iteration and Newton method is that the matrix must be computed and factored in each iteration. These computations can be prohibitively expensive for very large systems of equations if convergence is not reached within a reasonable number of iterations. Picard is a linearly convergent algorithm, whereas Newton normally converges quadratically, but often with a smaller radius of convergence. For example when the Navier-Stokes equations are solved with large Reynolds numbers (see chapter 6), both methods appear to be ineffective, see Engelman et al (1981). In optimization problems one has developed a new group of algorithms for the solution of non-linear equations known by the name quasi-Newton.

These methods are characterized by the fact that the Jacobian

matrix is not computed in each iteration explicitely, but a simple update in each step is computed which does not require a new LU-factorization. Although these methods take more iterations than Newton's method, a considerable saving in computing time is possible, since a new Jacobian matrix must be built up and factored each, say, 5 iterations.
The quasi-Newton method we shall treat is the one developed by Broyden (1965), using the efficient implementation of Engelman et al (1981).

Consider equation (5.6.23):

$$R(\underline{u}) = K(\underline{u})\underline{u} - \underline{f} \ . \qquad (5.6.26)$$

Newton's method (5.6.24) can be written as:

$$\underline{u}^{i+1} = \underline{u}^{i} - J^{-1}(\underline{u}^{i})R(\underline{u}^{i}) \ , \qquad (5.6.27)$$

with $J(\underline{u}^i)$ the Jacobian matrix (5.6.25).
Since we do not want to compute and factor J in each step, an approximation H_i of $J(\underline{u}^i)$ is defined satisfying:

$$H_i(\underline{u}^{i+1}-\underline{u}^{i}) = R(\underline{u}^{i+1}) - R(\underline{u}^{i}) \ . \qquad (5.6.28)$$

Define $\underline{p}_i = H_i^{-1}R_i \ , \quad \underline{y}_i = R(\underline{u}^{i+1}) - R(\underline{u}^{i}) \ , \qquad (5.6.29)$

$$\underline{\delta}_i = \underline{u}^{i+1} - \underline{u}^{i} \ .$$

Then Newton's method ($s_i = 1$, $H_i = J(\underline{u}^i)$), belongs to the general class of iteration methods:

$$\underline{u}^{i+1} = \underline{u}^{i} - s_i\underline{p}_i \ . \qquad (5.6.30)$$

(5.6.30) converges when a suitable value of s_i has been chosen. In order to fix H_i, n-1 equations are necessary apart from (5.6.28). Since the only available information of the Jacobian is in the direction of $\underline{u}^{i+1}-\underline{u}^{i}$, a possible choice is:

$$H_i\underline{q}_i = H_{i-1}\underline{q}_i \qquad \text{for all } \underline{q}_i \perp \underline{\delta}_i. \qquad (5.6.31)$$

One easily verifies that

$$H_i = H_{i-1} + \frac{(\underline{y}_i - H_{i-1}\underline{\delta}_i)\underline{\delta}_i^T}{\underline{\delta}_i^T\underline{\delta}_i} \tag{5.6.32}$$

satisfies (5.6.28) and (5.6.31). (Take the inner product with an arbitrary vector $\underline{v}$ and substitute $\underline{\delta}_i$ and $\underline{q}_i$).
In order to formulate the updates in terms of H_i^{-1}, the following formula, Householder (1964), will be used:

$$(A+\underline{x}\underline{y}^T)^{-1} = A^{-1} - \frac{A^{-1}\underline{x}\underline{y}^T A^{-1}}{1+\underline{y}^T A^{-1}\underline{x}} \; . \tag{5.6.33}$$

Substitution of (5.6.32) into (5.6.33) gives

$$H_i^{-1} = H_{i-1}^{-1} + \frac{(\underline{\delta}_i - H_{i-1}^{-1}\underline{y}_i)\underline{\delta}_i^T}{\underline{\delta}_i^T H_{i-1}^{-1}\underline{y}_i} H_{i-1}^{-1} \; . \tag{5.6.34}$$

In practical computations, where H_0 is a sparse matrix with a band structure, we do not actually calculate H_i^{-1}, but compute the LU-factorization of H_0. The vector $H_i^{-1}\underline{v}$, is computed by repeated application of formula (5.6.34). The following algorithm due to Engelman et al (1981), gives an efficient way of applying this process:

Start: give $\underline{u}^0$ and H_0 (for example H_0 is the Jacobian matrix $J(\underline{u}^0)$). The LU-factorization of H_0 is computed and stored.
Next $\underline{p}_0$ is found from $H_0\underline{p}_0 = R(\underline{u}_0)$ and a new approximation $\underline{u}^1$ from:

$$\underline{u}^1 = \underline{u}^0 - s_0\underline{p}_0.$$

Iteration: suppose we have found: (5.6.35)

$$\underline{u}^i, \quad \underline{p}_{i-1}, \quad s_{i-1},$$

$$\underline{\delta}_1,\ldots,\underline{\delta}_{i-1}\ ,\quad \underline{r}_1,\ldots,\underline{r}_{i-1}\ ,\quad \rho_1,\ldots,\rho_{i-1}.$$

(i) Calculate $\underline{q}_1$ from $H_0\underline{q}_1 = R(\underline{u}^i)$.

(ii) For j:=1,...,i-1 compute $\underline{q}_{j+1} = \underline{q}_j + \rho_j(\underline{\delta}_j-\underline{r}_j)\underline{\delta}_j^T\underline{q}_j$.

(iii) Form and store:

$$\underline{r}_i = \underline{q}_i - \underline{p}_{i-1}\ ,$$

$$\underline{\delta}_i = \underline{u}^i-\underline{u}^{i-1} = -\ s_{i-1}\underline{p}_{i-1}\ ,$$

$$\rho_i = \frac{1}{\underline{\delta}_i^T\underline{r}_i}\ .$$

(iv) Form

$$\underline{p}_i = \underline{q}_i + \rho_i(\underline{\delta}_i-\underline{r}_i)\underline{\delta}_i^T\underline{q}_i\ ,$$

$$\underline{u}^{i+1} = \underline{u}^i - s_i\underline{p}_i.$$

This algorithm requires the calculation and storage of two large vectors for each iteration, however, only one factorization of H_0 must be stored. One easily verifies that this algorithm is identical to Broyden's quasi-Newton method. In practical computations only limited updates of H_i are formed in this way (say 5 or 10), and then a new Jacobian matrix $J(\underline{u}^i)$ is used as initial guess.

Choice of s_i

A good choice for the parameter s_i may improve the rate of convergence of the algorithm (5.6.35). The simplest choice would be $s_i = 1$, making the algorithm a direct translation of the Newton method with approximate Jacobian. It is however, likely that a better convergence can be achieved by choosing s_i such that $R(\underline{u}^i-s_i\underline{p}_i)=0$ in the direction $\underline{p}_i$, i.e. $\underline{p}_i^T R(\underline{u}^i-s_i\underline{p}_i)=0$ A line search may be used to find an approximation for the value s where:

$$G(s) = \underline{p}_i^T R(\underline{u}^i - s\underline{p}_i) = 0 \; . \tag{5.6.36}$$

The point s must be found up to a modest accuracy only, since in general the computation of $R(\underline{u}^i - s\underline{p}_i)$, requiring the evaluation of a large vector, is very expensive. The following algorithm due to Matthies and Strang (1979) may be used to find an approximation of s by a line search.

Start: $s_0 = 0 \; , \quad G_a = G_0 = G(s_0) \; ,$ (5.6.37)

$$s_1 = 1 \; , \quad G_1 = G(s_1) \; ,$$

$$s_{max} = 16.$$

Find an interval included in $[s_0, s_1]$ with $s_1 \leq 16$ which contains the zero:

When $G_0 * G_1 > 0$ and $s_1 < s_{max}$ then:

$$s_0 = s_1 \; , \qquad s_1 = 2s_1 \; ,$$

$$G_0 = G_1 \; , \qquad G_1 = G(s_1).$$

Iteration: (5.6.38)

When $G_0 * G_1 < 0$ and $\left(|G_1| > \varepsilon |G_a| \text{ or } |s_1 - s_0| > \varepsilon \frac{s_1 + s_0}{2} \right)$

then: $s = s_1 - G_1 * (s_1 - s_0)/(G_1 - G_0) \; ,$

$G = G(s).$

When $G * G_1 > 0$ then $G_0 = G_0/2$

else $s_0 = s_1, \quad G_0 = G_1.$

$s_1 = s, \quad G_1 = G.$

It is remarked that the iteration step is the so-called Illinois algorithm. This is a modification of the regular falsi method, to locate the zero with tolerance ε. Not too many iterations may be performed, hence in practice the algorithm is stopped when more than 10 iterations are necessary. Practical computations have shown that a tolerance ε between 0.5 and 0.9

required the smallest amount of computing time.
The function G(0) has already been computed in the quasi Newton step, the function G(s) is necessary for the following iteration, hence the initial and end evaluation do not require extra computing time.

Also for the quasi-Newton method a good initial guess can be found by application of one Picard iteration.

PART II

APPLICATION OF THE FINITE ELEMENT METHOD TO THE NAVIER-STOKES EQUATIONS

Introduction

In the second part we shall treat various aspects of the FEM with regard to the construction of an approximate solution to the Navier-Stokes equations and the continuity equation. These equations must be solved simultaneously, because together they describe the flow. Unknowns in these equations are the velocity components and the pressure. In this part we limit ourselves to incompressible and isothermal flows. The equations describing the dynamics of fluid flow, then reduce to the Navier-Stokes equations and the continuity equation (see section 1.6).
In chapter 6 some alternative formulations of these equations will be given. Successively the velocity-pressure formulation, the stream function-vorticity formulation and the stream function formulation will be explained. Also the necessary initial and boundary conditions will be discussed. After a short evaluation we shall decide to use the velocity-pressure formulation. In the subsequent chapters the finite element approximations of these equations will be elaborated.
The procedure will always be that first the system of approximate equations corresponding to the Stokes equations and the continuity equation is derived, after which the non-linear convective term will be added to the system of equations. Hereafter the resulting discretization of the Navier-Stokes equations will be solved iteratively. In chapter 7 this is

carried out for the integrated method. For this method we start from the Navier-Stokes equations and the continuity equation, and apply the Galerkin method, using the basis functions for velocities and pressure as test functions. This leads to a system of equations in matrix form with velocity values and pressure values in the nodal points as unknowns; velocity and pressure are coupled simultaneously. Furthermore in this chapter some of the admissable elements are discussed. The chapter ends with a short evaluation of the integrated method, in which we conclude that the treatment of the continuity equation as a separate equation, which implies the simultaneous solution of velocity and pressure, results into a matrix with unfavourable properties. This is caused by the fact that in the Navier-Stokes equations the pressure appears as an unknown, while the pressure does not for the continuity equation.
Due to this fact the coefficient matrix contains zeros on the main diagonal, and hence partial pivoting must be applied during the solution process. The consequence is an inefficient computing process resulting in long computing times. In the next two chapters, two methods will be treated where the computation of velocity and pressure are decoupled and the continuity equation is taken into account implicitly in the matrix equations. Chapter 8 is concerned with the penalty method. In this method the continuity equation is considered as a constraint. The constrained problem is reduced to an unconstrained one by adding the continuity equation, multiplied by a large parameter, to the original equation. One can prove (see chapter 13) that for a sufficient large value of the penalty parameter, this approximation converges to the solution of the constrained problem. In this method it is easy to compute the pressure, once the velocities are known. Chapter 9 discusses the method of divergence-free elements. This method constructs an approximate solution of the Stokes equations with the help of divergence-free basis functions for the velocities. By this choice the continuity equation is automatically satisfied, and it is readily shown that the pressure disappears from the Stokes equations. As a consequence a system of equations is constructed that from a numerical point of view,

can be solved very efficiently. In this method also the pressure is calculated after the computation of the velocities. Finally chapter 10 is devoted to the instationary Navier-Stokes equations. These equations will be considered as consisting of a time-dependent part and a space part. The space part can be solved with the methods treated in the chapters 7, 8 and 9. For the time dependent part classical difference techniques will be utilised.

Chapter 6

Alternative formulations of Navier-Stokes equations

Introduction

In this chapter we are concerned with the basic equations of fluid dynamics: the Navier-Stokes equations and the continuity equation. The initial and boundary conditions of these equations will be discussed, using the practical example of a glass oven as reference. Some alternative formulations: the velocity-pressure formulation, the stream function-vorticity formulation and the stream function formulation will be considered. It is shown that for 3D problems the velocity-pressure formulation is the most attractive and hence it is concluded that the remaining chapters will only be concerned with the velocity-pressure or primitive variables formulation.

6.1 The basic equations of fluid dynamics

In this chapter a fluid flow is considered which has the following properties

i) The medium is incompressible.

ii) The medium has a Newtonian character.

iii) The medium properties are temperature-independent and uniform.

iv) The flow is non-turbulent.

For a 3D flow field in cartesian co-ordinate system, the

equations (1.6.1) and (1.6.5) then reduce to:

i) The continuity equation:

$$\frac{\partial u_1}{\partial x_1} + \frac{\partial u_2}{\partial x_2} + \frac{\partial u_3}{\partial x_3} = \operatorname{div} \underline{u} = 0 \ . \qquad (6.1.1)$$

ii) The Navier-Stokes equations:

$$\rho\left(\frac{\partial \underline{u}}{\partial t} + (\underline{u}\cdot\nabla)\underline{u}\right) - \mu \, \Delta \, \underline{u} + \nabla \, p = \rho \, \underline{f} \ , \qquad (6.1.2)$$

or componentwise:

$$\rho \frac{\partial u_i}{\partial t} - \mu\left(\frac{\partial^2 u_i}{\partial x_1^2} + \frac{\partial^2 u_i}{\partial x_2^2} + \frac{\partial^2 u_i}{\partial x_3^2}\right) + \rho\left(u_1 \frac{\partial u_i}{\partial x_1} + u_2 \frac{\partial u_i}{\partial x_2} + u_3 \frac{\partial u_i}{\partial x_3}\right)$$

$$+ \frac{\partial p}{\partial x_i} = \rho \, f_i \qquad i = 1,2,3. \qquad (6.1.3)$$

In these equations the following quantities are used:

$\underline{u} = (u_1,u_2,u_3)$: velocity vector with components u_1, u_2 and u_3 in respectively x_1, x_2 and x_3 direction,

p : pressure,

μ : viscosity of the fluid,

ρ : density of the fluid,

$\underline{f} = (f_1,f_2,f_3)$: body force per unit of mass.

The Navier-Stokes equations can be written in dimensionless form by scaling with characteristic quantities U (characteristic velocity) and L (characteristic length) in the following way:

$$x_i' = \frac{x_i}{L} \ , \quad \underline{u}' = \frac{\underline{u}}{U} \ , \quad p' = \frac{p}{\rho U^2} \ , \quad \underline{f}' = \frac{\underline{f}}{(U^2/L)} \ . \qquad (6.1.4)$$

substitution of these quantities in the equations and introducing the dimensionless Reynolds number:

$$Re = \frac{\rho \, U \, L}{\mu} \ , \qquad (6.1.5)$$

results in the dimensionless form of the equations in which the accents are dropped:

$$\frac{\partial \underline{u}}{\partial t} - \frac{1}{Re} \Delta \underline{u} + \underline{u}.\nabla \underline{u} + \nabla p = \underline{f} , \qquad (6.1.6)$$

$$\text{div } \underline{u} = 0 . \qquad (6.1.7)$$

This dimensionless form illustrates that for large Reynolds numbers the convective term dominates the viscous term. In that case the same phenomena can occur in the approximate solution as in the case of a dominant first derivative in the convection diffusion equation (see section 2.6). This subject will be returned to in section 6.3.2.

In the derivation of the Navier-Stokes equations, see section 1.6, the constitutive relation for a Newtonian fluid (1.6.4) and the continuity equation (1.6.1), are substituted into the momentum equation (1.6.2). The constitutive relation will be used later on in the formulation of the boundary conditions and reads for an incompressible and isotropic medium:

$$\underline{\underline{\sigma}} = - p \underline{\underline{I}} + \underline{\underline{d}} = - p \underline{\underline{I}} + 2 \mu \underline{\underline{e}} , \qquad (6.1.8)$$

with $\underline{\underline{\sigma}}$ the stress tensor,
$\underline{\underline{I}}$ the unit tensor,
$\underline{\underline{e}}$ the rate of strain tensor,
$\underline{\underline{d}}$ the deviatoric stress tensor.

In cartesian co-ordinates the matrix components of the stress tensor are:

$$\sigma_{ij} = - p \delta_{ij} + 2 \mu e_{ij} , \qquad i,j = 1,2,3 \qquad (6.1.9)$$

with matrices:

$$\sigma = \begin{bmatrix} \sigma_{11} & \sigma_{12} & \sigma_{13} \\ \sigma_{21} & \sigma_{22} & \sigma_{23} \\ \sigma_{31} & \sigma_{32} & \sigma_{33} \end{bmatrix} \qquad (6.1.10)$$

$$e = \begin{bmatrix} \frac{\partial u_1}{\partial x_1} & \frac{1}{2}(\frac{\partial u_1}{\partial x_2} + \frac{\partial u_2}{\partial x_1}) & \frac{1}{2}(\frac{\partial u_1}{\partial x_3} + \frac{\partial u_3}{\partial x_1}) \\ \frac{1}{2}(\frac{\partial u_2}{\partial x_1} + \frac{\partial u_1}{\partial x_2}) & \frac{\partial u_2}{\partial x_2} & \frac{1}{2}(\frac{\partial u_2}{\partial x_3} + \frac{\partial u_3}{\partial x_2}) \\ \frac{1}{2}(\frac{\partial u_3}{\partial x_1} + \frac{\partial u_1}{\partial x_3}) & \frac{1}{2}(\frac{\partial u_3}{\partial x_2} + \frac{\partial u_2}{\partial x_3}) & \frac{\partial u_3}{\partial x_3} \end{bmatrix} \quad (6.1.11)$$

Here the stress tensor components σ_{ij} represent the stress component in the x_i-direction exerted across a plane surface element normal to the x_j-direction. Mark that the stress tensor is symmetric which implies $\sigma_{ij} = \sigma_{ji}$.

The dimensionless form of the stress components can be derived by substituting:

$$\sigma'_{ij} = \frac{\sigma_{ij}}{\rho\, U^2} , \quad (6.1.12)$$

yielding:

$$\sigma_{ij} = - p\, \delta_{ij} + \frac{2}{Re} e_{ij} . \quad (6.1.13)$$

For flows in a region with cylinder co-ordinates r, ϕ and z (see fig. 6.1), the dimensionless relations can be derived analogously (Batchelor 1967)

$$\text{div } \underline{u} = 0 , \quad (6.1.14)$$

$$- \frac{1}{Re} \Delta \underline{u} + \underline{u}.\nabla \underline{u} + \nabla p = \underline{f} ,$$

or component wise:

$$\frac{1}{r} (\frac{\partial(ru_r)}{\partial r} + \frac{\partial u_\phi}{\partial \phi} + \frac{\partial(ru_z)}{\partial z}) = 0 , \quad (6.1.15)$$

$$- \frac{1}{Re} (\nabla^2 u_r - \frac{u_r}{r^2} - \frac{2}{r^2} \frac{\partial u_\phi}{\partial \phi}) + \frac{\partial p}{\partial r} + \underline{u}.\nabla u_r - \frac{u_\phi^2}{r} = f_r ,$$

$$- \frac{1}{Re} (\nabla^2 u_\phi + \frac{2}{r^2} \frac{\partial u_r}{\partial \phi} - \frac{u_\phi}{r^2}) + \frac{1}{r} \frac{\partial p}{\partial \phi} + \underline{u}.\nabla u_\phi + \frac{u_r u_\phi}{r} = f_\phi ,$$

$$- \frac{1}{Re} (\nabla^2 u_z) + \frac{\partial p}{\partial z} + \underline{u}.\nabla u_z = f_z ,$$

with $\nabla v = (\frac{\partial v}{\partial r}, \frac{1}{r} \frac{\partial v}{\partial \phi}, \frac{\partial v}{\partial z})^T$,

$$\nabla^2 v = \frac{1}{r} \frac{\partial}{\partial r} (r \frac{\partial v}{\partial r}) + \frac{1}{r^2} \frac{\partial^2 v}{\partial \phi^2} + \frac{\partial^2 v}{\partial z^2} .$$

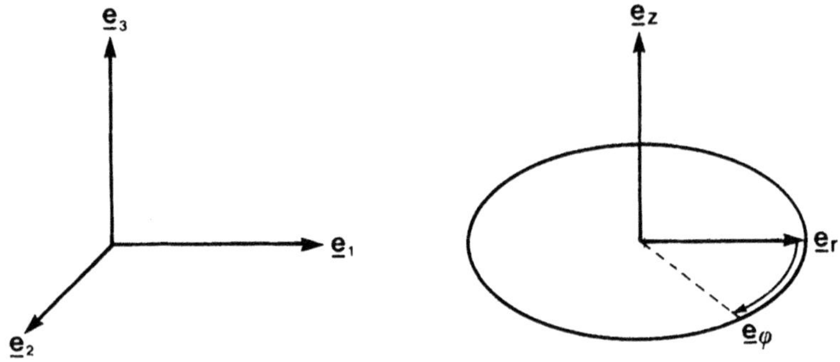

$\underline{x}=x_1\underline{e}_1+x_2\underline{e}_2+x_3\underline{e}_3=(x_1,x_2,x_3)$ $\underline{x}=r\underline{e}_r+\phi\underline{e}_\phi+z\underline{e}_z=(r,\phi,z)$

$\underline{u}=u_1\underline{e}_1+u_2\underline{e}_2+u_3\underline{e}_3=(u_1,u_2,u_3)$ $\underline{u}=u_r\underline{e}_r+u_\phi\underline{e}_\phi+u_z\underline{e}_z=(u_r,u_\phi,u_z)$

Fig. 6.1 a Cartesian and b axi-symmetric configurations with corresponding vector fields.

The constitutive relation in tensor notation (6.1.8) is also valid for cylindrical co-ordinates, where the matrix components in dimensionless form are given by:

$$\sigma_{rr} = - p + \frac{2}{Re} \frac{\partial u_r}{\partial r} , \tag{6.1.16}$$

$$\sigma_{\phi\phi} = - p + \frac{2}{Re} (\frac{u_r}{r} + \frac{1}{r} \frac{\partial u_\phi}{\partial \phi}) ,$$

$$\sigma_{zz} = - p + \frac{2}{Re} \frac{\partial u_z}{\partial z} ,$$

$$\sigma_{r\phi} = \frac{1}{Re} \left(r \frac{\partial}{\partial r} \left(\frac{u_\phi}{r}\right) + \frac{1}{r} \frac{\partial u_r}{\partial \phi}\right) = \sigma_{\phi r} ,$$

$$\sigma_{\phi z} = \frac{1}{Re} \left(\frac{1}{r} \frac{\partial u_z}{\partial \phi} + \frac{\partial u_\phi}{\partial z}\right) = \sigma_{z\phi} ,$$

$$\sigma_{rz} = \frac{1}{Re} \left(\frac{\partial u_r}{\partial z} + \frac{\partial u_z}{\partial r}\right) = \sigma_{zr} .$$

Notice that both in the system of equations (6.1.15), as in the expressions (6.1.16), terms appear that contain a factor $1/r$. So the co-ordinate transform causes a singularity in $r = 0$. Since, in a physical flow no singularity can be present, we have to take special care at $r = 0$. We shall return to this in section 6.3.

For axi-symmetric flow fields the flow is constant in the azimuthal direction which implies that the derivatives $\partial/\partial\phi$ are zero. A special case of axi-symmetric flow fields is the one in which the azimuthal component of velocity u_ϕ is zero everywhere and the velocity vector lies in a plane through the axis of symmetry. In the following chapters we shall mainly restrict ourselves to the case $u_\phi = 0$.

6.2 Alternative formulations

In this section we shall derive some alternative formulations for the equations as given in section 6.1. In the sequel we shall use the same dimensionless quantities as introduced by eq. (6.1.4) and here also the accents are dropped.

For an incompressible medium a velocity vector potential $\underline{\psi}$ can be introduced (see section 1.2) according to

$$\underline{u} = \text{curl } \underline{\psi} , \tag{6.2.1}$$

since div curl $\underline{\psi} = 0$.

For a 2D flow (cartesian co-ordinates) an applicable vector potential is:

$$\underline{\psi} = (0,0,\psi) \tag{6.2.2}$$

where the stream function ψ satisfies:

$$\frac{\partial \psi}{\partial x_1} = - u_2 , \qquad \frac{\partial \psi}{\partial x_2} = u_1 . \tag{6.2.3}$$

It is easy to verify that the continuity equation is always satisfied exactly. It is clear from (6.2.3) that the stream function can be computed from the velocity up to an additive constant, while the difference between the values of ψ at two points in the flow field is equal to the flux of fluid volume across any curve joining those two points. Moreover, it follows that ψ is constant along a streamline of the flow field and, for the steady case, also along a pathline of the fluid particles.
Another vector which plays an important role in fluid mechanics is:

$$\underline{\omega} = \text{curl } \underline{u} , \tag{6.2.4}$$

which for a 2D flow reduces to

$$\underline{\omega} = (0,0,\omega) . \tag{6.2.5}$$

The vector $\underline{\omega}$, and also the scalar ω in the 2D flow field, is termed the local vorticity of the fluid. Its physical meaning is that $\underline{\omega}$ appears to be twice the effective local angular velocity of the fluid, see Batchelor (1967). Hence, in case $\underline{\omega} = 0$ the fluid is commonly called irrotational.
For 2D cartesian co-ordinates ω reads:

$$\omega = \frac{\partial u_2}{\partial x_1} - \frac{\partial u_1}{\partial x_2} . \tag{6.2.6}$$

In 2D flow problems the stream function and vorticity can be introduced in the Navier-Stokes equations:

$$\frac{\partial u_1}{\partial t} - \frac{1}{Re}\left(\frac{\partial^2 u_1}{\partial x_1^2} + \frac{\partial^2 u_1}{\partial x_2^2}\right) + u_1 \frac{\partial u_1}{\partial x_1} + u_2 \frac{\partial u_1}{\partial x_2} + \frac{\partial p}{\partial x_1} = f_1, \quad (6.2.7)$$

$$\frac{\partial u_2}{\partial t} - \frac{1}{Re}\left(\frac{\partial^2 u_2}{\partial x_1^2} + \frac{\partial^2 u_2}{\partial x_2^2}\right) + u_1 \frac{\partial u_2}{\partial x_1} + u_2 \frac{\partial u_2}{\partial x_2} + \frac{\partial p}{\partial x_2} = f_2. \quad (6.2.8)$$

The pressure term will be eliminated by differentiating (6.2.7) to x_2 and (6.2.8) to x_1. After subtraction of the two equations and use of the continuity equation (6.1.1) the following relation remains:

$$-\frac{\partial \omega}{\partial t} + \frac{1}{Re}\Delta\omega - \left(\frac{\partial \psi}{\partial x_2}\frac{\partial \omega}{\partial x_1} - \frac{\partial \psi}{\partial x_1}\frac{\partial \omega}{\partial x_2}\right) = \frac{\partial f_1}{\partial x_2} - \frac{\partial f_2}{\partial x_1}. \quad (6.2.9)$$

The definition of the vorticity leads to the vector relation:

$$\underline{\omega} = \text{curl } \underline{u} = \text{curl curl } \underline{\psi} = \nabla(\nabla \underline{\psi}) - \Delta \underline{\psi} ,$$

which reduces for the 2D case to:

$$\omega = - \Delta\psi . \quad (6.2.10)$$

The equations (6.2.9) and (6.2.10) are equivalent to the original system of continuity equation and Navier-Stokes equations and are called the stream function-vorticity formulation of that system. The unknowns are ω and ψ.

The second alternative formulation follows by substituting (6.2.10) into the rewritten Navier-Stokes equation (6.2.9). Hence:

$$\frac{\partial \Delta\psi}{\partial t} - \frac{1}{Re}\Delta^2\psi + \frac{\partial \psi}{\partial x_2}\frac{\partial \Delta\psi}{\partial x_1} - \frac{\partial \psi}{\partial x_1}\frac{\partial \Delta\psi}{\partial x_2} = \frac{\partial f_1}{\partial x_2} - \frac{\partial f_2}{\partial x_1}. \quad (6.2.11)$$

This fourth order PDE is also equivalent to the original system and is called the stream function formulation of the system. The only unknown is ψ.

Also in the axi-symmetric case such a procedure is possible. A possible vector potential satisfying $\underline{u}$ = curl $\underline{\psi}$ is now:

$$\underline{\psi} = (0,\frac{\psi}{r},0) \ . \qquad (6.2.12)$$

The scalar ψ is called the Stokes stream function and is again interpretable as a measure of volume flux, see Batchelor (1967). The stream function ψ then satisfies:

$$\frac{1}{r}\frac{\partial\psi}{\partial r} = u_z \ , \qquad -\frac{1}{r}\frac{\partial\psi}{\partial z} = u_r \ , \qquad (6.2.13)$$

while for the vorticity a vector $\underline{\omega} = (0,\omega,0)$ is found with

$$\omega = \frac{\partial u_r}{\partial z} - \frac{\partial u_z}{\partial r} \ . \qquad (6.2.14)$$

It is remarked that the azimuthal component of velocity u_ϕ does not enter into the continuity equation, due to the assumed axial symmetry and hence it cannot be obtained from ψ. Let us now assume $u_\phi = 0$, then the Navier-Stokes equations reduce analogously to the 2D case to:

$$\frac{\partial\omega}{\partial t} - \frac{1}{Re}\left\{\frac{\partial}{\partial r}\frac{1}{r}\frac{\partial r\omega}{\partial r} + \frac{\partial^2\omega}{\partial z^2}\right\} + \frac{1}{r}\left(\frac{\partial\psi}{\partial z}\frac{\partial\omega}{\partial r} - \frac{\partial\psi}{\partial r}\frac{\partial\omega}{\partial z}\right) - \frac{1}{r}\frac{\partial\psi}{\partial z}\frac{\omega}{r} =$$

$$= -\frac{\partial f_1}{\partial z} + \frac{\partial f_2}{\partial r} \ , \qquad (6.2.15)$$

while the vector relation $\underline{\omega} = \nabla(\nabla\underline{\psi}) - \Delta\underline{\psi}$ reduces to

$$\omega = -\frac{1}{r}\left(\frac{\partial^2\psi}{\partial r^2} - \frac{1}{r}\frac{\partial\psi}{\partial r} + \frac{\partial^2\psi}{\partial z^2}\right). \qquad (6.2.16)$$

It will be clear that from (6.2.15), (6.2.16) the stream function formulation of the Navier-Stokes equation can be derived. This formulation will not be treated in the sequel.

From both formulations a solution for the stream function and if needed the vorticity is found. If one is interested in the velocities, then the stream function solution has to be differentiated in the 2D case with respect to x_1 or x_2. In order to compute the pressure as well one has to solve

$$\Delta p = \frac{\partial f_1}{\partial x_1} + \frac{\partial f_2}{\partial x_2} + 2\left\{\frac{\partial^2 \psi}{\partial x_1^2}\frac{\partial^2 \psi}{\partial x_2^2} - \left(\frac{\partial^2 \psi}{\partial x_1 \partial x_2}\right)^2\right\} \qquad (6.2.17)$$

(cartesian co-ordinates).
(6.2.17) can be found by taking the divergence of the Navier-Stokes equations (6.2.7), (6.2.8) and substitution of the relations (6.1.1) and (6.2.3).

6.3 Initial and boundary conditions

6.3.1 Introduction

For the solution of the instationary Navier-Stokes equations, both initial and boundary conditions are required. The initial condition can simply be given by prescribing an initial velocity field at $t = 0$. Of course this velocity field must satisfy the continuity equation.
The problem of imposing the correct boundary conditions is important in practical flow computations. Although from a theoretical point of view it may be clear what kind of boundary conditions must be prescribed, in practical computations these boundary conditions are not known, or are known very inaccurately. Therefore one is often forced to prescribe boundary conditions containing (large) errors. At the end of this chapter we shall return to this important question but first the boundary conditions are given which are possible from the theoretical point of view.

For a second order elliptic PDE it is necessary and sufficient to give exactly one boundary condition for each dependent variable. Fourth order elliptic PDEs require exactly two boundary conditions. Although it may be possible in some cases to prove existence and uniqueness of the solution of an elliptic second order PDE if for example one gives two boundary conditions along one part of the boundary and none along another part, such a problem would be ill-posed in the sense

that a small perturbation of the boundary condition would yield a large perturbation of the solution. The system of Navier-Stokes equations and continuity equation in velocity-pressure formulation essentially consists of three (or two in 2D) second order elliptic PDEs for the velocity unknowns, coupled by the continuity equation. Application of the above statements to this system of differential equations thus requires exactly three (resp. two) boundary conditions for the velocity or the stresses along the boundary (see Chapter 11). Furthermore these boundary conditions must be given in independent directions (see Chapter 7). So it is not allowed to give a normal velocity component and a normal stress component in the same point of the boundary. One can prove, Ladyzhenskaya (1969), that in the case of incompressible flows, no boundary conditions for the pressure are necessary, emphasizing the fact that the pressure plays a special role in the solution process of the Navier-Stokes equations. Boundary conditions for the pressure can only be given implicitly through the normal stress components. That no boundary conditions for the pressure are necessary can be made plausible, if we realise ourselves that it is possible to eliminate the pressure from the Navier-Stokes equations with the aid of the continuity equation, as will be shown in chapters 8 and 9.
For the unique solvability of the system of equations in the stream function-vorticity or the stream function formulation (in $\mathbb{R}^2$ only), two boundary conditions are necessary along the boundary. If one wants to compute the pressure from the Poisson equation (6.2.17), a value for the pressure on the boundary must be imposed as well. These boundary conditions can be expressed in terms of the velocities.
The various boundary conditions and the possible formulations will be the subject of the next sections.

6.3.2 Velocity-pressure formulation

For the explanation of the various types of boundary conditions, we shall use the specific example of fig. 6.2.

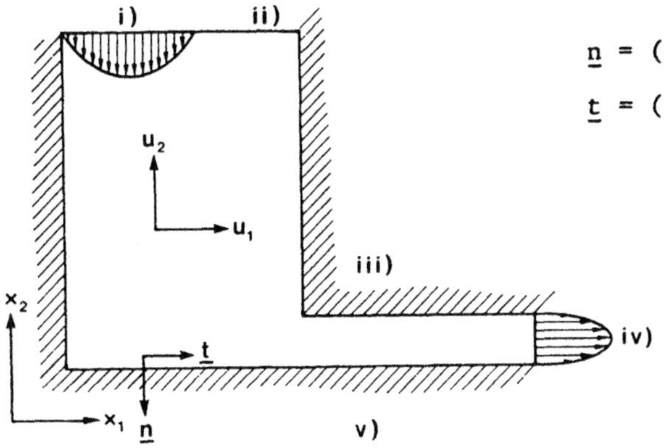

$\underline{n} = (n_1, n_2)$

$\underline{t} = (t_1, t_2)$

Fig. 6.2 Glass oven with boundaries:
i) inflow
ii) free surface
iii) fixed wall
iv) outflow
v) fixed wall

The example represents a very simplified 2D model of a glass oven with an inflow and a free surface at the top and an outflow through a long horizontal pipe at the right-hand side. The fluid flow in this case is regarded to be stationary and isothermal.

The normal ($\underline{n}$) and tangential ($\underline{t}$) unit vectors at the boundaries are given in the figure. On each of the sides two boundary conditions are necessary. First we shall give them for the specific example and afterwards a generalization is formulated:

i) inflow: the velocity is given, for example $u_1 = 0$, $u_2 = u_2(x_1)$.

ii) free surface: the surface is a streamline, i.e. $u_2 = 0$. Furthermore, the stress component in the surface is zero. So

$$\sigma_{12} = \frac{1}{Re}\left(\frac{\partial u_2}{\partial x_1} + \frac{\partial u_1}{\partial x_2}\right) = 0 .$$

Using $u_2 = 0$ $\left(\frac{\partial u_2}{\partial x_1} = 0\right)$, this condition reduces to $\frac{\partial u_1}{\partial x_2} = 0$.

iii),v) fixed walls: along the walls both the tangential (no-slip condition) and the normal velocity components must be zero. Hence

$$u_1 = u_2 = 0.$$

iv) outflow: when the flow is fully developed, the velocity can be prescribed by:

$$u_1 = u_1(x_2) , \quad u_2 = 0.$$

As we have seen in chapter 2 for the discretization of the convection-diffusion equation, such a Dirichlet boundary condition could lead to numerical oscillations in the solution of the Navier-Stokes equations. In chapter 2 it was recommended to choose Neumann boundary conditions, in such a case. A possible alternative is then:

$$\frac{\partial u_1}{\partial x_1} = 0 , \quad u_2 = 0 ,$$

or expressed in terms of the stresses:

$$\sigma_{11} = d_{11} - p = -p + \frac{2}{Re}\frac{\partial u_1}{\partial x_1} = -p , \quad u_2 = 0.$$

In the last case the stress component normal to the outflow surface is prescribed. As the flow is assumed to be fully developed, the pressure can be considered to be constant along the outflow boundary.

Because in the Navier-Stokes equations only derivatives of the pressure are present, the pressure is fixed up to an additive constant and hence it can be given in one point. At the outflow one could choose for example $p = 0$, and then the boundary conditions become:

$$\sigma_{11} = 0 , \quad u_2 = 0.$$

With the aid of the normal and tangential unit vectors of fig. 6.2 these boundary conditions can be formulated generally:

i) inflow: $\underline{u}$ given: Dirichlet boundary condition.

ii) free surface: $\underline{u}.\underline{n}$ given: Dirichlet boundary condition.
$\sigma_{nt} = 0$ or $\underline{n}.\text{grad } u_t = 0$ with $u_t = \underline{u}.\underline{t}$: Neumann boundary condition.

iii) and v) fixed walls: $\underline{u} = 0$: Dirichlet boundary condition.

iv) outflow: $\underline{u}.\underline{t} = 0$: Dirichlet boundary condition.

$\sigma_{nn} = 0$: Neumann boundary condition.

In this 2D example we see that on each boundary always two boundary conditions for velocity or stress are given; one in the normal and one in the tangential direction. There are no explicit boundary conditions for the pressure.

It will be clear that also for axi-symmetric flows the same type of boundary conditions can be given. Due to the presence of a symmetry axis at r = 0, it is necessary to give symmetry conditions at that axis when it is a part of the region:

$$u_r = 0 , \quad \frac{\partial u_z}{\partial r} = 0 , \quad u_\phi = 0. \tag{6.3.1}$$

The conditions $u_r = 0$ and $u_\phi = 0$ for $r = 0$ are necessary because of the apparent singularity as we found in the system of equations (6.1.15).

By the choice of the above mentioned boundary conditions the solution of the system of Navier-Stokes equations and continuity equation is unique, for Re small enough, see Chapter 11. However, this does not hold for every case where two boundary conditions are given. When constructing boundary conditions one should take into account the following two general points:

i) When Dirichlet boundary conditions are prescribed in the normal directions of the boundaries Γ of the region Ω:

$$u_n = g_0, \quad \text{with} \quad \underline{x} \in \Gamma , \tag{6.3.2}$$

there exists only a solution of the system of Navier-Stokes equations and continuity equation when g_0 satisfies the compatibility condition (see section 1.8):

$$\int_\Gamma g_0 \, d\Gamma = 0 . \tag{6.3.3}$$

ii) As we have seen, the solution of the pressure from the Navier-Stokes equation is determined up to an additive constant, when nowhere normal stresses have been prescribed on the boundary. One way to fix this constant is to demand that:

$$\int_\Omega p \, d\Omega = 0 . \tag{6.3.4}$$

Of course this condition can be discretized, but in numerical computations for the sake of simplicity one usually prescribes the pressure in one point.
When a normal stress is prescribed on a part of the boundary the constant is fixed implicitly as we have seen in the outflow condition of the glass oven example.

6.3.3 Stream function-vorticity formulation

In the preceding section all boundary conditions have been formulated in terms of velocity and pressure. For the solution of the system in stream function-vorticity or stream function form, those boundary conditions must be expressed in terms of the stream function ψ. We shall illustrate this "translation" for the 2-D case, using the same example from fig. 6.2:

i) inflow: $u_1 = 0$, $u_2 = u_2(x_1)$.

These conditions can be rewritten as:

$$\frac{\partial\psi}{\partial x_1} = -u_2(x_1) \quad \text{and} \quad \frac{\partial\psi}{\partial x_2} = 0 ,$$

hence defining the left vertex as x = 0:

$$\psi(x_1) = \int_0^{x_1} -u_2(x)dx + \psi_0 \quad \text{and} \quad \frac{\partial\psi}{\partial x_2} = 0 .$$

ii) free surface: $u_2 = 0, \frac{\partial u_1}{\partial x_2} = 0$.

Rewriting gives, using x_2 is constant along the surface:

$$\psi = \text{constant} , \quad \frac{\partial^2\psi}{\partial x_2^2} = 0 .$$

iii), v) fixed walls: $u_1 = u_2 = 0$.

These conditions are equivalent to the conditions:

$$\psi = \text{constant and} \quad \frac{\partial\psi}{\partial n} = 0 .$$

iv) outflow: $\frac{\partial u_1}{\partial x_1} = 0, u_2 = 0$ or also $u_2 = 0, \frac{\partial u_2}{\partial x_1} = 0.$

Rewriting this we get

$$\frac{\partial\psi}{\partial x_1} = 0 , \quad \frac{\partial^2\psi}{\partial x_1 \partial x_2} = 0 \quad \text{or} \quad \frac{\partial^2\psi}{\partial x_1^2} = 0 .$$

These boundary conditions can also be generalized with the normal and tangential unit vector as sketched in fig. 6.2, and we shall use the fact that the stream function is continuous along the boundary. Since the stream function is determined up to an additive constant, the stream function value in the left vertex of boundary i) is made equal to zero, so:

$$\psi_0 = 0 .$$

The boundary conditions are then:

i) inflow ($n_1 = 0, n_2 = 1$):

$$\psi(x_1) = \int_0^{x_1} - v(x)dx \ , \quad \frac{\partial\psi}{\partial n} = \underline{n}.\nabla\psi = n_1 \frac{\partial\psi}{\partial x_1} + n_2 \frac{\partial\psi}{\partial x_2} = 0 \ .$$

ii) free surface ($n_1 = 0, n_2 = 1$):

$\psi = \psi_1$ with ψ_1 the value in the right vertex of boundary i).

$$\frac{\partial^2\psi}{\partial n^2} = \underline{n}.\nabla(\frac{\partial\psi}{\partial n}) =$$

$$= n_1(n_1 \frac{\partial^2\psi}{\partial x_1^2} + n_2 \frac{\partial^2\psi}{\partial x_1 \partial x_2}) + n_2(n_1 \frac{\partial^2\psi}{\partial x_2 \partial x_1} + n_2 \frac{\partial^2\psi}{\partial x_2^2}) = 0.$$

iii), v) fixed walls:

$$\psi = \psi_1 \ , \quad \frac{\partial\psi}{\partial n} = 0 \ .$$

iv) outflow ($n_1 = 1, n_2 = 0$):

$$\frac{\partial\psi}{\partial n} = 0 \ , \quad \frac{\partial^2\psi}{\partial n^2} = 0 \ .$$

It is evident that also in this case we have two boundary conditions for ψ on all boundaries. That is sufficient for the solution of the system in stream function formulation, as well as in the stream function-vorticity formulation. For the solution of the system in stream function-vorticity form it would be attractive to have boundary conditions for stream function <u>and</u> vorticity on each boundary. In practical applications there are generally only two boundary conditions for the stream function. This obstructs the numerical solution considerably. Besides that, when at the inflow not the velocity but the pressure (through the normal stress) is given, the value of ψ_1 is unknown and a special treatment of the boundary condition at boundaries ii) and iii) is necessary.

When in this formulation of the system of equations the pressure must be computed as well, a boundary condition for the

pressure on the whole boundary is necessary. This boundary condition can be derived by taking the inner product of the Navier-Stokes equations with the outward unit normal on the boundary.
Hence:

$$\frac{\partial p}{\partial n} = \underline{n}.\nabla p = \underline{f}.\underline{n} + \frac{1}{Re}(\Delta\underline{u}).\underline{n} - (\underline{u}.\nabla\underline{u}).\underline{n} \ . \qquad (6.3.5)$$

It turns out that this pressure boundary condition contains third derivatives of the computed solution of the stream function ψ near the boundary.

6.3.4 Some practical remarks concerning the boundary conditions

In the preceding sections the necessary boundary conditions for the solution of the Navier-Stokes equations have been derived and some possibilities for a practical problem have been given. In practice, however, it is not so easy to find the correct values of the boundary conditions. In reality boundary conditions can only be given accurately for fixed or moving walls, free surfaces and fully developed pipe flows. For inflow and outflow we have to use either measured values with the corresponding inaccuracies, or to use the well known and accurate boundary conditions that belong to a fully developed flow. Therefore one could incorporate inlet and outlet pipes, if possible, in order to get more accurate boundary conditions. When the measured velocity components have to be prescribed at the outlet, the measurement inaccuracies, even if they are small may lead to numerical problems as demonstrated in chapter 2 for the convection diffusion equation. Consequently an unnatural mesh refinement or an (inaccurate) upwinding technique should be used to suppress the numerical oscillations found. In Segal (1982) and also in chapter 2 it has been shown that Neumann boundary conditions for the convection diffusion equation, do not exhibit this behaviour. Therefore it is preferable to use Neumann conditions or prescribed stresses at outflows. Such conditions may be the vanishing tangential velocity and the given normal stress component, which is usually sufficient to suppress the oscillations. When the practical situation differs too much from the fully developed flow the tangential stress could be prescribed instead of the tangential velocity. In the case of an inflow not the problem of oscillations but of accuracy is important. The conditions at inflow fully predict the solution in the whole region, unless there is a pipe forcing the velocity to a fully developed flow. Therefore in the case of inaccurate data at inflow one could also prescribe stresses instead of velocities, in the hope that errors in stresses at inflow have a smaller effect on the solution (Segal (1982)), or as an alternative one can always put an artificial pipe at the inflow for the computational region. The reader must always keep in mind that the solution can never be more accurate then the boundary conditions. Besides that, the addition of in- and ouflow pipes may be helpful but it has to be remarked that in both cases the situation, in a physical sense, could be changed.

6.4 Evaluation of the various formulations

For a 2D flow we have seen that there are three formulations for the Navier-Stokes equations and the continuity equation. Dropping the convective terms, these formulations read:

i) The velocity-pressure formulation:

$$\frac{\partial \underline{u}}{\partial t} - \frac{1}{Re} \Delta \underline{u} + \nabla p = \underline{f} , \tag{6.4.1}$$

$$\text{div } \underline{u} = 0 .$$

This system consists of 3 equations with 3 unknowns.

ii) The stream function-vorticity formulation:

$$\frac{\partial \omega}{\partial t} - \frac{1}{Re} \Delta \omega = - \frac{\partial f_1}{\partial x_2} + \frac{\partial f_2}{\partial x_1} , \tag{6.4.2}$$

$$- \Delta \psi = \omega .$$

Here we have 2 equations with 2 unknowns.

iii) The stream function formulation:

$$\frac{\partial \Delta\psi}{\partial t} - \frac{1}{Re}\,\Delta^2\psi = \frac{\partial f_1}{\partial x_2} - \frac{\partial f_2}{\partial x_1}\,. \tag{6.4.3}$$

The system has been reduced to one equation with one unknown.

We have seen that a similar set of equations can be derived for the axi-symmetric case with $u_\phi = 0$. In both cases we can formulate adequate boundary conditions. We shall now compare the advantages and disadvantages of the various formulations.

i) The velocity-pressure formulation.

The advantage of this method is that we are working directly with the physical most relevant quantities: pressure and velocity. Furthermore the method can be applied both in 2D and 3D. Numerically, however, problems arise because the pressure does not appear in the continuity equation. Usually this problem is solved by elimination of the pressure.

In fact the role of the pressure is a very special one. As will be shown in chapters 8 and 9 the pressure is closely related to the continuity equation: when the continuity equation is dropped, for example by elimination, then the pressure terms disappear also. Furthermore the pressure is fixed up to an additive constant, and no boundary conditions for the pressure are defined explicitly. These boundary conditions appear only implicitly in expressions for the stresses on the boundary. In conclusion, the pressure unknowns cannot be treated in the same way as the velocity unknowns.

ii) The stream function-vorticity formulation.

The main advantage of this formulation is that there are no problems with the continuity equation; this equation is satisfied exactly without the introduction of numerical problems. Moreover the number of unknowns has been reduced to two. For some problems, the calculated stream function and vorticity are the interesting quantities. A disadvantage, however, is that this method can be applied in 2D or axi-symmetric cases only, due to the definition of the stream function. Because in many problems only boundary conditions for the stream function are available and not for the vorticity, these equations must be solved simultaneously. From a numerical point of view this is unfavourable; practical computations have shown that the ψ can be solved accurately, but that the ω is very inaccurate. Since the velocity must be computed by numerical differentiation of the computed ψ, which is an inaccurate process, the resulting velocity is less accurate than in the velocity-pressure formulation. The pressure computation exhibits this behaviour more strongly, since in this case second order derivatives (for the differential equation) and third order derivatives (for the boundary conditions) of the computed stream function are required.

iii) The stream function formulation.

The stream function formulation does not have only the advantages of the stream function-vorticity formulation, but also it consists of only one equation with one unknown. However, this results in a fourth order PDE. The construction of an approximate solution by the FEM is a heavy task, since for conforming elements, the basis functions must have continuous first derivatives over the element boundaries, see chapters 5 and 13. This requires high order elements (at least fourth degree polynomials, or complicated third degree composition elements), and implies a large amount of computing time. Therefore the FEM is generally not applied in this formulation. The discretization of the stream function formulation by finite differences is less troublesome so that this formulation is a very popular one in FDM. Of course the problems related to the accuracy of velocity and pressure and the restriction to 2D remain the same for FDM as for FEM.

Since it is our aim to derive the equations for the 3D case and as we are primary interested in an accurate approximation of velocity and pressure we shall limit ourselves to the velocity-pressure formulation. In the next chapters several methods to solve the resulting equations will be treated. Firstly the integrated method is treated, using the complete system of equations, secondly two methods are derived from which both the continuity equation and the pressure have been eliminated. The chapters 7, 8 and 9 are limited to the steady case, the instationary Navier-Stokes equations are the subject of chapter 10.

Chapter 7

The integrated method

Introduction

This chapter is devoted to the numerical solution of the basic equations of fluid dynamics by the FEM. The direct method or integrated method will be discussed and problems related to the continuity equation will be investigated. It is shown that the approximation of the pressure and the velocity are strongly related and only a limited number of elements may be used to solve the Navier-Stokes equations. Finally the structure of the resulting system of equations is considered. This structure is such that partial pivoting is necessary which increases the computing time and the computer memory required.

7.1 General approach

In the integrated method, the Navier-Stokes equations and continuity equation are discretized by Galerkin's method (see Section 3.6). The non-linear convective terms are treated in 7.3; in this section we limit ourselves to the linear Stokes equations:

$$-\frac{1}{Re}\,\Delta\, u_1 + \frac{\partial p}{\partial x_1} = f_1\ , \qquad (7.1.1)$$

$$-\frac{1}{Re}\,\Delta\, u_2 + \frac{\partial p}{\partial x_2} = f_2\ ,$$

$$-\frac{1}{Re}\,\Delta\, u_3 + \frac{\partial p}{\partial x_3} = f_3 \;,$$

$$\text{div}\; \underline{u} = 0 \;,$$

defined on the 3D domain Ω with boundary Γ.

For simplicity Dirichlet boundary conditions are assumed:

$$u_1 = g_1(\underline{x}) \;, \quad u_2 = g_2(\underline{x}) \;, \quad u_3 = g_3(\underline{x}) \;; \qquad \underline{x} \in \Gamma, \qquad (7.1.2)$$

$$\int_\Omega p \; d\Omega = 0 \;. \qquad (7.1.3)$$

g_1, g_2 and g_3 must satisfy the compatibility condition (6.3.3).

The general approach in the weighted residual method is to multiply these equations by arbitrary test functions (see Section 3.6), integrate over the domain Ω, apply partial integration if necessary, and finally substitute the boundary conditions. If we choose the basis functions of the approximation as test functions we end up with the Galerkin equations. As test functions in first instance v_1, v_2, v_3 and q are chosen. Because of the boundary conditions for u_1, u_2 and u_3 we assume that:

$$v_1 = v_2 = v_3 = 0 \quad \text{on} \quad \Gamma \;. \qquad (7.1.4)$$

The resulting integral equations are:

$$\int_\Omega \left(-\frac{1}{Re}\,\Delta\, u_i + \frac{\partial p}{\partial x_i}\right) v_i \; d\Omega = \int_\Omega f_i \; v_i \; d\Omega \;, \quad i=1,2,3 \;, \qquad (7.1.5)$$

$$\int_\Omega q \; \text{div}\; \underline{u} \; d\Omega = 0 \;. \qquad (7.1.6)$$

We now apply partial integration to the equations (7.1.5), and use the relations:

$$v_i \Delta u_i = \text{div}(v_i \nabla u_i) - \nabla u_i \cdot \nabla v_i \;,$$

$$v_i \frac{\partial p}{\partial x_i} = \frac{\partial}{\partial x_i}(p v_i) - p \frac{\partial v_i}{\partial x_i} \;,$$

together with the Gauss' theorem,

$$\int_\Omega \operatorname{div} \underline{a}\, d\Omega = \int_\Gamma \underline{a}\cdot\underline{n}\, d\Gamma \quad \text{with } \underline{n} \text{ the outward normal.}$$

Choice of respectively $\underline{a} = v_i \nabla u_i$ and $\underline{a} = (pv_1,0,0)$, $\underline{a} = (0,pv_2,0)$, $\underline{a} = (0,0,pv_3)$, transforms the equations (7.1.5), (7.1.6) into:

$$\int_\Omega \left(\frac{1}{Re} \nabla u_i \cdot \nabla v_i - p \frac{\partial v_i}{\partial x_i}\right) d\Omega = \int_\Omega f_i v_i \, d\Omega + \int_\Gamma \frac{1}{Re}(\nabla u_i \cdot \underline{n}) v_i \, d\Gamma +$$

$$- \int_\Gamma p v_i n_i \, d\Gamma , \qquad i=1,2,3 , \tag{7.1.7}$$

$$\int_\Omega q \operatorname{div} \underline{u}\, d\Omega = 0 . \tag{7.1.8}$$

with n_i the components of the outward normal $\underline{n}$.
Substitution of the boundary conditions for v_i gives:

$$\int_\Omega \left\{\frac{1}{Re} \nabla u_i \cdot \nabla v_i - p \frac{\partial v_i}{\partial x_i}\right\} d\Omega = \int_\Omega f_i v_i \, d\Omega , \quad i=1,2,3 , \tag{7.1.9}$$

$$\int_\Omega q \operatorname{div} \underline{u}\, d\Omega = 0 . \tag{7.1.10}$$

For the weak formulation of problem (7.1.1) - (7.1.3) the function spaces $\underline{V}_0, \underline{V}_g$ and Q must be introduced.
Let $\underline{V}_0$ be the space of vector functions $\underline{v} = (v_1,v_2,v_3)$ satisfying (7.1.4) and $\underline{V}_g$ the space of vector functions $\underline{u} = (u_1,u_2,u_3)$ satisfying (7.1.2). Functions in $\underline{V}_0$ and $\underline{V}_g$ must be so smooth that the integrals in (7.1.9) exist. Let furthermore Q be the space of functions satisfying (7.1.3) that are so smooth that the integral (7.1.10) exists. Then the weak formulation of (7.1.1)-(7.1.3) reads:

Find $\underline{u} \in \underline{V}_g$ and $p \in Q$ such that

$$\int_\Omega \left\{\frac{1}{Re} \nabla u_i \cdot \nabla v_i - p \frac{\partial v_i}{\partial x_i}\right\} d\Omega = \int_\Omega f_i v_i d\Omega , \quad i=1,2,3 , \tag{7.1.11}$$

$$\int_{\Omega} q \operatorname{div} \underline{u} \, d\Omega = 0 \ , \tag{7.1.12}$$

for all $\underline{v} \in \underline{V}_0$ and all $q \in Q$.

For the construction of the approximation of the solution, u_1, u_2, u_3 and p are written as linear combinations of basis functions:

$$u_i = u_{i0}(\underline{x}) + \sum_{j=1}^{\infty} u_{ij}\phi_j(\underline{x}) \ , \quad i=1,2,3 \ , \tag{7.1.13}$$

$$p = \sum_{j=1}^{\infty} p_j \psi_j(\underline{x}) \ . \tag{7.1.14}$$

In (7.1.13), (7.1.14) ϕ_j are the basis functions for the velocity components and ψ_j the basis functions for the pressure. The basis functions ϕ_j must be chosen such that the functions $(\phi_j,0,0)$, $(0,\phi_j,0)$, $(0,0,\phi_j)$ $(j=1(1)\infty)$ span the complete space $\underline{V}_0$ and ψ_j such that the space Q is spanned. Basis functions $\phi_j(\underline{x})$ that are piecewise continuously differentiable and continuous over the domain Ω satisfy the smoothness requirements of the spaces $\underline{V}_g$ and $\underline{V}_0$. For the basis functions $\psi_j(\underline{x})$ no continuity is required.
The functions $u_{i0}(\underline{x})$ must be chosen such that:

$$u_{i0}(\underline{x}) = g_i(\underline{x}) \ , \quad \underline{x} \in \Gamma \ . \tag{7.1.15}$$

The basis functions ϕ_i are substituted subsequently for the test functions v_i in (7.1.11), and the test function q in (7.1.12) is replaced by ψ_i. If we limit ourselves to approximate solutions, constructed by a finite number of basis functions i.e.

$$\tilde{u}_i(\underline{x}) = \tilde{u}_{i0}(\underline{x}) + \sum_{j=1}^{N} \tilde{u}_{ij}\phi_j(\underline{x}) \ , \quad i=1,2,3 \ , \tag{7.1.16}$$

$$\tilde{p}(\underline{x}) = \sum_{j=1}^{M} \tilde{p}_j \psi_j(\underline{x}) \ , \tag{7.1.17}$$

as for example defined by the FEM, then the following system of Galerkin equations remains:

$$\int_\Omega \frac{1}{Re} \nabla\tilde{u}_1 \cdot \nabla\phi_i \, d\Omega - \int_\Omega \tilde{p} \frac{\partial \phi_i}{\partial x_1} d\Omega = \int_\Omega f_1 \phi_i \, d\Omega \,, \tag{7.1.18}$$

$$\int_\Omega \frac{1}{Re} \nabla\tilde{u}_2 \cdot \nabla\phi_i \, d\Omega - \int_\Omega \tilde{p} \frac{\partial \phi_i}{\partial x_2} d\Omega = \int_\Omega f_2 \phi_i \, d\Omega \,,$$

$$\int_\Omega \frac{1}{Re} \nabla\tilde{u}_3 \cdot \nabla\phi_i \, d\Omega - \int_\Omega \tilde{p} \frac{\partial \phi_i}{\partial x_3} d\Omega = \int_\Omega f_3 \phi_i \, d\Omega \,, \quad i=1,2,\ldots,N \,,$$

$$\int_\Omega \psi_i \, \mathrm{div}\, \tilde{\underline{u}} \, d\Omega = 0 \,, \quad i=1,2,\ldots,M \,. \tag{7.1.19}$$

(7.1.18), (7.1.19) form a system of 3N+M equations with 3N+M unknowns; in the case of the FEM these are the approximations for the nodal point values of the velocity components and the pressure.

Remarks

i) For convenience in the next chapters u_{ij} and p_j will be used instead of $\tilde{u}_{ij}$ and $\tilde{p}_j$.

ii) In practical computations the basis functions ψ_i are not constructed such that:

$$\int_\Omega \psi_i \, d\Omega = 0 \,, \tag{7.1.20}$$

but one of the pressure parameters is fixed. There is no essential difference between both approaches.

7.2 Practical elaboration

7.2.1 Complicated boundary conditions

The treatment of other boundary conditions than the Dirichlet boundary conditions of section 7.1, is roughly the same as the treatment in the sections 3.2 and 3.6. This will be explained on the basis of the 2D example of section 6.3.2. The boundary conditions on the five boundaries were given by:

i) inflow: $(n_1=0,\ n_2=1)$: $u_1 = 0\ ,\quad u_2 = u_2(x_1)\ .$ (7.2.1)

ii) free surface $(n_1=0,\ n_2=1)$: $\dfrac{\partial u_1}{\partial x_2} = 0,\ u_2 = 0\ .$

iii), v) fixed walls : $\underline{u} = 0\ .$

iv) outflow $(n_1=1,\ n_2=0)$: $\dfrac{\partial u_1}{\partial x_1} = 0,\ u_2 = 0,\ p = 0\ .$

If we assume that the test functions v_1 and v_2 satisfy homogeneous Dirichlet boundary conditions, whenever u_1 and u_2 satisfy Dirichlet boundary conditions, i.e:

$$u_1 = g_1(\underline{x}) \iff v_1 = 0\ , \tag{7.2.2}$$

$$u_2 = g_2(\underline{x}) \iff v_2 = 0\ , \qquad \underline{x} \in \Gamma\ ,$$

then all boundary integrals:

$$\int_\Gamma (\nabla u_1\cdot\underline{n})v_1\, d\Gamma\ , \quad \int_\Gamma p v_1 n_1\, d\Gamma\ , \tag{7.2.3}$$

$$\int_\Gamma (\nabla u_2\cdot\underline{n})v_2\, d\Gamma\ , \quad \int_\Gamma p v_2 n_2\, d\Gamma\ ,$$

in (7.1.7) vanish on all boundaries. For this specific example these conditions result in the same type of Galerkin equations as was found for the Dirichlet boundary conditions. The only difference is that some basis functions corresponding to boundary points where no Dirichlet conditions are prescribed, must be used as test functions and that generally the number of equations for the two velocity components is not equal. In the case of inhomogeneous Neumann and Robbins conditions, line integrals must be added to the system of equations in exactly the same way as demonstrated in section 5.3.

In the case that stresses on boundaries are prescribed, it is generally easier to start with the stress tensor formulation (1.6.2) before applying the Galerkin method. Multiplication of (1.6.2) by test functions $\underline{v}$, integration over the domain Ω and application of partial integration (Green's formula) leads in the stationary case to:

$$\int_{\Omega} (\underline{\underline{\sigma}}.\nabla\underline{v}+\rho(\underline{u}.\nabla)\underline{u}.\underline{v}-\rho\underline{f}.\underline{v})\, d\Omega = \int_{\Gamma} (\sigma_n\underline{v}.\underline{n}+\sigma_t\underline{v}.\underline{t})\, d\Gamma \,. \quad (7.2.4)$$

In order to treat the boundary integral adequately, it is necessary that all terms in the boundary integral are related to boundary conditions. For example in the term $\sigma_n\underline{v}.\underline{n}$ it is necessary that either $\underline{v}.\underline{n} = 0$ (i.e. the normal component of the velocity is prescribed) or something is known about σ_n (σ_n known, or σ_n is a combination of velocities). In the same way either $\underline{v}.\underline{t}$ must be zero (i.e. tangential component of the velocity is prescribed) or σ_t is given.
This demonstrates that boundary conditions must be given in independent directions.

7.2.2 Necessary conditions for the elements

According to the remarks in section 3.5.1, the basis functions for velocity and pressure must satisfy the following requirements:

i) In the Galerkin equations first order derivatives (7.2.5) of the velocities occur. That means that these basis functions must be continuously differentiable in each element and continuous in the whole domain Ω.

ii) The basis functions for the pressure are not differentiated in the Galerkin equations. These basis functions have to be continuous in each element, but in the domain Ω they are allowed to be discontinuous. They must be integrable.

So the conditions for the basis functions of the pressure are weaker than the conditions for the velocity basis functions. An element approximation satisfying the requirements (7.2.5) is called conforming (see Section 5.5).

Another important point concerns the relation between the velocity basis functions and the pressure basis functions. Consider for example the system of approximate equations for the 2D continuity equation:

$$\int_\Omega \left(\frac{\partial \tilde{u}_1}{\partial x_1} + \frac{\partial \tilde{u}_2}{\partial x_2}\right) \psi_i \, d\Omega = 0 \ , \qquad i = 1,2,\ldots,M \ . \tag{7.2.6}$$

Substitution of the approximate solution (7.1.15) yields:

$$\sum_{j=1}^{N} \int_\Omega \left\{u_{1j} \frac{\partial \phi_j}{\partial x_1} + u_{2j} \frac{\partial \phi_j}{\partial x_2} + \frac{\partial \tilde{u}_{10}}{\partial x_1} + \frac{\partial \tilde{u}_{20}}{\partial x_2}\right\} \psi_i \, d\Omega = 0. \tag{7.2.7}$$

$$i = 1,2,\ldots,M$$

We see that there are M equations with 2N unknowns. Essentially M and N are the number of nodal points in the region Ω. However, due to the influence of the boundary conditions, the number of equations and the number of degrees of freedom is smaller. To see this, consider the following boundary conditions:

i) the pressure is prescribed in one point. This implies that the number of basis functions is decreased by one, and

$$\tilde{p}(\underline{x}) = p_c + \sum_{j=1}^{P-1} p_j \psi_j(\underline{x}) \ , \tag{7.2.8}$$

with p_c the additive constant.

ii) the velocity is prescribed on the boundary Γ. Suppose that the number of nodal points on the boundary is equal to L. The approximate solutions can be written as:

$$\tilde{u}_1(\underline{x}) = \tilde{u}_{10}(\underline{x}) + \sum_{j=1}^{K-L} u_{1j} \phi_j(\underline{x}) \ , \tag{7.2.9}$$

$$\tilde{u}_2(\underline{x}) = \tilde{u}_{20}(\underline{x}) + \sum_{j=1}^{K-L} u_{2j}\phi_j(\underline{x}) \ .$$

Due to the boundary conditions, which are much more severe for the velocity than for the pressure, there remains a system of P-1 equations with 2(K-L) unknowns. Here P is the number of nodal points for the pressure, K the number of nodal points for the velocity and L the number of nodal points on the boundary where Dirichlet boundary conditions have been prescribed. Since we do not want that the velocity approximation is completely described by the continuity equation, we demand:

$$P - 1 < 2(K-L) \iff M < 2N \ , \qquad (7.2.10)$$

for any number of elements.

(7.2.10) restricts the choice of elements considerably.
A generally accepted requirement is the following:

The pressure must be approximated by interpolation polynomials that are at least one degree less than the polynomials for the velocity. (7.2.11)

That this demand is useful, but not generally applicable, is explained in the next two examples. Consider the two elements and corresponding regions Ω, consisting of 8 elements, given in fig. 7.1.
The interpolation functions and basis functions for the velocity are elementwise linear for both elements (see Section 4.2). Furthermore the pressure is elementwise constant in both cases. The velocity is continuous in the region Ω for the element of fig. 7.1a, whereas it is discontinuous for the element of fig. 7.1b. This can be easily checked by applying the interpolation formulae of section 4.4 at the vertices of each element. The pressure is in both cases discontinuous in the domain Ω, but elementwise integrable. The element with the vertices as nodal points is conforming according to definition (7.2.5), the element in fig. 7.1b is non-conforming. Both elements satisfy the requirement (7.2.11), the velocity is

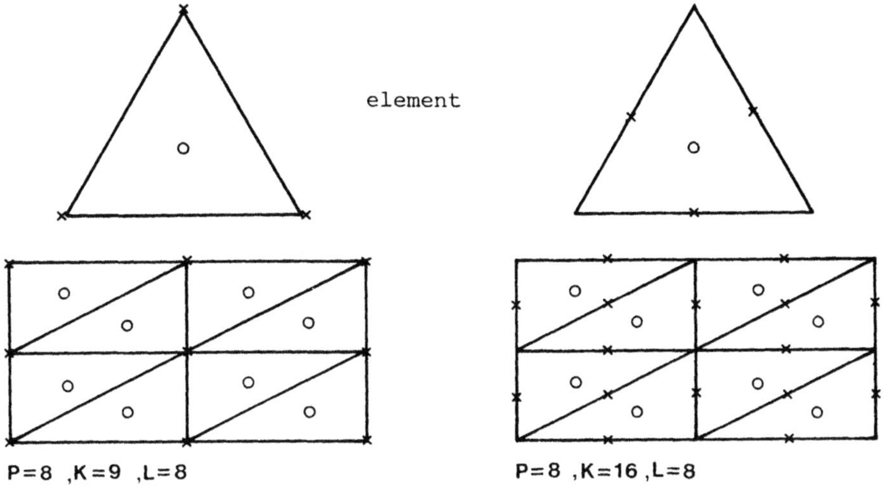

Region Ω with
P: pressure nodal points
K: velocity nodal points
L: velocity nodal points
on the outer boundary

Fig. 7.1 Triangular elements with three nodal points for the velocity (x) and one nodal point for the pressure (0) :
a. vertices are the velocity nodal points: conforming element,
b. mid-points of the sides are the velocity nodal points: non-conforming element.

linearly approximated, the pressure by a constant in each element. Nevertheless, the conforming element is not applicable since the number of free degrees of freedom for the velocity (2(K-L)=2) is much smaller than the number of equations that follows from the discretization of the continuity equation (P-1=7). This is not the case for the non-conforming element (2(K-L)=16), however, due to the non-conformity, it has a very limited practical value. Of course the requirement (7.2.10) could be satisfied by increasing the number of elements, however, one cannot expect that an element that produces singular matrices for a small number of elements gives regular matrices for a large number of elements. (7.2.11) gives just an indication of what one should demand. In the literature, see for example Fortin (1981) and also in chapter 14, an exact formulation of the condition that elements must satisfy in order to be admissable is mentioned. This condition is known under the name Brezzi-Babuška condition (or BB condition). However, the BB condition is rather abstract and in practice it is very difficult to verify whether the BB condition is satisfied or not. Fortin (1981) has given a simple method to check the BB condition on a number of elements.

This method is based on the following statement:
an element satisfies the BB condition, whenever, given a continuous differentiable vector field $\underline{u}$, one can explicitly build a discrete vector field $\tilde{\underline{u}}$ such that

$$\int_{\Omega} \psi_i \,\mathrm{div}\, \tilde{\underline{u}} \, d\Omega = \int_{\Omega} \psi_i \,\mathrm{div}\, \underline{u} \, d\Omega \qquad (7.2.12)$$

for all basis functions ψ_i.

In section 7.2.3.3 it will be shown how such a vector field can be constructed for the non-conforming triangle of fig. 7.1. Another possibility to check the BB condition is to construct explicitly divergence-free basis functions, see chapter 9. The preceding examples show that nodal points in the centre of the sides increase the number of free velocity degrees of freedom, and furthermore that the demand (7.2.11) is not always sufficient. In most practical applications a quadratic velocity approximation and a linear pressure approximation is used. Such elements are treated in the next section. These results can be formulated in the following "engineering" statement (Fortin 1981):

Midside velocity points in two dimensions and centroid velocity points on surfaces in three dimensions make it possible to control the amount of flow through a side (2D) and through a surface (3D) of an element, without altering the amount of flow through other sides or surfaces. Hence such nodal points make it easier to satisfy the continuity equation.

It is remarked that in the literature frequently elements are used, that do not satisfy the BB condition. Such elements cannot be used with the direct method, however the penalty function method (see chapter 8), permits the use of these elements. For a discussion concerning this point, the user is referred to section 8.2.2.

7.2.3 Examples of admissible elements

7.2.3.1 Introduction

We shall now discuss some of the most important elements and recall some of their properties. To this end, the elaboration of the basis functions as given in Chapter 4 is used, as well as the accuracy properties described in Section 5.5. First the 2D elements are treated and secondly some 3D ones. The axi-symmetric elements are the same as the 2D elements, however, the numerical integration must be carried out more accurately. The elements can be divided into two groups: elements with continuous pressure (the Taylor-Hood family) and elements with discontinuous pressure (the Crouzeix-Raviart family). Examples of both families will be discussed. For the accuracy errors mentioned in this section, the reader is referred to expression (5.5.5) as well as Chapter 13.

7.2.3.2 Triangular elements (Taylor-Hood (1973))

These elements are characterized by the fact that the pressure is continuous in the region Ω. An example is given in fig. 7.2;

the velocity is quadratic, the pressure linear. Due to the choice of the nodal points, both velocity and pressure are continuous over the element boundaries. For example for the pressure, the linear approximation as well as the use of the vertices fix the pressure along a boundary, once the nodal point values are given. Since for adjacent elements, the joint side has the same nodal points and the same linear approximation, the pressure is continuous on this boundary in both elements, and therefore it is continuous in Ω.

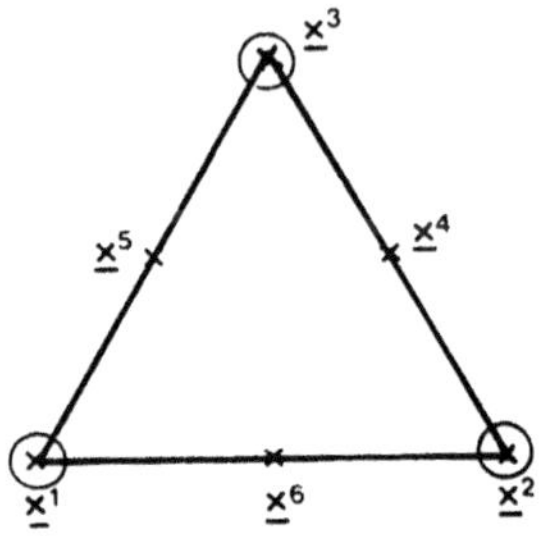

Velocity: quadratic (6 nodal points:x)
Pressure: linear (3 nodal points:0)

Accuracy velocity: $O(h^3)$
Accuracy pressure: $O(h^2)$

Fig. 7.2 Taylor-Hood element (P_2-P_1)

The basis functions for this element have been given in Section 4.2. For the sake of completeness they are repeated:

$$\tilde{u}_i = \sum_{j=1}^{6} u_{ij}\phi_j , \tag{7.2.13}$$

with

$$\phi_j = \lambda_j(2\lambda_j-1) , \qquad j = 1,2,3 ,$$

$$\phi_4 = 4\lambda_2\lambda_3, \quad \phi_5 = 4\lambda_1\lambda_3, \quad \phi_6 = 4\lambda_1\lambda_2,$$

and $\tilde{p} = \sum_{j=1}^{3} p_j\psi_j, \quad$ with $\psi_j = \lambda_j, \quad j = 1,2,3.$

u_{ij} and p_j are the nodal point values of velocity and pressure and λ_j the linear polynomials for a triangular element. In the same way as shown in Chapter 5 the element matrix and element vector corresponding to the Galerkin equations of the Stokes equation can be derived. In order to maintain the accuracy of the approximation, the numerical integration rule:

$$\int_e f(\underline{x})\, d\Omega = \frac{|e|}{3} \left(f(\underline{x}^4) + f(\underline{x}^5) + f(\underline{x}^6)\right),$$

with $|e|$ the area of the element, can be used. For axisymmetric elements this integration rule is not sufficiently accurate; a 6 or 7 point Gauss rule must be applied.

One can show, Segal (1979), that this element is admissible if at least 3 elements are used. With 3 elements and Dirichlet boundary conditions for the velocity along the whole boundary we have: $P-1 = 4$, $2(K-L) = 4$ and the system of equations can be solved formally. The element has proven to be a good, and numerically useful element, in practical computations, however, it requires a large amount of computing time. For that reason other elements have been developed.

7.2.3.3 Triangular elements (Crouzeix-Raviart (1973))

These elements are characterized by a discontinuous pressure; discontinuous on element boundaries. For output purposes (printing, plotting etc.) these discontinuous pressures are averaged in vertices for all the adjoining elements, see fig. 7.3. We shall discuss some of the Crouzeix-Raviart elements.

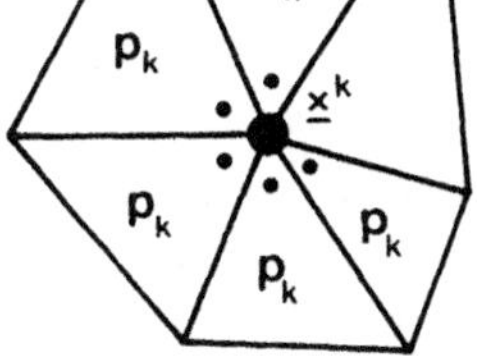

Fig. 7.3
Averaging in each nodal point $\underline{x}_k$ over all elements containing $\underline{x}_k$, in order to get a continuous pressure for Crouzeix-Raviart elements.

a. The most simple Crouzeix-Raviart element (see Section 4.2.3) is shown in fig. 7.4. It is the non-conforming element, already mentioned in Section 7.2.2, where the velocity is approximated linearly and the pressure is approximated by a constant.

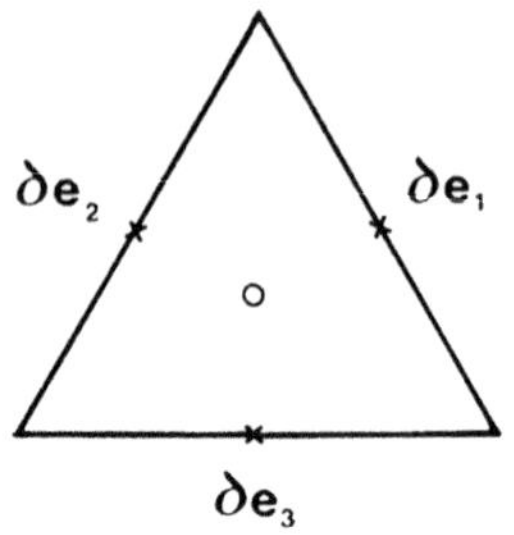

Velocity: linear (3 nodal points: x)
Pressure: constant (1 nodal point:0)
Accuracy velocity: $O(h^2)$
Accuracy pressure: $O(h)$

Fig. 7.4 Crouzeix-Raviart element (P_1-P_0)

We shall show that this element is admissible by building explicitly a vector field $\tilde{\underline{u}}$ such that (7.2.12) is satisfied, i.e.:

$$\int_e \operatorname{div} \underline{u}\, d\Omega = \int_e \operatorname{div} \tilde{\underline{u}}\, d\Omega = \int_{\partial e} \tilde{\underline{u}}\cdot\underline{n}\, d\Gamma , \qquad (7.2.14)$$

given the continuous vector field $\underline{u}$.

To that end we define $\tilde{\underline{u}}$ in the midside point of an element in the following way:

$$\int_{\partial e_k} \underline{u}\, d\Gamma = \int_{\partial e_k} \tilde{\underline{u}}\, d\Gamma = |\partial e_k|\, \underline{u}_k , \qquad (7.2.15)$$

with $|\partial e_k|$ the length of side ∂e_k and $\underline{u}_k$ the velocity in the midside point on side ∂e_k. Definition (7.2.15) causes no contradictions, since the value of $\underline{u}_k$ in adjacent elements is the same. Furthermore (7.2.15) is such that (7.2.14) is satisfied. This element has only a limited applicability and is rather inaccurate. It can not be used for axi-symmetric co-ordinates. The main interest of this element is theoretical; it shows that midpoints on sides make it possible to satisfy the BB condition, or otherwise stated: midside points on elements enable us to control the amount of flow through a side.

Remark: the n-dimensional extension of this element is treated in section 12.3.6.

b. The simplest conforming Crouzeix-Raviart element is shown in fig. 7.5. The velocity is approximated quadratically, the pressure by a constant. Due to the bad approximation of the pressure, which affects the velocity approximation through the continuity equation, the theoretical accuracy of the velocity is only $O(h^2)$, which has been confirmed by numerical experiments, see Segal (1979).

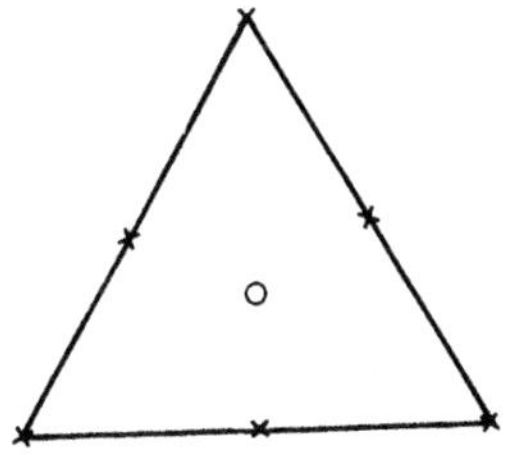

Velocity: quadratic (6 nodal points: x)
Pressure: constant (1 nodal point: 0)
Accuracy velocity: $O(h^2)$
Accuracy pressure: $O(h)$

Fig. 7.5 Crouzeix-Raviart element (P_2-P_0)

Because of the rough approximation of the pressure this element is as accurate as the non-conforming P_1-P_0 element and is hardly used. Due to the large number of degrees of freedom for the velocity it is an admissible element, which can be proven simply. See for example Crouzeix and Raviart (1973).

c. Fig. 7.6 shows the trivial extension of the P_2-P_0 element. It resembles very much the Taylor-Hood element, the velocity is quadratic, the pressure linear.

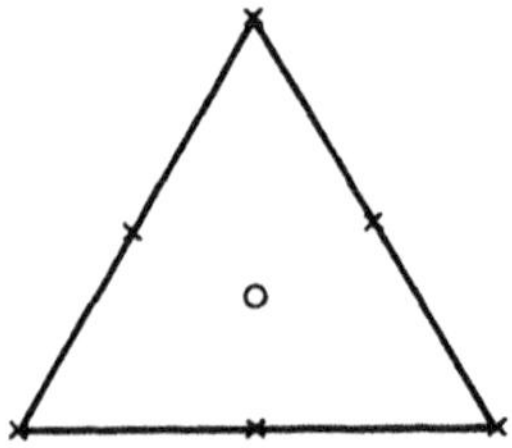

Velocity: quadratic (6 nodal points: x)
Pressure: linear (1 nodal point: 0 including 2 derivatives)
Accuracy velocity } Not admissible
Accuracy pressure }

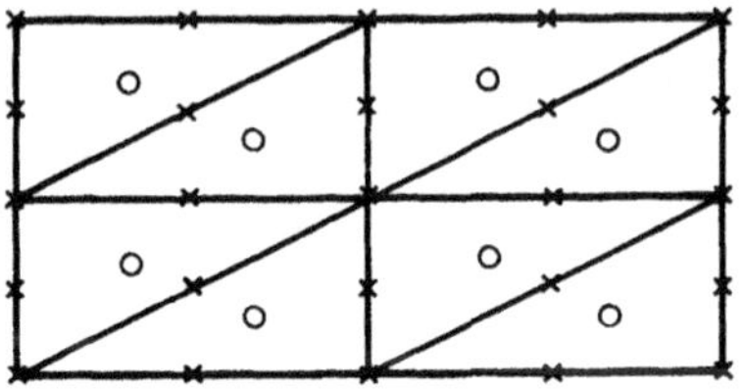

$2(K-L) < P - 1$

Fig. 7.6 Crouzeix-Raviart element (P_2-P_1) with region Ω.

Because of the choice to define the pressure polynomial by its value and the value of its derivatives in the centroid instead of its value in the three vertices, this element, unlike the Taylor-Hood element, is not applicable.

d. An extension of the P_2-P_1 element can be constructed by taking an extended (or enriched) quadratic approximation, using

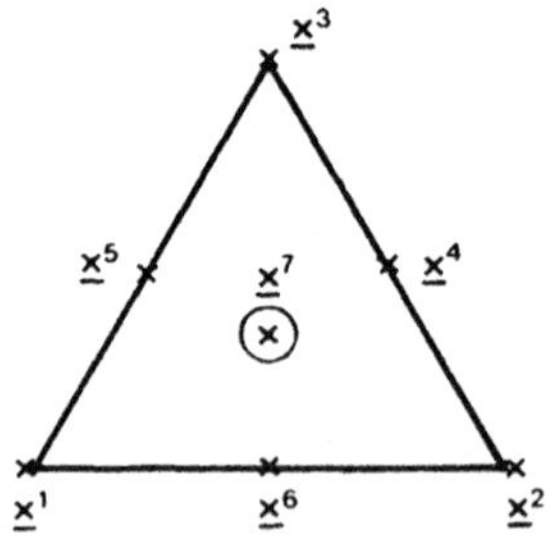

Velocity: enriched quadratic (7 nodal points: x)
Pressure: linear (1 nodal point: 0 including 2 derivatives)
Accuracy velocity: $O(h^3)$
Accuracy pressure: $O(h^2)$

Fig. 7.7 Crouzeix-Raviart element (P_2^+-P_1)

the centroid as an additional nodal point. This element is shown in fig. 7.7.

The extension increases the number of velocity degrees of freedom in the region Ω in fig. 7.6 by 16. It will be clear that this makes the element admissible. We repeat the basis functions (see also Section 4.2.5):

$$\tilde{u}_i = \sum_{j=1}^{7} u_{ij}\phi_j \,, \tag{7.2.16}$$

with $\phi_j = \lambda_j(2\lambda_j-1)+3\lambda_1\lambda_2\lambda_3$, $\quad j = 1,2,3$,

$\phi_4 = 4\lambda_2\lambda_3-12\lambda_1\lambda_2\lambda_3$, $\phi_5 = 4\lambda_1\lambda_3-12\lambda_1\lambda_2\lambda_3$, $\phi_6 = 4\lambda_1\lambda_2-12\lambda_1\lambda_3\lambda_3$,

$\phi_7 = 27\lambda_1\lambda_2\lambda_3$,

and $\tilde{p} = p_7\psi_1 + \frac{\partial p}{\partial x_1}(\underline{x}^7)\psi_2 + \frac{\partial p}{\partial x_2}(\underline{x}^7)\psi_3$,

with $\psi_1 = 1$,

$\psi_2 = x_1 - x_1^7$,

$\psi_3 = x_2 - x_2^7$.

u_{ij} are nodal point values for the velocity and p_7 is the value of the pressure in the centroid. λ_i is the linear polynomial for a triangular element. In order to maintain accuracy, the numerical integration rule (see Section 5.4.1):

$$\int_e g \, d\Omega = \frac{|e|}{60}\{3\sum_{i=1}^{3} g(\underline{x}^i) + 8\sum_{i=4}^{6} g(\underline{x}^i) + 27g(\underline{x}^7)\}, \tag{7.2.17}$$

with $|e|$ the area of the element, may be used to compute the element matrix and element vector. For axisymmetric co-ordinates this rule is not accurate enough and a 6 point or 7 point Gauss integration rule is necessary.

The $P_2^+ - P_1$ element is conforming; the velocity polynomials are continuous over the element boundaries, the pressure polynomials are integrable.

We shall show that this element is admissible by explicitly constructing a vector field $\tilde{\underline{u}}$ such that (7.2.12) is satisfied, i.e.:

$$\int_e \operatorname{div} \underline{u}\, d\Omega = \int_e \operatorname{div} \tilde{\underline{u}}\, d\Omega = \int \tilde{\underline{u}} \cdot \underline{n}\, d\Gamma \; , \qquad (7.2.18)$$

and

$$\int_e (x_k - x_k^7) \operatorname{div} \underline{u}\, d\Omega = \int_e (x_k - x_k^7) \operatorname{div} \tilde{\underline{u}}\, dx, \quad k=1,2 \; ; \qquad (7.2.19)$$

given the continuous vector field $\underline{u}$.
In order to satisfy (7.2.18) we define $\tilde{\underline{u}}$ such that $\tilde{\underline{u}}(\underline{x}^i)=\underline{u}(\underline{x}^i)$, i=1,2,3, and furthermore $\tilde{\underline{u}}_4$, $\tilde{\underline{u}}_5$ and $\tilde{\underline{u}}_6$ such that

$$\int_{\partial e_k} \underline{u}\, ds = \int_{\partial e_k} \tilde{\underline{u}}\, ds \; , \qquad (7.2.20)$$

with ∂e_k a side of the triangle, (k=1,2,3).

Given $\tilde{\underline{u}}_1, \ldots, \tilde{\underline{u}}_6$ by (7.2.20), $\tilde{u}_7$ can be chosen such that (7.2.18) is satisfied. This will be shown in Section 8.3.2.

This element is a useful element. However, compared to the Taylor-Hood element, there are more degrees of freedom for the same order of accuracy. Hence the element requires a larger amount of computing time without improvement of the results. In chapter 8 it will be demonstrated, that a modification of this element makes it more useful. Because of the special structure of the basis functions for the pressure, it is possible to eliminate the velocity components in the centroid. Besides that, the gradient of the pressure disappears, and therefore the number of equations as well as the number of degrees of freedom reduces with four in each element. The resulting element is sketched in fig. 7.8.

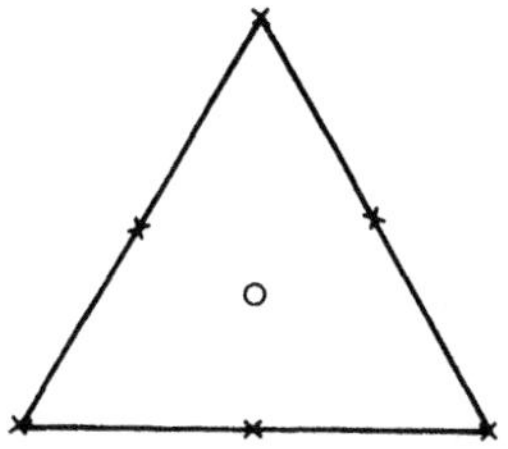

Velocity: enriched quadratic
(6 nodal points: x)

Pressure: linear
(1 nodal point: 0)

Accuracy velocity: $O(h^3)$
Accuracy pressure: $O(h^2)$

Fig. 7.8 Modified Crouzeix-Raviart element ($P_2^+ - P_1$)

The modification of the original element maintains its accuracy, however, the number of equations and degrees of freedom is reduced. There are exactly the same number of velocity degrees of freedom as in the Taylor-Hood element, but less pressure unknowns (one instead of three in each element). Since the number of internal vertices of the triangles is generally smaller than the number of centroids, the total amount of degrees of freedom is somewhat larger than of the Taylor-Hood element. Hence the Taylor-Hood element is a fraction cheaper than the Crouzeix-Raviart element. The Crouzeix-Raviart type elements, however, enable us to use the more economic methods that will be treated in the chapters 8 and 9, and it is there where its main interest lies. Another advantage of the Crouzeix-Raviart element is that it satisfies the continuity equation elementwise. This is easily shown by considering the Galerkin equations corresponding to the continuity equation:

$$\int_\Omega \left(\frac{\partial \tilde{u}_1}{\partial x_1} + \frac{\partial \tilde{u}_2}{\partial x_2}\right) \psi_i d\Omega = 0 \ , \qquad i = 1,2,\ldots,M \ . \qquad (7.2.21)$$

If we restrict ourselves to the constant term in each element, (7.2.21) can be written as:

$$\int_e \left(\frac{\partial \tilde{u}_1}{\partial x_1} + \frac{\partial \tilde{u}_2}{\partial x_2}\right) d\Omega = 0 = \int_{\partial e} \underline{\tilde{u}} \cdot \underline{n} \ d\Gamma \ , \qquad (7.2.22)$$

since $\psi_i = 1$ inside the element and $\psi_i = 0$ outside the element. (7.2.22) states that the mass produced or consumed in each element is zero, and so, by choosing a discontinuous pressure, the conservation of mass (continuity equation) is satisfied elementwise. This is in contrast with the Taylor-Hood element where the pressure is approximated continuously over Ω and where elements are not mass-conservative. The Taylor-Hood element only conservates mass over the whole domain Ω. From a physical point of view, elementwise conservation of mass is preferable, it is however not clear whether the accuracy of the approximating solution is influenced by this property or not.

7.2.3.4 Quadrilateral elements

The derivation of basis functions, element matrices and vectors is more complicated for quadrilateral elements than for triangular elements, except in the case that the boundaries of the elements are curved. This is because always, even for straight sides an isoparametric transformation is needed, see Section 4.5. However, in general applications, quadrilateral elements are favourable above triangular elements, except when the region is such that it cannot be approximated easily by quadrilaterals. This is because of the following reasons:

i) Compared to triangular elements, only one half of the number of elements is needed. This reduces the computation time for the construction of matrices and vectors.

ii) If we divide a quadrilateral into two triangles, the

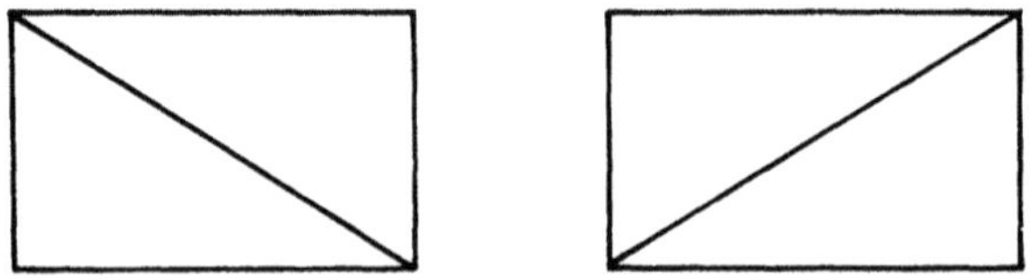

Fig. 7.9 Possible division of quadrilaterals into 2 triangles; mark the different diagonal directions.

diagonal of the quadrilateral has one of two possible directions. Numerical computations for convective-dominant flows, have shown, at least for rectangular grids, that the solution is sensitive to the direction of the diagonal. This sensitivity reduces when the size of the elements decreases. Quadrilateral elements do not exhibit this behaviour; their results are more "symmetric" for symmetrical problems.

In the same way as for triangular elements, we can distinguish between elements with continuous pressure (Taylor-Hood family) and with discontinuous pressure (Crouzeix-Raviart family).

The trivial extension of the P_2-P_1 Taylor-Hood element is the Q_2-Q_1 9-node quadrilateral element. Sometimes also the Q_2-Q_1 8-node (serendipity) element is used, where the central point is neglected. The 9-node element is sketched in fig. 7.10.

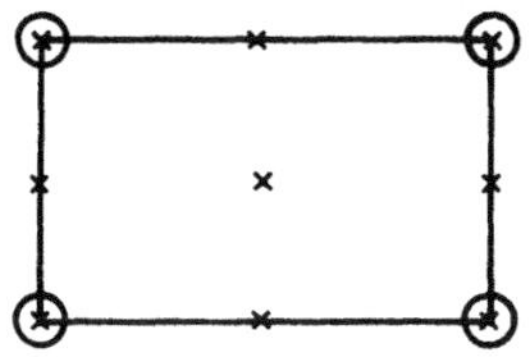

Velocity: bi-quadratic
(9 nodal points: x)
Pressure: bi-linear
(4 nodal points: 0)

Accuracy velocity: $O(h^3)$
Accuracy pressure: $O(h^2)$

Fig. 7.10 Taylor-Hood element (Q_2-Q_1)

The basis functions for this element are bi-quadratic polynomials for the velocity and bi-linear polynomials for the pressure, both defined on the reference element (see Section 4.5). For the derivation of the element matrix and element vector corresponding to the Galerkin equations of the Stokes equation, the transformation of the reference element to the actual element must be applied, see Section 4.5. In order to maintain the accuracy of the approximation, a 9 point Newton-

Cotes integration rule or a 3×3 point Gauss rule may be applied (see Section 5.4.2). For axisymmetric elements, the Newton-Cotes rule is not accurate enough, and only the 3×3 Gauss rule is applicable. This element is admissible if at least 2 elements are used.

The quadrilateral counterpart of the enriched quadratic triangle is the Q_2-P_1 9-node quadrilateral. It is probably the best one known for incompressible computations Fortin (1981). This element is sketched in fig. 7.11.

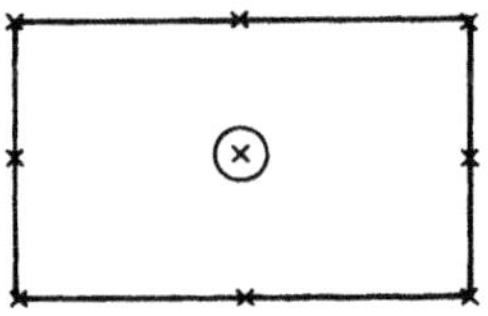

Velocity: bi-quadratic
(9 nodal points: x)
Pressure: linear
(1 nodal point: 0,
including 2 derivatives)

Accuracy velocity: $O(h^3)$
Accuracy pressure: $O(h^2)$

Fig. 7.11 Crouzeix-Raviart quadrilateral (Q_2-P_1)

The basis functions for this element are bi-quadratic polynomials on the reference element for the velocity and linear (discontinuous) polynomials for the pressure. Like for the enriched triangle, the velocity components and the pressure gradient in the centroid may be eliminated, reducing the number of velocity degrees of freedom to 16 and the number of pressure degrees of freedom to 1.

7.2.3.5 3D elements

The finite element literature on 3D elements for incompressible flows is not large at this moment. In this section we only mention some of the admissible elements. For a derivation of the element matrices and element vectors the reader is referred to Fortin (1981).

The "enriched" trilinear velocity-constant pressure element ($Q_1^+-P_0$) is probably the simplest admissible 3D element. The velocity is approximated by a trilinear approximation, based on the vertices of the hexahedron, see fig. 7.12. In order to maintain control over the amount of fluid flow, on each face a mid face point has been introduced with one degree of freedom, i.e. the normal velocity component. The corresponding basis function is a "bubble" Q_2 function that is zero in all other nodal points. For example on the reference cube ((0,1)×(0,1)×(0,1)), this bubble can be written for the x_3=1 face:

$$b(x_1,x_2,x_3) = \tfrac{1}{2}(x_3+1)x_1^2x_2^2$$

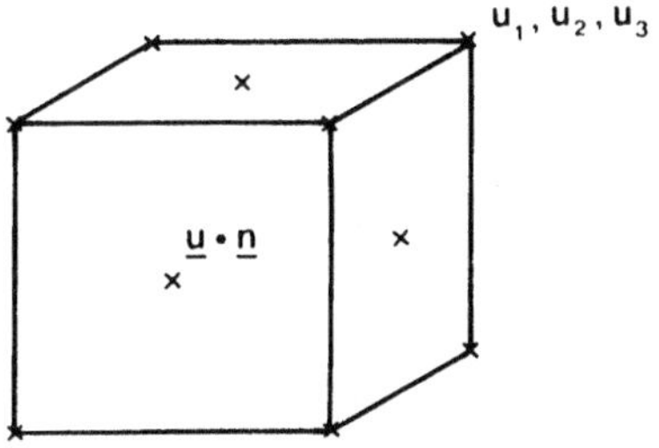

Velocity: trilinear + bubble function (14 nodal points: x)
Pressure: constant

Accuracy velocity: $O(h^2)$
Accuracy pressure: $O(h)$

Fig. 7.12 The $Q_1^+-P_0$ element

The full triquadratic-linear pressure element ($Q_2^{(27)}-P_1$) has a velocity that is approximated by a full triquadratic approximation based on the 27 nodes of fig. 7.13 (see Section

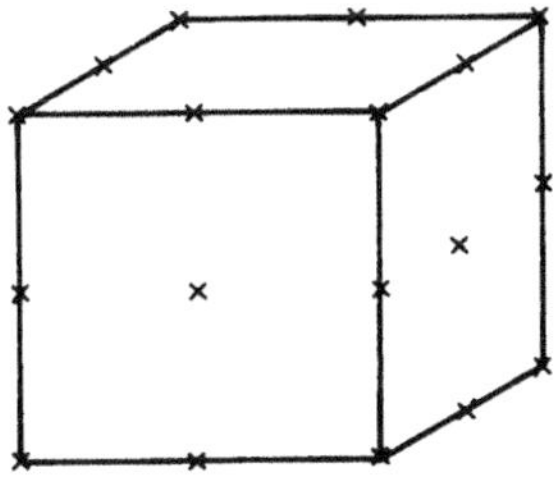

Velocity: triquadratic (27 nodal points: x)
Pressure: linear
Accuracy velocity: $O(h^3)$
Accuracy pressure: $O(h^2)$

Fig. 7.13 The Q_2-P_1 element

4.6.2). The pressure is approximated linearly, and is discontinuous. The internal velocity components as well as the gradient of the pressure may be eliminated in the same way as in the case of the discontinuous 2D elements.

7.2.4 The structure of the equations

In this Section the structure of the system of equations will be examined closer. To this end, consider the Galerkin equations (7.1.18), (7.1.19) and substitute the approximations (7.1.16) and (7.1.17) into them. The resulting system of equations becomes:

$$\sum_{j=1}^{N} u_{1j} \int_\Omega \frac{1}{Re} \nabla\phi_j \cdot \nabla\phi_i d\Omega - \sum_{j=1}^{M} p_j \int_\Omega \psi_j \frac{\partial\phi_i}{\partial x_1} d\Omega = \int_\Omega f_1 \phi_j d\Omega, \quad (7.2.23)$$

$$\sum_{j=1}^{N} u_{2j} \int_\Omega \frac{1}{Re} \nabla\phi_j \cdot \nabla\phi_i d\Omega - \sum_{j=1}^{M} p_j \int_\Omega \psi_j \frac{\partial\phi_i}{\partial x_2} d\Omega = \int_\Omega f_2 \phi_i d\Omega,$$

$$\sum_{j=1}^{N} u_{3j} \int_\Omega \frac{1}{Re} \nabla\phi_j \cdot \nabla\phi_i d\Omega - \sum_{j=1}^{M} p_j \int_\Omega \psi_j \frac{\partial\phi_i}{\partial x_3} d\Omega = \int_\Omega f_3 \phi_i d\Omega,$$

$$i=1,2,\ldots,N,$$

$$-\sum_{j=1}^{N} \int_\Omega \left(u_{1j} \frac{\partial\phi_j}{\partial x_1} + u_{2j} \frac{\partial\phi_j}{\partial x_2} + u_{3j} \frac{\partial\phi_j}{\partial x_3}\right) \psi_i d\Omega = 0, \quad i=1,2,\ldots,M. \quad (7.2.24)$$

Or in matrix-vector notation:

$$\begin{bmatrix} S_{u_1u_1} & 0 & 0 & L^T_{u_1p} \\ 0 & S_{u_2u_2} & 0 & L^T_{u_2p} \\ 0 & 0 & S_{u_3u_3} & L^T_{u_3p} \\ L_{pu_1} & L_{pu_2} & L_{pu_3} & 0 \end{bmatrix} \begin{bmatrix} \underline{u}_1 \\ \underline{u}_2 \\ \underline{u}_3 \\ \underline{p} \end{bmatrix} = \begin{bmatrix} \underline{f}_1 \\ \underline{f}_2 \\ \underline{f}_3 \\ \underline{0} \end{bmatrix} \quad (7.2.25)$$

with submatrices:

$$S_{u_k u_k}(i,j) = \frac{1}{Re} \int_\Omega \nabla\phi_i . \nabla\phi_j \, d\Omega ,$$

$$L_{pu_k}(i,j) = \int_\Omega \psi_i \frac{\partial \phi_j}{\partial x_k} \, d\Omega = L^T_{u_k p}(j,i) ,$$

unknowns:

$$\underline{u}_k^T = [u_{k1},\ldots,u_{kN}] , \qquad \underline{p}^T = [p_1,\ldots,p_M] ,$$

and right-hand-side vectors:

$$f_k(i) = \int_\Omega f_k \phi_i \, d\Omega .$$

The integrals can be computed efficiently elementwise in the same way as is shown in Section 5.4 using numerical integration rules. For example in the case of the 2D P_2-P_1 Taylor-Hood element, see fig. 7.2, this results in a 6×6 element coefficient matrix for each of the diffusion matrices $S_{u_k u_k}$. The velocity pressure element matrices $L_{u_{kp}}$ are 6×3 matrices. Also the element right-hand-side vectors can be computed by numerical integration. Assembly of the element matrices and element vectors yields the large coefficient matrix and right-hand-side vector. Because of the complicated formulae involved this will not be elaborated, but it will be clear that the matrices $S_{u_k u_k}$ are identical to the stiffness matrices corresponding to the Poisson equation, which have been treated in Chapter 5. The incorporation of different types of boundary conditions is not really a problem; in exactly the same way as shown in Sections 3.6, 5.1,...,5.3 this results into extra terms in the right-hand-side vector or the coefficient matrix. Hence the system of equations can be computed completely, and solved numerically by a Gaussian elimination method.

If we consider the system of equations (7.1.18) - (7.1.19) in the sequence as indicated, i.e. first all velocity components u_1, u_2 and u_3 and then the pressure p, it will be clear that the profile of the large matrix is very big (see fig. 7.14).

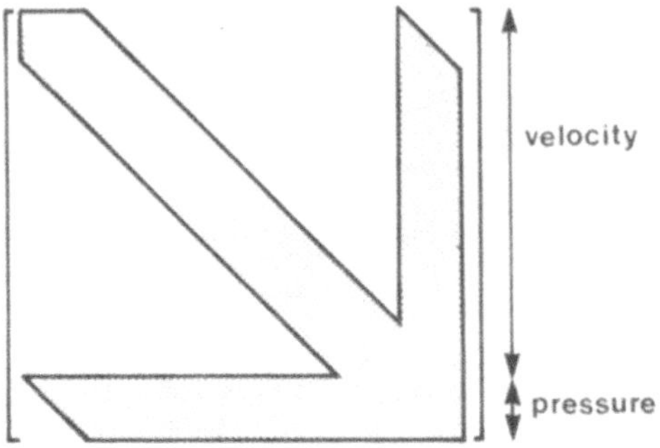

Fig. 7.14 Profile of the large matrix.

Even profile methods appear to be very time consuming for this type of profiles. In order to avoid such big profiles, in practice the sequence of the unknowns is chosen nodal pointwise, i.e. first all unknowns of the first nodal point, then for the second, etc. In this way a band matrix arises which has a much smaller profile and which can be solved more efficiently. In this matrix the degrees of freedom and corresponding rows where Dirichlet boundary conditions are prescribed are eliminated. For the example of fig. 7.15, this yields the following sequence of unknowns:

$$u_{11}, u_{21}, p_1, u_{12}, u_{22}, u_{13}, u_{23}, p_3, u_{14}, u_{24}, u_{15}, u_{25}, p_5, u_{16}, u_{26} \text{ etc.},$$

or due to the boundary conditions (suppose $u_1=u_2=0$ on the outer boundary, $p=p_1$ in nodal point 1):

$$p_3, p_5, u_{27}, v_{17} \quad \text{etc.}$$

a

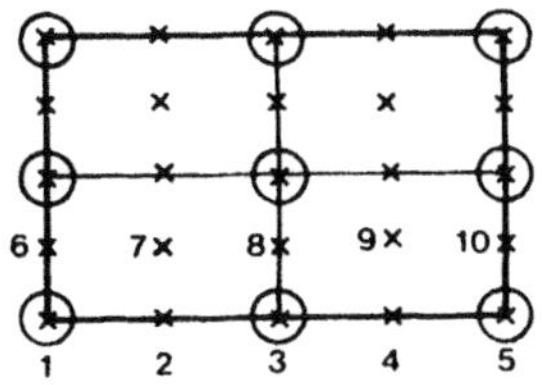

b Degrees of freedom: $u_{11}, u_{12}, u_{13}, \ldots; u_{21}, u_{22}, u_{23}, \ldots$
$p_1, p_3, \ldots$
c Degrees of freedom: $u_{11}, u_{21}, p_1, u_{12}, u_{22}, u_{13}, u_{23}, p_3, \ldots$

Fig. 7.15 Renumbering of unknowns:
a the region Ω,
b sequence of unknowns component-wise,
c sequence of unknowns nodal point wise.

The resulting system of equations has a much smaller profile than the one for the original system of equations. Due to renumbering, however, it is possible that the first diagonal elements of the matrix are equal to zero. This is for example the case in fig. 7.15c where, because of the boundary conditions, the first degrees of freedom are pressures, which do not appear in the continuity equation. In order to prevent zeros on the main diagonal partial pivoting must be applied. Unfortunately, partial pivoting reorders the sequence of the equations and increases the profile or bandwidth. Therefore a large amount of extra computing time and computer memory is required. However, it still remains cheaper than application of the numbering of fig. 7.15. In practice partial pivoting is always applied for the systems (7.1.18) - (7.1.19).

Our main concern in the preceding sections has been the solution of flows in cartesian co-ordinate systems. For axi-symmetric flows almost the same procedure can be followed. The element shapes are exactly the same as in section 7.2.3, the only difference is that more accurate numerical integration rules are necessary, because of the non-constant Jacobian in the surface element of the integration. Another important point is the symmetry axis r=0, which can cause apparent singularities, see section 6.1. Boundary conditions for r=0, see section 6.3.2, are therefore inevitable.

7.3 The Navier-Stokes equations

7.3.1 Introduction

In this section the convective term in the Navier-Stokes equations will be considered. The Navier-Stokes equations in vector notation can be written as (6.1.6):

$$-\frac{1}{Re}\Delta\underline{u} + \underline{u}.\nabla\underline{u} + \nabla p = \underline{f} , \qquad (7.3.1)$$

or component-wise (in cartesian co-ordinates):

$$-\frac{1}{Re}\Delta u_1 + u_1\frac{\partial u_1}{\partial x_1} + u_2\frac{\partial u_1}{\partial x_2} + u_3\frac{\partial u_1}{\partial x_3} + \frac{\partial p}{\partial x_1} = f_1 , \qquad (7.3.2)$$

$$-\frac{1}{Re}\Delta u_2 + u_1\frac{\partial u_2}{\partial x_1} + u_2\frac{\partial u_2}{\partial x_2} + u_3\frac{\partial u_2}{\partial x_3} + \frac{\partial p}{\partial x_2} = f_2 ,$$

$$-\frac{1}{Re}\Delta u_3 + u_1\frac{\partial u_3}{\partial x_1} + u_2\frac{\partial u_3}{\partial x_2} + u_3\frac{\partial u_3}{\partial x_3} + \frac{\partial p}{\partial x_3} = f_3 .$$

The continuity equation is:

$$\frac{\partial u_1}{\partial x_1} + \frac{\partial u_2}{\partial x_2} + \frac{\partial u_3}{\partial x_3} = 0 . \qquad (7.3.3)$$

For simplicity Dirichlet boundary conditions are assumed:

$$u_1 = g_1(\underline{x}) , \quad u_2 = g_2(\underline{x}) , \quad u_3 = g_3(\underline{x}) , \quad \underline{x} \in \Gamma ,$$

and $p(\underline{x}^0) = p_c$ for some point $\underline{x}^0 \in \Omega$.

g_1, g_2 and g_3 satisfy the compatibility condition (6.3.3).

For system (7.3.2), (7.3.3) one can derive the corresponding Galerkin equations, in a completely analogous way as shown in Section 7.1:

$$\int_\Omega [\frac{1}{Re} \nabla\tilde{u}_1 \cdot \nabla\phi_i + (\tilde{u}_1 \frac{\partial\tilde{u}_1}{\partial x_1} + \tilde{u}_2 \frac{\partial\tilde{u}_1}{\partial x_2} + \tilde{u}_3 \frac{\partial\tilde{u}_1}{\partial x_3})\phi_i - p \frac{\partial\phi_i}{\partial x_1}]d\Omega = \int_\Omega f_1\phi_i d\Omega$$

$$\int_\Omega [\frac{1}{Re} \nabla\tilde{u}_2 \cdot \nabla\phi_i + (\tilde{u}_1 \frac{\partial\tilde{u}_2}{\partial x_1} + \tilde{u}_2 \frac{\partial\tilde{u}_2}{\partial x_2} + \tilde{u}_3 \frac{\partial\tilde{u}_2}{\partial x_3})\phi_i - p \frac{\partial\phi_i}{\partial x_2}]d\Omega = \int_\Omega f_2\phi_i d\Omega$$

$$\int_\Omega [\frac{1}{Re} \nabla\tilde{u}_3 \cdot \nabla\phi_i + (\tilde{u}_1 \frac{\partial\tilde{u}_3}{\partial x_1} + \tilde{u}_2 \frac{\partial\tilde{u}_3}{\partial x_2} + \tilde{u}_3 \frac{\partial\tilde{u}_3}{\partial x_3})\phi_i - p \frac{\partial\phi_i}{\partial x_3}]d\Omega = \int_\Omega f_3\phi_i d\Omega$$

$$i = 1,2,\ldots,N,$$

$$\int_\Omega [\frac{\partial\tilde{u}_1}{\partial x_1} + \frac{\partial u_2}{\partial x_2} + \frac{\partial u_3}{\partial x_3}]\ \psi_i\ d\Omega = 0\ , \qquad i = 1,2,\ldots,M\ . \qquad (7.3.4)$$

(7.3.4) contains, due to the convection, non-linear terms of the type:

$$\tilde{u}_1 \frac{\partial\tilde{u}_1}{\partial x_1} = \sum_{j=1}^{N} \sum_{k=1}^{N} u_{1j}u_{1k}\phi_j \frac{\partial\phi_k}{\partial x_1}\ , \qquad (7.3.5)$$

with u_{1j} and u_{1k} as unknowns.

Since only linear systems of equations can be solved directly, equations (7.3.4) have to be linearized and combined with some iteration process. In the next sections three commonly used procedures for the linearization of the Navier-Stokes equations will be discussed, and after that the structure of the system of equations will be considered.

7.3.2 The Picard iteration

In the Picard iteration method (also called successive substitution), the velocity computed in the preceding iteration is substituted into the convective term. This means that the linearized Navier-Stokes equations after the nth iteration can be written as:

$$-\frac{1}{Re}\Delta \underline{u}^n + \nabla p^n = \underline{f} - \underline{u}^{n-1}.\nabla \underline{u}^{n-1} , \tag{7.3.6}$$

with the continuity equation:

$$-\operatorname{div} \underline{u}^n = 0 . \tag{7.3.7}$$

A commonly used initial guess is the solution of the (linear) Stokes equations. These linearized Navier-Stokes equations can be discretized into a system of linear equations by Galerkin's method. The advantage of this method is that in each iteration only the right-hand-side vector changes, but that the large coefficient matrix is kept unchanged. From a numerical point of view this seems very attractive. However, even for moderate values of Re, this iteration process appears to converge very slowly and for larger values of Re it even diverges. Such a behaviour can be expected if one realizes that for larger values of Re the convective terms dominate the diffusive term. Picard-like iterations that take the importance of the convective terms into account, are for example:

i) Choose $\underline{u}$ at the preceding and $\nabla \underline{u}$ at the new level.
ii) Choose $\nabla \underline{u}$ at the preceding and $\underline{u}$ at the new level.

In these Picard iterations, the Navier-Stokes equations are linearized such that the coefficient matrix is changed in each step. In case ii), the Galerkin equations become:

$$\int_\Omega \{\frac{1}{Re} \nabla \tilde{u}_k^n.\nabla \phi_i + (\tilde{u}_1^n \frac{\partial \tilde{u}_k^{n-1}}{\partial x_1} + \tilde{u}_2^n \frac{\partial \tilde{u}_k^{n-1}}{\partial x_2} + \tilde{u}_3^n \frac{\partial \tilde{u}_k^{n-1}}{\partial x_3})\phi_i \tag{7.3.8}$$

$$- p^n \frac{\partial \phi_i}{\partial x_k}\} d\Omega = \int_\Omega f_k \phi_i \, d\Omega , \qquad i = 1,2,\ldots,N , \quad k = 1,2,3 .$$

Practical computations have shown that method i) is superior to method ii), in the sense that both its convergence region and its rate of convergence is larger. Although method ii) converges faster than method (7.3.6) it still is a slow converging process. The convergence speed of a Picard-type iteration is always linear.

7.3.3 Newton and Quasi-Newton methods

Newton's method and quasi-Newton methods are usually applied directly to the system of non-linear equations (7.3.4). These methods are treated in Section 5.6. In this section we shall derive a simple method to linearize equation (7.3.1) before application of the Galerkin procedure. This method can be regarded as an alternative way to derive Newton's method. To that end we write the solution at the new level as the sum of the preceding level and a correction:

$$\underline{u}^n = \underline{u}^{n-1} + \delta\underline{u}^{n-1} .$$

When the nth iteration $\underline{u}^n$ is in the neighbourhood of $\underline{u}$, then $\delta\underline{u}$ is small. The convective terms can be written as:

$$\begin{aligned}\underline{u}^n.\nabla\underline{u}^n &= (\underline{u}^{n-1}+\delta\underline{u}^{n-1}).\nabla(\underline{u}^{n-1}+\delta\underline{u}^{n-1}) = \\ &= \underline{u}^{n-1}.\nabla\underline{u}^n + (\underline{u}^n-\underline{u}^{n-1}).\nabla(\underline{u}^{n-1}+\delta\underline{u}^{n-1}) \\ &= \underline{u}^{n-1}.\nabla\underline{u}^n + \underline{u}^n.\nabla\underline{u}^{n-1} - \underline{u}^{n-1}.\nabla\underline{u}^{n-1} + \delta\underline{u}^{n-1}.\nabla\delta\underline{u}^{n-1} .\end{aligned} \tag{7.3.9}$$

Neglecting the quadratic term in $\delta\underline{u}$, the linearized form of (7.3.1) reads:

$$-\frac{1}{Re}\,\Delta\underline{u}^n + \underline{u}^n.\nabla\underline{u}^{n-1} + \underline{u}^{n-1}.\nabla\underline{u}^n + \nabla p^n = \underline{f} + \underline{u}^{n-1}.\nabla\underline{u}^{n-1} ,$$

$$\text{div}\,\underline{u}^n = 0 . \tag{7.3.10}$$

(7.3.10) is called the Newton linearization of the Navier-Stokes equations and continuity equation. One can easily verify that Galerkin's method applied to (7.3.10) gives exactly the same system of equations as application of Newton's method (5.6.24) to the system of equations (7.3.1) would do.
The main advantage of Newton's method is that it converges quadratically. For the initial guess one can use either the solution of the Stokes equations, or the result of some Picard iterations. For large values of Re it is possible that Newton's method does not converge due to a bad initial guess. Engelman et al (1981) report that the result of one Picard iteration is usually a good starting value for the Newton process.
When Newton's method does not converge, a so-called continuation method could be used. That is, start with a small Reynolds number, compute the solution and use this solution as initial guess for a larger Reynolds number. In this way it is possible to reach the ultimate Reynolds number through a number of continuation steps.
A very promising alternative for Newton's methods is a quasi-Newton method as treated in section 5.6.3.3. Engelman et al (1981) report savings of about 50% compared to Newton's method for Re in the range of 100 to 1000, where the initial guess is formed by one Picard iteration.

7.3.4 The structure of the system of equations

The linearized Navier-Stokes equations, as derived in Sections 7.3.2 and 7.3.3, can be discretized by Galerkin's method, completely analogous to the Stokes equations. This results into a system of linear equations that can be solved in each iteration step. In Section 5.6.3.3 it has been shown how this system of equations can be solved efficiently when the quasi-Newton method is used. Let us consider the discretization of the Newton linearization of the convective terms more closely. Analogous to the Picard iteration they are given by:

$$\int_\Omega \sum_{\alpha=1}^{3} \{\tilde{u}_\alpha^n \frac{\partial \tilde{u}_k^{n-1}}{\partial x_\alpha} + \tilde{u}_\alpha^{n-1} \frac{\partial \tilde{u}_k^{n}}{\partial x_\alpha} - \tilde{u}_\alpha^{n-1} \frac{\partial \tilde{u}_k^{n-1}}{\partial x_\alpha}\} \phi_i d\Omega \ , \qquad (7.3.11)$$

$$k = 1,2,3 \ ,$$
$$i = 1,2,\ldots,N \ .$$

Substitution of the approximate solutions:

$$\tilde{u}_k^n = \sum_{j=1}^{N} u_{kj}^n \phi_j \ , \qquad (7.3.12)$$

results into the following terms:

$$\sum_{j=1}^{N} [u_{1j}^n \{\int_\Omega (\phi_j \frac{\partial \tilde{u}_1^{n-1}}{\partial x_1} + \tilde{u}_1^{n-1} \frac{\partial \phi_j}{\partial x_1} + \tilde{u}_2^{n-1} \frac{\partial \phi_j}{\partial x_2} + \tilde{u}_3^{n-1} \frac{\partial \phi_j}{\partial x_3}) \phi_i d\Omega\}$$

$$+ u_{2j}^n \{\int_\Omega \phi_j \frac{\partial \tilde{u}_1^{n-1}}{\partial x_2} \phi_i d\Omega\} + u_{3j}^n \{\int_\Omega \phi_j \frac{\partial \tilde{u}_1^{n-1}}{\partial x_3} \phi_i d\Omega\}], \ i=1,\ldots,N, \qquad (7.3.13)$$

and comparable terms for the other components.

In matrix form we get for the convective term:

$$\begin{bmatrix} N_{u_1u_1} & N_{u_1u_2} & N_{u_1u_3} & 0 \\ N_{u_2u_1} & N_{u_2u_2} & N_{u_2u_3} & 0 \\ N_{u_3u_1} & N_{u_3u_2} & N_{u_3u_3} & 0 \\ 0 & 0 & 0 & 0 \end{bmatrix} \begin{bmatrix} \underline{u}_1^n \\ \underline{u}_2^n \\ \underline{u}_3^n \\ \underline{p}^n \end{bmatrix} - \begin{bmatrix} \underline{f}_4 \\ \underline{f}_5 \\ \underline{f}_6 \\ \underline{0} \end{bmatrix} \qquad (7.3.14)$$

with submatrices:

$$N_{u_k u_k}(i,j) = \int_\Omega (\phi_j \frac{\partial \tilde{u}_k^{n-1}}{\partial x_k} + \tilde{u}_1^{n-1} \frac{\partial \phi_j}{\partial x_1} + \tilde{u}_2^{n-1} \frac{\partial \phi_j}{\partial x_2} + \tilde{u}_3^{n-1} \frac{\partial \phi_j}{\partial x_3}) \phi_i d\Omega ,$$

(7.3.15)

$$N_{u_k u_\ell}(i,j) = \int_\Omega \phi_j \frac{\partial \tilde{u}_k^{n-1}}{\partial x_\ell} \phi_i d\Omega , \qquad i,j = 1,2,\ldots,N , \quad k,\ell = 1,2,3 ,$$

with unknowns:

$$(\underline{u}_k^n)^T = [u_{k1}^n, u_{k2}^n, \ldots, u_{kN}^n] , \qquad k = 1,2,3 , \qquad (7.3.16)$$

and right-hand-side components:

$$\underline{f}_{3+k}(i) = \int_\Omega (\tilde{u}_1^{n-1} \frac{\partial \tilde{u}_k^{n-1}}{\partial x_1} + \tilde{u}_2^{n-1} \frac{\partial \tilde{u}_k^{n-1}}{\partial x_2} + \tilde{u}_3^{n-1} \frac{\partial \tilde{u}_k^{n-1}}{\partial x_3}) \phi_i d\Omega ,$$

$$k = 1,2,3 \quad , \quad i = 1,2,\ldots,N . \qquad (7.3.17)$$

(7.3.14) combined with (7.2.25) yields the system of linear equations to be solved in each iteration step:

$$\begin{bmatrix} S_{u_1u_1}+N_{u_1u_1} & N_{u_1u_2} & N_{u_1u_3} & L^T_{u_1p} \\ N_{u_2u_1} & S_{u_2u_2}+N_{u_2u_2} & N_{u_2u_3} & L^T_{u_2p} \\ N_{u_3u_1} & N_{u_3u_2} & S_{u_3u_3}+N_{u_3u_3} & L^T_{u_3p} \\ L_{pu_1} & L_{pu_2} & L_{pu_3} & 0 \end{bmatrix} \begin{bmatrix} \underline{u}_1 \\ \underline{u}_2 \\ \underline{u}_3 \\ \underline{p} \end{bmatrix} = \begin{bmatrix} \underline{f}_1+\underline{f}_4 \\ \underline{f}_2+\underline{f}_5 \\ \underline{f}_3+\underline{f}_6 \\ 0 \end{bmatrix}$$

In this system of equations the integrals can also be computed elementwise. After a renumbering of the unknowns (see fig. 7.15) this system can be solved, using partial pivoting.

7.4 Evaluation of the integrated method

With the integrated method, the Navier-Stokes equations and the continuity equation:

$$- \frac{1}{Re} \Delta \underline{u} + \underline{u}.\nabla \underline{u} + \nabla p = \underline{f} \ , \qquad (7.4.1)$$

$$\text{div } \underline{u} = 0 \ , \qquad (7.4.2)$$

have been transformed into a system of linear equations. Formally this system can be written in matrix-vector notation as:

$$S\underline{u} + N(\underline{u})\underline{u} + L^T\underline{p} = \underline{F} \ , \qquad (7.4.3)$$

$$L\underline{u} = \underline{0} \ ,$$

with:

S : diffusion matrix, components: $S_{u_k u_k}$,

N : convection matrix, components: $N_{u_k u_\ell}$,

L : continuity matrix, components: L_{pu_k},

$\underline{u}$: velocity vector, components: $\underline{u}_k^n$,

$\underline{p}$: pressure vector, components: $\underline{p}^n$,

$\underline{F}$: right-hand-side vector, components: $\underline{f}_k + \underline{f}_{k+3}$.

The main advantage of the integrated method is that velocity and pressure are computed directly, without numerical differentiations. However, an important disadvantage is that the system of equations becomes very large, since in most nodal points there are three degrees of freedom (in $\mathbb{R}^2$; in $\mathbb{R}^3$ four), and hence a lot of computing is required. Moreover, since there are no pressure degrees of freedom in the continuity equation, the large matrix contains zeros on the main diagonal, which makes partial pivoting necessary. This fact also causes an increase of computing time. Using a suitable numbering it is possible to avoid partial pivoting without increasing the bandwidth considerably, but in practice such a numbering is hard to find and difficult to program. Although the integrated method is direct and clear to understand, it is not fitted to complicated practical problems. In the next two chapters some methods will be introduced that are more efficient from the numerical point of view. These methods have in common that the pressure and the continuity equation are eliminated, thus reducing the number of degrees of freedom.

Chapter 8

The penalty function method

Introduction

The numerical solution of the Navier-Stokes equations by the integrated method, takes a lot of computing time and computer memory, due to the presence of the incompressibility constraint. In this Chapter the penalty method is discussed. In the penalty method the continuity equation is perturbed with a small parameter times the pressure. Next the pressure is eliminated, thus uncoupling the momentum equations and the continuity equation. The main advantage of the penalty method is the large reduction of the system of equations as well as the fact that partial pivoting is not longer necessary. Compared to the integrated method, the reduction in computing time and memory is considerable.

8.1 General approach

The system of Stokes equations and continuity equation:

$$-\frac{1}{Re}\Delta \underline{u} + \nabla p = \underline{f} , \qquad (8.1.1)$$

$$\text{div } \underline{u} = 0 ,$$

will be considered with boundary conditions:

$$\underline{u} = \underline{0} \quad \text{on } \Gamma \text{ boundary of } \Omega. \qquad (8.1.2)$$

The pressure p is defined up to an additive constant. The idea of the penalty method is to perturb the continuity equation with a small term containing the pressure:

$$\varepsilon p + \operatorname{div} \underline{u} = 0 \ , \tag{8.1.3}$$

with ε a small parameter. One can consider this perturbation as the introduction of a slight artificial compressibility. This will be more obvious in chapter 10, where instationary Navier-Stokes equations are considered. Replacing the continuity equation in (8.1.1) by relation (8.1.3) results into:

$$- \frac{1}{Re} \Delta \underline{u} + \nabla p = \underline{f} \ , \tag{8.1.4}$$

$$p = - \tau \operatorname{div} \underline{u} \ , \quad (\tau = 1/\varepsilon) \ ,$$

$$\underline{u}\big|_{\Gamma} = 0.$$

From (8.1.4) it follows that:

$$\int_{\Omega} p \, d\Omega = \int_{\Omega} - \tau \operatorname{div} \underline{u} \, d\Omega. \tag{8.1.5}$$

This means that after $\underline{u}$ has been computed, the additive constant for the pressure is determined by relation (8.1.5). System of equations (8.1.4) is called the penalty function formulation and τ the penalty parameter. Approximation (8.1.3) can be justified because the continuity equation is satisfied approximately for $\varepsilon \to 0$, due to the finite value of the pressure. Furthermore (8.1.3) is not in contradiction with the boundary conditions. Moreover, one can show that the solution of the penalty function formulation approaches the solution of the unperturbed system (8.1.1), for increasing values of τ. We shall give a sketch of the proof for the boundary conditions (8.1.2), a general proof can be found in Temam (1977). See also Chapter 14.

Let $\underline{u}^{\tau} = (u_1^{\tau}, u_2^{\tau}, u_3^{\tau})$ and p^{τ} be the solution of (8.1.4). Subtraction of $- \frac{1}{Re} \Delta \underline{u} + \nabla p = \underline{f}$, and

$$- \frac{1}{Re} \Delta \underline{u}^{\tau} + \nabla p^{\tau} = \underline{f} \ ,$$

yields:

$$-\frac{1}{Re}\,\Delta(\underline{u}^{\tau}-\underline{u}) + \nabla(p^{\tau}-p) = 0. \qquad (8.1.6)$$

If we take the inner product of (8.1.6) with $(\underline{u}^{\tau}-\underline{u})$ and integrate over the domain Ω then:

$$-\frac{1}{Re}\int_{\Omega}(\underline{u}^{\tau}-\underline{u}).\Delta(\underline{u}^{\tau}-\underline{u})d\Omega + \int_{\Omega}(\underline{u}^{\tau}-\underline{u}).\nabla(p^{\tau}-p)d\Omega = 0. \qquad (8.1.7)$$

Application of Gauss' divergence theorem, gives in the same way as in Section 7.1:

$$\frac{1}{Re}\int_{\Omega}\nabla(\underline{u}^{\tau}-\underline{u}).\nabla(\underline{u}^{\tau}-\underline{u})d\Omega - \int_{\Omega}(p^{\tau}-p)\,\mathrm{div}(\underline{u}^{\tau}-\underline{u})d\Omega \qquad (8.1.8)$$

$$-\frac{1}{Re}\int_{\Gamma}(\underline{u}^{\tau}-\underline{u}).\nabla(\underline{u}^{\tau}-\underline{u}).\underline{n}\;d\Gamma + \int_{\Gamma}(p^{\tau}-p)(\underline{u}^{\tau}-\underline{u}).\underline{n}\;d\Gamma = 0.$$

Substitution of $\mathrm{div}\;\underline{u} = 0$ and $\mathrm{div}\;\underline{u}^{\tau} = -\frac{1}{\tau}\,p^{\tau}$, as well as boundary condition (8.1.2) yields:

$$\frac{1}{Re}\int_{\Omega}\nabla(\underline{u}^{\tau}-\underline{u}).\nabla(\underline{u}^{\tau}-\underline{u})d\Omega + \frac{1}{\tau}\int_{\Omega}(p^{\tau}-p)p^{\tau}d\Omega = 0. \qquad (8.1.9)$$

From the inequality:

$$(a-b)^2 \geq 0 \iff 2ab \leq a^2 + b^2 ,$$

and hence

$$\int_{\Omega} ab\;d\Omega \leq \frac{1}{2}\int_{\Omega} a^2\;d\Omega + \frac{1}{2}\int_{\Omega} b^2\;d\Omega , \qquad (8.1.10)$$

or

$$\frac{1}{\tau}\int_{\Omega} pp^{\tau}\;d\Omega \leq \frac{1}{2\tau}\int_{\Omega} p^2\;d\Omega + \frac{1}{2\tau}\int_{\Omega}(p^{\tau})^2\;d\Omega , \qquad (8.1.11)$$

it follows that:

$$\frac{1}{Re}\int_{\Omega}\nabla(\underline{u}^{\tau}-\underline{u}).\nabla(\underline{u}^{\tau}-\underline{u})d\Omega + \frac{1}{2\tau}\int_{\Omega}(p^{\tau})^2 d\Omega \leq \frac{1}{2\tau}\int_{\Omega} p^2 d\Omega. \qquad (8.1.12)$$

All integrals in (8.1.12) are positive. Since p is finite we have also $\int_\Omega p^2 d\Omega$ finite and it follows that

$$\int_\Omega (p^\tau)^2 \, d\Omega \text{ is finite.}$$

So when $\tau \to \infty$, then the pressure integrals vanish and

$$\int_\Omega \nabla(\underline{u}^\tau - \underline{u}) . \nabla(\underline{u}^\tau - \underline{u}) d\Omega \to 0 \ ,$$

or

$$\nabla(\underline{u}^\tau - \underline{u}) \to \underline{0} \ .$$

Using now the fact that $\underline{u}^\tau$ and $\underline{u}$ coincide on the boundary, we find that $\underline{u}^\tau \to \underline{u}$. (8.1.13)

From (8.1.6) we see that $\nabla(p^\tau - p) \to 0$. (8.1.14)

Hence $p^\tau \to p + c$ where c is some constant. (8.1.15)

Since the pressure p is defined up to an additive constant, whereas the constant in p^τ is prescribed by the relation (8.1.5), (8.1.15) completes the proof. Hence we have proved the convergence of the solution of the penalty function formulation to the solution of the unperturbed equation when $\tau \to \infty$.

This simple proof is only valid for the Stokes equations. For a general proof in the case of the Navier-Stokes equations the reader is referred to Pelissier (1975).

The discretization of the penalty function formulation (8.1.4):

$$-\frac{1}{Re} \Delta\underline{u} + \nabla p = \underline{f} \ ,$$

$$p = -\tau \operatorname{div} \underline{u} \ ,$$

$$\underline{u}\big|_\Gamma = 0 \ ,$$

is completely analogous to the integrated method, see Section 7.1. By approximating the velocity and pressure unknowns by:

$$\tilde{u}_k(\underline{x}) = \sum_{j=1}^{N} u_{kj}\phi_j(\underline{x}) \ , \qquad k = 1,2,3 \ , \tag{8.1.16}$$

$$\tilde{p}(\underline{x}) = \sum_{j=1}^{M} p_j\psi_j(\underline{x}) \ ,$$

the following system of Galerkin equations can be derived:

$$\int_\Omega \frac{1}{Re} \nabla\tilde{u}_k \cdot \nabla\phi_i d\Omega - \int_\Omega \tilde{p} \frac{\partial\phi_i}{\partial x_k} d\Omega = \int_\Omega f_k\phi_i d\Omega \ , \tag{8.1.17}$$

$$i = 1,2,\ldots,N \ ,$$
$$k = 1,2,3 \ ,$$

$$\int_\Omega \tilde{p} \ \psi_i d\Omega = - \tau \int_\Omega \psi_i \ \text{div} \ \tilde{u} \ d\Omega \ , \qquad i = 1,2,\ldots,M. \tag{8.1.18}$$

Using the nomenclature as defined in sections 7.2.4 and 7.4, (8.1.17) and (8.1.18) can be written in matrix-vector notation as:

$$S \ \underline{u} + L^T\underline{p} = \underline{F} \ , \tag{8.1.19}$$

$$D \ \underline{p} = \tau \ L \ \underline{u} \ ,$$

with pressure mass matrix:

$$D(i,j) = \int_\Omega \psi_i\psi_j \ d\Omega \ , \qquad i,j = 1,2,\ldots,M. \tag{8.1.20}$$

From (8.1.19) we can eliminate the pressure, resulting into:

$$\underline{p} = \tau \ D^{-1}L \ \underline{u} \ , \tag{8.1.21}$$

$$S \ \underline{u} + \tau \ L^TD^{-1}L \ \underline{u} = \underline{F} \ , \tag{8.1.22}$$

and analogously for the Navier-Stokes equations:

$$S \ \underline{u} + N(\underline{u})\underline{u} + \tau \ L^TD^{-1}L \ \underline{u} = \underline{F} \ . \tag{8.1.23}$$

The main advantage of this construction is that the computation of the velocity and pressure are uncoupled. First the velocity $\underline{u}$ is computed from (8.1.22) or (8.1.23) and then the pressure is computed from (8.1.21). When the pressure mass matrix is a diagonal matrix, which is the case for some of the admissible elements, the computation of the pressure is a straightforward matrix-vector multiplication, and can be carried out very efficiently.

Before studying the penalty function formulation in more detail we shall give some alternative formulations of the penalty method in the next section.

8.2 Alternative formulations of the penalty function method

8.2.1 Minimization formulation

Originally the penalty function method has been formulated on the basis of a minimization problem, see Chapter 3. We shall give a short outline. We start with the Stokes equations (8.1.1):

$$\left.\begin{aligned} -\frac{1}{Re}\,\Delta\underline{u} + \nabla p &= \underline{f} \\ \operatorname{div}\,\underline{u} &= 0 \end{aligned}\right\} \quad \text{on } \Omega \text{ and } \underline{u}\big|_{\Gamma} = 0 \,. \tag{8.2.1}$$

Taking the inner product of the momentum equations with an arbitrary function $\underline{v}$, integrating over the domain Ω and applying Gauss' theorem (see Section 7.1) gives:

$$\int_{\Omega} \frac{1}{Re} \nabla\underline{u}.\nabla\underline{v}\, d\Omega - \int_{\Omega} p \operatorname{div}\,\underline{v}\, d\Omega = \int_{\Omega} \underline{f}.\underline{v}\, d\Omega \quad \text{for all } \underline{v} \in \underline{V}_0. \tag{8.2.2}$$

In (8.2.2) it is assumed that $\underline{v}\big|_{\Gamma} = 0$. If we restrict ourselves to functions $\underline{v}$ satisfying:

$$\operatorname{div}\,\underline{v} = 0 \quad \text{and} \quad \underline{v}\big|_{\Gamma} = 0 \,, \tag{8.2.3}$$

then (8.2.2) reduces to:

$$\int_{\Omega} \frac{1}{Re} \nabla\underline{u}.\nabla\underline{v}\, d\Omega = \int_{\Omega} \underline{f}.\underline{v}\, d\Omega \,, \quad \text{for all } \underline{v} \in \underline{V}_0 \,, \tag{8.2.4}$$

with constraints $\text{div}\ \underline{u} = \text{div}\ \underline{v} = 0$,

$$\underline{u}|_\Gamma = \underline{v}|_\Gamma = 0.$$

Without constraints, (8.2.4) is equivalent to the following minimization problem (compare with Section 3.2.2):

> Find $\underline{w} = \underline{u}(\underline{x})$, $\underline{x} \in \Omega$, with $\underline{u}|_\Gamma = \underline{0}$ minimizing the following functional:
>
> $$J(\underline{w}) = \frac{1}{2Re} \int_\Omega |\nabla\underline{w}|^2 d\Omega - \int_\Omega \underline{f}\cdot\underline{w}\ d\Omega. \qquad (8.2.5)$$

In fact one easily verifies that problem (8.2.1) is equivalent to the minimization problem (8.2.5) under the constraint $\text{div}\ \underline{u} = 0$. See for example Temam (1977). The pressure p turns out to be the Lagrangian multiplier for the constrained minimization problem, Segal (1979), emphasizing the special role of the pressure.

Problem (8.2.5) can be discretized by the Ritz method, when we introduce the approximation:

$$\tilde{u}_k = \sum_{j=1}^{N} u_{kj}\ \phi_j\ , \qquad k = 1,2,3\ . \qquad (8.2.6)$$

Substitution of (8.2.6) into (8.2.5) yields:

> Find $\underline{w} = \underline{u} = (u_{11}, u_{12}, \ldots, u_{3N})$ minimizing the following functional:
>
> $$J(\underline{w}) = \tfrac{1}{2}\ \underline{w}^T S\ \underline{w} - \underline{w}^T \underline{F}. \qquad (8.2.7)$$

The constraint $\text{div}\ \underline{u} = 0$ must be discretized in the same way as in Section 7.1, (7.1.6):

$$\int_\Omega q\ \text{div}\ \underline{u}\ d\Omega = 0\ , \quad \text{for all } q \in Q\ , \qquad (8.2.8)$$

or (7.1.19):

$$\int_\Omega \psi_i\ \text{div}\ \tilde{\underline{u}}\ d\Omega = 0\ , \quad \text{for all basis functions } \psi_i. \qquad (8.2.9)$$

With the notation of Sections 7.2.4 and 7.4:

$$L\,\underline{u} = \underline{0}\ . \tag{8.2.10}$$

The minimization problem:

> Find $\underline{w} = \underline{u} = (u_{11},\ldots,u_{3N})$ minimizing the following functional:
>
> $$J(\underline{w}) = \tfrac{1}{2}\,\underline{w}^T S\,\underline{w} - \underline{w}^T \underline{F}\ ,$$
>
> with the constraint
>
> $$L\,\underline{u} = 0\ ,$$

can be solved with standard techniques from the theory of constrained minimization, Gill and Murray (1974). One of these techniques is the penalty function method, where one tries to satisfy the constraint by adding a large penalty term to the minimization problem, of the following form:

$$\tfrac{1}{2}\tau\,|L\underline{u}|^2 = \tfrac{1}{2}\tau(L\underline{u})^T\cdot(L\underline{u}) = \tfrac{1}{2}\tau\underline{u}^T L^T L\underline{u}\ , \tag{8.2.11}$$

with τ large. The penalty is always positive. By the choice of a quadratic term $|L\underline{u}|^2$ and τ large, we force the term $L\underline{u}$ to zero in the following minimization problem:

Find $\underline{w} = \underline{u}$ minimizing the following functional:

$$J(\underline{w}) = \tfrac{1}{2}\underline{w}^T S\,\underline{w} + \tfrac{1}{2}\tau\,\underline{w}^T L^T L\,\underline{w} - \underline{w}^T\underline{F}\ . \tag{8.2.12}$$

This is the so-called minimization formulation of the penalty function problem. This method is frequently used in mechanics. We see that also in this set of equations pressure and velocity are uncoupled. Also for this system one can prove that the solution converges to the solution of (8.2.1) when $\tau \to \infty$, and that the pressure can be computed from

$$\underline{p} = \tau\,L\,\underline{u}\ . \tag{8.2.13}$$

One easily verifies that the solution of (8.2.12) satisfies:

$$S \underline{u} + \tau L^T L \underline{u} = \underline{F} . \qquad (8.2.14)$$

Comparing (8.2.14) with (8.1.22), we see that both expressions are identical when D is a diagonal matrix with a constant diagonal. This is for example the case when the Crouzeix-Raviart elements are used and all elements have the same area.

For the Navier-Stokes equations, there exists no equivalent minimization problem. Therefore the same penalty is introduced in the discrete equations:

$$S \underline{u} + N(\underline{u})\underline{u} + \tau L^T L \underline{u} = \underline{F} . \qquad (8.2.15)$$

There is no proof for the justification of this approach. Practical computations show that the results are the same as those obtained by the integrated method and hence the usage of the method has at least some justification.

8.2.2 The continuous penalty function method

The penalty function method as introduced in Section 8.1 will be called discrete penalty function method, since the pressure is eliminated after the discretization of the continuity equation and the momentum equations. It is also possible to eliminate the pressure from (8.1.3) before the discretization. Let

$$p = -\tau \text{ div } \underline{u} , \quad (\tau = 1/\varepsilon) , \qquad (8.2.16)$$

and substitute this expression in (8.1.1):

$$- \frac{1}{Re} \Delta \underline{u} - \tau \text{ grad } (\text{div } \underline{u}) = \underline{f} . \qquad (8.2.17)$$

Application of the standard Galerkin method to equation (8.2.17) results in the following system of linear equations:

$$\int_\Omega \frac{1}{Re} \nabla \tilde{u}_k \cdot \nabla \phi_i d\Omega + \tau \int_\Omega \text{div } \tilde{\underline{u}} \frac{\partial \phi_i}{\partial x_k} d\Omega = \int_\Omega f_k \phi_i d\Omega , \qquad (8.2.18)$$

$$i = 1,2,\ldots,N , \quad k = 1,2,3 ,$$

or in matrix notation:

$$S\,\underline{u} + \tau\,A\,\underline{u} = \underline{F}. \tag{8.2.19}$$

At first sight it seems as if (8.2.19) is more attractive than (8.1.22), since it is not necessary to form the complicated matrix $L^TD^{-1}L$. However, a closer examination of (8.1.22) and (8.2.19) shows that (8.2.19) has some disadvantages which are not present in (8.1.22). From (8.1.22) it follows that if $\tau\to\infty$:

$$L^TD^{-1}L\,\underline{u} = \underline{0}, \tag{8.2.20}$$

and it is easy to show that, if the element is an admissible one (in the sense of 7.2.2), (8.2.20) is equivalent with:

$$L\,\underline{u} = \underline{0}. \tag{8.2.21}$$

From (8.2.19) it follows that if $\tau \to \infty$:

$$A\,\underline{u} = 0\ , \tag{8.2.22}$$

or: $$\int_\Omega \operatorname{div} \tilde{\underline{u}} \operatorname{div} \underline{\phi}_i \, d\Omega = 0\ , \tag{8.2.23}$$

for all basis functions $\underline{\phi}_i$. We see that (8.2.23) is equivalent to the discretized form of (7.1.10), where div $\underline{\phi}_i$ plays the role of the test function q. In fact (8.2.23) can be written as:

$$\int_\Omega \psi_i \operatorname{div} \tilde{\underline{u}} \, d\Omega = 0\ , \tag{8.2.24}$$

with ψ_i the divergence of an arbitrary discretized velocity field. From the discussion in 7.2.2 and 7.2.3, it is clear that we may not use any basis function ψ_i for the pressure, since otherwise (7.1.19) leads to an overdetermined system of equations. When ψ_i is the divergence of a discretized velocity field, this situation may arise.

Consider for example the Taylor-Hood triangle of 7.2.3.2. The velocity is quadratic, so that the divergence is a linear polynomial. This linear polynomial is discontinuous over the element boundaries, and therefore (8.2.24) is comparable with the quadratic Crouzeix-Raviart element with linear pressure. In 7.2.3.3 it has been shown that this element is not admissible, which implies that we can expect poor results from this approach. The pressure computed from (8.2.16) showed indeed unrealistic wiggles when using this element (Sani et al (1981)). This is not caused by a singularity of the matrix, since the matrix $S + \tau A$ is positive definite. These wiggles are known as spurious modes or checkerboard modes for the pressure. This is because on a rectangular grid, pressures on the "black" fields seem to be independent of the pressures on the "white" fields. These spurious modes show that we are dealing with overdetermined systems of equations. In the literature these checkerboard modes are suppressed, either by computing the matrix A inaccurately (reduced integration) or by filtering the spurious modes.

With the reduced integration technique (see Malkus et al (1978)), the integrals (8.2.23) are computed so inaccurate, that the solution does not exhibit spurious pressure modes anymore. In fact, the numerical integration forces the term div $\underline{\tilde{u}}$ to a lower degree polynomial and in this way the element becomes admissible. From Section 7.2.3.3 it is known that for the Crouzeix-Raviart triangle, the discontinuous pressure must be a constant polynomial in each element. The result of these manipulations is an element with quadratic velocity and constant pressure approximation, which is an inaccurate element (see Section 7.2.3.3).

Filtering of the spurious pressure modes is treated in Sani et al (1981). This process can be considered as a smoothing technique, where the wiggles are smoothed out. Especially on non-rectangular grids a good filtering technique is hard to find.

The observations in this Section lead to the conclusion, that the discrete penalty function method of Section 8.1 is superior to the continuous penalty function method.

8.2.3 Iterative penalty function method

A different approach to construct an approximate solution for the continuity equation and the Stokes equations is to use iterative methods. In these methods the pressure at the preceding iteration is used, which uncouples the velocity and pressure computation. As an example the Powell and Hestenes method (Gill and Murray (1974)), will be treated. For this purpose we start with the system of discretized Stokes equations in matrix form:

$$S\,\underline{u} + L^T\underline{p} = \underline{F}\ , \qquad (8.2.25)$$

$$L\,\underline{u} = \underline{0}\ .$$

The constraint in (8.2.25) is imposed by a penalty term in the same way as in Section 8.2.1, i.e.

$$S\,\underline{u} + L^T\underline{p} + \tau\, L^T L\,\underline{u} = \underline{F}\ , \qquad (8.2.26)$$

with τ a large parameter. Velocity and pressure are uncoupled by choosing the pressure at the preceding level in the iteration process, i.e. in the ith iteration level:

$$S\,\underline{u}^i + L^T\underline{p}^{i-1} + \tau\, L^T L\,\underline{u}^i = \underline{F}\ . \qquad (8.2.27)$$

An estimate of the pressure can be computed from:

$$\underline{p}^i = \underline{p}^{i-1} + \tau\, L\,\underline{u}^i\ . \qquad (8.2.28)$$

As initial guess for $\underline{p}$ one chooses $\underline{p}^0 = \underline{0}$.

This iteration process is repeated until convergence has been achieved; in that case the continuity equation is satisfied. One can show, Segal (1979), that this process converges fast for large values of τ. It will be clear that the first iteration is identical to the discrete penalty function method of Section 8.2.1. In general one iteration is sufficient, but for example when τ is too small, more than one iteration may be necessary.

In the literature alternative iterative methods as for example the Uzawa scheme (Girault, Raviart (1979), Thomasset (1981)) are mentioned. These methods, however, converge very slowly and are not popular anymore.
In the next sections we shall restrict ourselves to the discrete penalty function method.

8.3 Practical consequences

8.3.1 Element conditions

In the same way as for the integrated method, see Section 7.2.2, the basis functions for velocity and pressure have to satisfy some requirements in order to be conforming. Also in this case, the basis functions for the velocity must be piecewise continuously differentiable and continuous over the element boundaries. The pressure approximation must be continuous in each element. Since the continuity equation is not used explicitly it seems as if the basis functions for velocity and pressure may be chosen independently. From Section 8.1, however, it is clear that the solution of the penalty function method approaches the solution of the integrated method. Therefore only elements satisfying the BB condition may be used. Only when the computation of the pressure matrix D and the penalty matrix $L^TD^{-1}L$ (see (8.1.21), (8.1.22)) can be carried out elementwise, the construction of the matrix $L^TD^{-1}L$ is simple. This is only the case when we deal with discontinuous pressure elements, like the Crouzeix-Raviart elements. The inversion of D is simplest if D is a diagonal matrix. The matrix D can be approximated by a diagonal matrix, if it is lumped, that is, all off-diagonal matrix elements are moved to and added to the main diagonal. This can for example be achieved by the application of a Newton-Cotes integration rule for the computation of $D(i,j) = \int_\Omega \psi_i \psi_j \, d\Omega$. Lumping is not necessary when only the basis function of the pressure in the centroid in each element is used as test function. Due to the special form of the pressure approximation, D is then a diagonal matrix with the areas of the elements on the main diagonal. In the next sections it will be shown how the number of test functions derived from the basis functions for the pressure in the case of the Crouzeix-Raviart $P_2^+-P_1$ element can be reduced to one, by the elimination of the gradient of the pressure. This modified Crouzeix-Raviart element is therefore an important example of an element with a diagonal pressure mass matrix D.

From the preceding remarks it will be clear that the penalty function method is usually combined with discontinuous pressure elements (Crouzeix-Raviart elements). Of course the penalty function method can also be applied to the Taylor-Hood elements, but for these elements the method is inefficient and most of its advantages are lost. This is caused by the structure of the matrices L and D, which is essentially different from that of the Crouzeix-Raviart type elements, because the components of L and D are not determined by one finite element only. Due to this difference, the bandwidth of the matrix $\tau L^T D^{-1} L$ is increased by a factor 2, even if the matrix D is lumped. In practice the penalty function method is never combined with Taylor-Hood type elements.

Finally it is remarked that from Section 8.2.1 it is clear that the matrix D does not play an important role in the accuracy of the process; D may be replaced by the identity matrix. Fortin (1983) states that D may be replaced by any positive definite matrix; D may be considered as a weighting matrix. The computation of L, however, must be carried out accurately as is shown in Section 8.2.2.

8.3.2 The modified $P_2^+-P_1$ Crouzeix-Raviart element

8.3.2.1 Introduction

In this section the $P_2^+-P_1$ Crouzeix-Raviart element is

considered. It is shown how the velocities and the gradient of the pressure in the centroid can be eliminated as unknowns. The accuracy of the element is not affected by this procedure. In fact the process can be applied for all discontinuous pressure elements in IR^2 and IR^3. The importance of this method is that the number of velocity and pressure unknowns is reduced; important for the integrated method, see Section 7.2.3.3, but also for the other approaches. Furthermore the pressure matrix D gets a simple structure (important for the penalty function method, see Section 8.3.1). Finally the divergence-free elements can be derived more easily in this way, see Section 9.2.3.

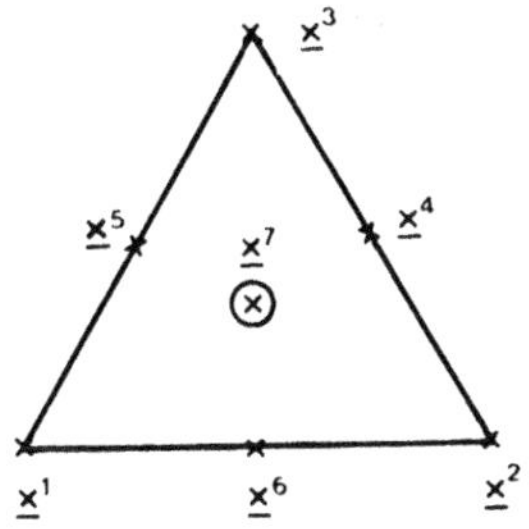

Velocity: quadratic
(7 nodal points: x)

Pressure: linear
(1 nodal point: 0
+ 2 derivatives)

Approximation $\tilde{u}_k = \sum_{i=1}^{7} u_{ki}\phi_i$

$$\tilde{p} = \sum_{i=1}^{3} p_i\psi_i$$

Fig. 8.1 Crouzeix-Raviart element $(P_2^+-P_1)$

Consider the Crouzeix-Raviart element of figure 8.1. According to Section 7.2.3.3 the approximate pressure can be written as:

$$\tilde{p} = p_{123}\psi_1 + \frac{\partial p}{\partial x_1}(\underline{x}^7)\psi_2 + \frac{\partial p}{\partial x_2}(\underline{x}^7)\psi_3 , \qquad (8.3.1)$$

with $\psi_1 = 1$ within the element, $\psi_1 = 0$ outside the element,

$\psi_2 = x_1-x_1^7$ within the element, $\psi_2 = 0$ outside the element,

$\psi_3 = x_2-x_2^7$ within the element, $\psi_3 = 0$ outside the element.

The pressure gradient is constant over the element, since

$$\frac{\partial \tilde{p}}{\partial x_k} = \frac{\partial p}{\partial x_k}(\underline{x}^7). \tag{8.3.2}$$

8.3.2.2 Elimination of the velocities in the centroid

Consider the discretized continuity equation (7.1.19):

$$L\underline{u} = 0 \iff \int_\Omega \left(\frac{\partial \tilde{u}_1}{\partial x_1} + \frac{\partial \tilde{u}_2}{\partial x_2}\right)\psi_i \, d\Omega = 0 \ , \ i = 1,2,\ldots,M. \tag{8.3.3}$$

With the expressions (8.3.1), (8.3.3) can be written as:

$$\int_e \left(\frac{\partial \tilde{u}_1}{\partial x_1} + \frac{\partial \tilde{u}_2}{\partial x_2}\right) d\Omega = 0 \ , \tag{8.3.4}$$

$$\int_e (x_1 - x_1^7)\left(\frac{\partial \tilde{u}_1}{\partial x_1} + \frac{\partial \tilde{u}_2}{\partial x_2}\right) d\Omega = 0 \ ,$$

$$\int_e (x_2 - x_2^7)\left(\frac{\partial \tilde{u}_1}{\partial x_1} + \frac{\partial \tilde{u}_2}{\partial x_2}\right) d\Omega = 0 \ ,$$

for each element e.

The first relation implies, as is shown in Section 7.2.3.3, that the continuity equation is satisfied elementwise in the mean. The other two equations can be transformed using Gauss' theorem:

$$\int_\Omega a \frac{\partial b}{\partial x_k} d\Omega = \int_\Gamma ab \, n_k d\Gamma - \int_\Omega \frac{\partial a}{\partial x_k} b \, d\Omega \ , \tag{8.3.5}$$

to

$$\int_{\partial e} (x_1 - x_1^7)(\tilde{u}_1 n_1 + \tilde{u}_2 n_2) d\Gamma - \int_e \tilde{u}_1 \, d\Omega = 0 \ , \tag{8.3.6}$$

$$\int_{\partial e} (x_2 - x_2^7)(\tilde{u}_1 n_1 + \tilde{u}_2 n_2) d\Gamma - \int_e \tilde{u}_2 \, d\Omega = 0 \, , \tag{8.3.7}$$

with ∂e the boundary of the element e and $\underline{n}$ the outward normal. The values of the velocity components u_{17} and u_{27} in the centroid give no contribution to the boundary integrals since their corresponding basis functions are zero on the boundary of the element. Therefore the boundary integrals can be written as:

$$\int_{\partial e} (x_k - x_k^7) \tilde{\underline{u}} \cdot \underline{n} \, d\Gamma = G_k(u_{11}, u_{12}, \ldots, u_{16}, u_{21}, \ldots, u_{26}), \quad k = 1,2. \tag{8.3.8}$$

For the surface integrals we have:

$$\int_e \tilde{u}_k \, d\Omega = \int_e \left(\sum_{i=1}^{6} u_{ki} \phi_i \right) d\Omega + u_{k7} \int_e \phi_7 d\Omega, \quad k = 1,2. \tag{8.3.9}$$

Since $\int_e \phi_7 d\Omega \neq 0$, the velocity components in the centroid can explicitly be expressed in the velocity components on the boundaries:

$$u_{k7} = \left[G_k - \sum_{i=1}^{6} u_{ki} \int_e \phi_i d\Omega \right] / \int_e \phi_7 d\Omega, \quad k = 1,2. \tag{8.3.10}$$

Substitution of (8.3.10) into the velocity approximation:

$$\tilde{u}_k = \sum_{i=1}^{6} u_{ki} \phi_i + u_{k7} \phi_7, \qquad k = 1,2 \, , \tag{8.3.11}$$

gives:

$$\tilde{u}_k = \sum_{i=1}^{6} u_{ki} \Phi_i + G_k \Phi_7, \qquad k = 1,2 \, , \tag{8.3.12}$$

with Φ_i $(i=1,2,\ldots,7)$ functions of the basis functions ϕ_i. Although G_1 and G_2 are functions of the nodal point velocities u_{ki}, we still have:

$$\tilde{u}_k(\underline{x}_j) = u_{kj} \quad \text{and} \quad \Phi_i(\underline{x}_j) = \delta_{ij}, \; i,j = 1,2,\ldots,6. \tag{8.3.13}$$

In vector notation, the velocity approximation satisfies:

$$\underline{\tilde{u}} = \sum_{i=1}^{6} (u_{1i}\underline{\Phi}_{1i} + u_{2i}\underline{\Phi}_{2i}). \qquad (8.3.14)$$

8.3.2.3 Elimination of the pressure derivatives

A similar procedure is possible for the elimination of the pressure derivatives:

$$\frac{\partial p}{\partial x_k}(\underline{x}^7).$$

To that end, we consider the Galerkin approximation to the Stokes equations at element level, with the basis function ϕ_7 as test function. Since ϕ_7 is zero outside the element, the integrals over Ω can be replaced by integrals over the element e:

$$\int_e \frac{1}{Re} \nabla\tilde{u}_k \cdot \nabla\phi_7 d\Omega - \int_e \tilde{p} \frac{\partial\phi_7}{\partial x_k} d\Omega = \int_e f_k\phi_7 d\Omega, \quad k = 1,2. \qquad (8.3.15)$$

Substitution of (8.3.1) into the pressure term gives:

$$\int_e \tilde{p} \frac{\partial\phi_7}{\partial x_k} d\Omega = p_7 \int_e \frac{\partial\phi_7}{\partial x_k} d\Omega + \frac{\partial p}{\partial x_1}(\underline{x}^7) \int_e (x_1 - x_1^7) \frac{\partial\phi_7}{\partial x_k} d\Omega$$

$$+ \frac{\partial p}{\partial x_2}(\underline{x}^7) \int_e (x_2 - x_2^7) \frac{\partial\phi_7}{\partial x_k} d\Omega, \quad k = 1,2. \qquad (8.3.16)$$

Application of the Gauss' theorem and using the fact that $\phi_7 = 0$ at the boundary ∂e of the element gives:

$$\int_e \tilde{p} \frac{\partial\phi_7}{\partial x_k} d\Omega = -\frac{\partial p}{\partial x_k}(\underline{x}^7) \int_e \phi_7 \, d\Omega, \qquad k = 1,2. \qquad (8.3.17)$$

From (8.3.17) and (8.3.15) it follows that

$$\frac{\partial p}{\partial x_k}(\underline{x}^7) = [-\int_e \frac{1}{Re} \nabla\tilde{u}_k \cdot \nabla\phi_7 d\Omega + \int_e f_k\phi_7 d\Omega] / \int_e \phi_7 d\Omega, \qquad (8.3.18)$$

$$k = 1,2.$$

So these pressure derivatives are completely determined by the velocity approximations in each element, and can be computed once the velocity is known. The relations (8.3.18) can be written as:

$$\frac{\partial p}{\partial x_1}(\underline{x}^7) = G_3(u_{11}\ldots u_{16}, u_{21}\ldots u_{26}), \tag{8.3.19}$$

$$\frac{\partial p}{\partial x_2}(\underline{x}^7) = G_4(u_{11}\ldots u_{16}, u_{21}\ldots u_{26}). \tag{8.3.20}$$

The pressure approximation (8.3.1) becomes:

$$\tilde{p} = p_7\psi_1 + G_3(u_{11}\ldots u_{26})\psi_2 + G_4(u_{11}\ldots u_{26})\psi_3. \tag{8.3.21}$$

8.3.2.4 Construction of matrices

Finally we shall consider the structure of the matrices S and L after elimination of the velocities and pressure derivatives in the centroid. From a numerical point of view it is attractive to construct these matrices from the original (non-eliminated) ones. To that end the unknowns in the original system:

$$S\,\underline{u} + L^T\underline{p} = \underline{F}\,, \tag{8.3.22}$$

$$L\,\underline{u} = \underline{0}\,,$$

will be written as:

$$\underline{u} = \begin{bmatrix} \hat{\underline{u}} \\ \underline{u}_z \end{bmatrix}, \quad \underline{p} = \begin{bmatrix} \hat{\underline{p}} \\ \underline{p}_\nabla \end{bmatrix}, \tag{8.3.23}$$

with $\hat{\underline{u}}$: velocity components in all nodal points on boundaries of elements,

$\underline{u}_z$: velocity components in centroids of elements,

$\hat{\underline{p}}$: pressure values in centroids of elements,

$\underline{p}_\nabla$: pressure gradients in centroids of elements.

With (8.3.23), (8.3.22) can be splitted into:

$$[S_1\ S_2]\begin{bmatrix}\hat{\underline{u}}\\ \underline{u}_z\end{bmatrix} + [L_1^T\ L_2^T]\begin{bmatrix}\hat{\underline{p}}\\ \underline{p}_\nabla\end{bmatrix} = \underline{F}\ , \tag{8.3.24}$$

$$L_1\underline{u} = \underline{0} \quad \text{or} \quad L_{11}\hat{\underline{u}} + L_{12}\underline{u}_z = \underline{0}\ , \tag{8.3.25}$$

$$L_2\underline{u} = \underline{0} \quad \text{or} \quad L_{21}\hat{\underline{u}} + L_{22}\underline{u}_z = \underline{0}\ . \tag{8.3.26}$$

In (8.3.10) it is shown that the velocity components in the centroid, can be expressed explicitly in the velocity components on the boundary of the elements. In this way the unknowns $\underline{u}_z$ have been eliminated. This elimination process can formally be written with the aid of a transformation matrix R_z:

$$\underline{u} = R_z\hat{\underline{u}} \quad \text{with} \quad R_z = \begin{bmatrix} I \\ R_0 \end{bmatrix} \quad \text{and} \quad \underline{u}_z = R_0\hat{\underline{u}}. \tag{8.3.27}$$

From (8.3.26) it follows that $\underline{u}_z = -L_{22}^{-1}L_{21}\hat{\underline{u}}$, hence

$$R_0 = -L_{22}^{-1}L_{21}. \tag{8.3.28}$$

Therefore:

$$L_2R_z = L_{21}I + L_{22}R_0 = L_{21}I - L_{22}L_{22}^{-1}L_{21} = 0, \tag{8.3.29}$$

so that $R_z^TL_2^T = 0$.

Substitution of (8.3.27) in (8.3.24) yields:

$$S_1\hat{\underline{u}} + S_2R_0\hat{\underline{u}} + L_1^T\hat{\underline{p}} + L_2^T\underline{p}_\nabla = \underline{F}\ , \tag{8.3.30}$$

and after premultiplication with R_z^T, using (8.3.29):

$$R_z^TS_1\hat{\underline{u}} + R_z^TS_2R_0\hat{\underline{u}} + R_z^TL_1^T\hat{\underline{p}} = R_z^T\underline{F}\ . \tag{8.3.31}$$

Or:

$$(R_z^TS_1 + R_z^TS_2R_0)\hat{\underline{u}} + R_z^TL_1^T\hat{\underline{p}} = R_z^T\underline{F}\ , \tag{8.3.32}$$

with continuity equation

$$L_1\underline{u} = L_1 R_z \hat{\underline{u}} = \underline{0} \ . \tag{8.3.33}$$

The result of the elimination process is:

$$\hat{S}\,\hat{\underline{u}} + \hat{L}^T\hat{\underline{p}} = \hat{\underline{F}} \ , \tag{8.3.34}$$

$$\hat{L}\,\hat{\underline{u}} = \underline{0} \ ,$$

with $\hat{S} = R_z^T S\, R_z = R_z^T S_1 + R_z^T S_2 R_0$,

$\hat{L} = L_1 R_z$ and $\hat{\underline{F}} = R_z^T \underline{F}$.

8.3.2.5 Concluding remarks

The procedure in this section resulted into the construction of an element in which the approximations for velocity and pressure are determined by six nodal points for the velocity and one nodal point for the pressure. This element is drawn in fig. 8.2 and is called the modified Crouzeix-Raviart element.

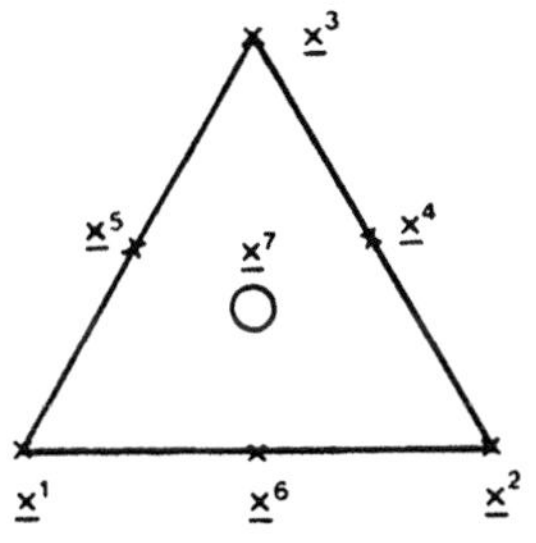

Velocity: extended quadratic (6 nodal points: x)

Pressure: linear approximation (1 nodal point: 0)

Approximation:

$$\underline{\tilde{u}} = \sum_{i=1}^{6} (u_{1i}\underline{\Phi}_{1i} + u_{2i}\underline{\Phi}_{2i})$$

$$\tilde{p} = p_7\psi_1 + G_3(u_{11}\ldots u_{26})\psi_2 + G_4(u_{11}\ldots u_{26})\psi_3$$

Fig. 8.2: Modified Crouzeix-Raviart element ($P_2^+ - P_1$).

Since the whole elimination process can be considered as a part of the Gaussian elimination process, the accuracy of the original $P_2^+-P_1$ element is retained.

The main advantage of this element is that the number of unknowns in each element is reduced with 4. The number of equations in each element is also decreased with 4, since 2 of the 14 Galerkin equations corresponding to the momentum equations and 2 of the 3 Galerkin equations corresponding to the continuity equation, have been used. In this way the number of equations is reduced considerably without affecting the accuracy. This reduction is illustrated in fig. 8.3. The resulting system of equations can be written as:

$$\hat{S}\,\hat{\underline{u}} + \hat{L}^T\hat{\underline{p}} = \hat{\underline{F}}\ , \qquad (8.3.35)$$

$$\hat{L}\,\hat{\underline{u}} = \underline{0}\ ,$$

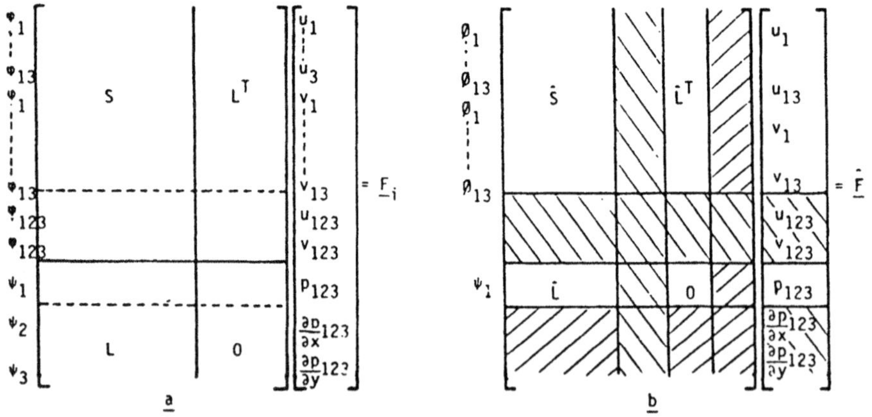

Fig. 8.3 Visualisation of the modification process for one $P_2^+-P_1$ element $\underline{a}$ before and $\underline{b}$ after modification. The shaded parts indicate the rows and columns that are eliminated.

where, if the region Ω consists of one element, $\hat{\underline{u}}$ and $\hat{\underline{p}}$ are equal to:

$$\hat{\underline{u}}^T = [u_{11}, u_{12} \cdots u_{16}, u_{21}, u_{22} \cdots u_{26}] ,$$

$$\hat{\underline{p}} = [p_7] .$$

This element is generally applicable. As mentioned in Chapter 7 it can be used for the integrated method. It is however, pre-eminently suitable for the penalty function method and as will be clear in Chapter 9, it forms the basis for the divergence-free element. The fact that this element is discussed in this chapter, does certainly not imply that it is only useful in the penalty function method.

8.3.3 The structure of the equations

Consider the penalty function equations (8.1.22):

$$S\,\underline{u} + \tau\, L^T D^{-1} L\,\underline{u} = \underline{F} . \qquad (8.3.36)$$

As is shown in (7.2.4), the matrix S corresponds to a double discretization of the Laplacian operator (see Section 3.5) and is therefore positive definite. The matrix components of $L_{u_k p}$ corresponding to the velocity components u_{ki} are equal to

$$L_{u_k p}(i,j) = \int_\Omega \frac{\partial \phi_j}{\partial x_k} \psi_i \, d\Omega . \qquad (8.3.37)$$

For a Crouzeix-Raviart element with

$$\psi_k = 1 \quad \text{in element} \quad e_k , \qquad (8.3.38)$$

$$= 0 \quad \text{outside} \quad e_k ,$$

(8.3.37) is equal to $\int_{e_i} \frac{\partial \phi_j}{\partial x_k} d\Omega$.

Furthermore when $D_{ii} = |e_i|$ is the area of element e_i then the components of $L^T D^{-1} L$ satisfy:

$$L^T_{u_k p} D^{-1} L_{p u_\ell}(i,j) = \sum_{\theta=1}^{M} \frac{1}{|e_\theta|} \int_{e_\theta} \frac{\partial \phi_i}{\partial x_k}\, d\Omega \int_{e_\theta} \frac{\partial \phi_j}{\partial x_\ell}\, d\Omega \,, \qquad (8.3.39)$$

$$i,j = 1,2,\ldots,N, \quad k,\ell = 1,2.$$

This shows that the matrix $L^T D^{-1} L$ is symmetrical and furthermore the matrix is positive.
For the matrix equation:

$$(S + \tau\, L^T D^{-1} L)\underline{u} = \underline{F} \,, \qquad (8.3.40)$$

this means that the matrix $S + \tau\, L^T D^{-1} L$ is positive definite, since S is positive definite (see Section 3.6). This property ensures that the Gaussian elimination process can be carried out efficiently.

Another important property of the $L^T D^{-1} L$ matrix is that this matrix can be computed elementwise for a Crouzeix-Raviart element. This is due to the fact that the elements of this matrix consist of the sum of quantities that are defined elementwise. As we have already seen in Section 5.1 this is also the case for the matrix S, which makes the construction of $L^T D^{-1} L$ very simple. For the Taylor-Hood elements this is not valid anymore. In practice the penalty function method is therefore only combined with the use of Crouzeix-Raviart elements.

Finally a remark is in order concerning the magnitude of the penalty function parameter τ. The matrix L is not a square matrix but a rectangular matrix (M×2N). Therefore the matrix L has at most M independent rows. The same is true for the matrix $\tau\, L^T D^{-1} L$ and so this matrix is singular. As a consequence, in the penalty function method a singular matrix multiplied by a large parameter is added to the regular matrix S. When τ is large enough, the resulting matrix will also be singular if we compute with a finite number of digits. If τ is somewhat

smaller, the resulting matrix is not singular, but has a poor condition, and therefore loss of decimals will occur in the Gaussian elimination process. So the penalty parameter τ must be chosen such that it is not too large because of the poor condition of the large matrix. On the other hand it may be not too small, because otherwise the approximation of the continuity equation is not accurate enough. In practice one usually chooses τ in the range of $O(10^3)$ to $O(10^9)$ depending on the number of decimals available on the computer.
Although the examples in this section are restricted to 2D, the whole theory remains valid for 3D without any essential modification.

8.4 Evaluation of the penalty function method

The discrete penalty function method is a very efficient method for the construction of an approximate solution of the continuity equation and the Navier-Stokes equations. The main advantage of this method compared to the integrated method is that velocity computation and pressure computation have been uncoupled. This leads to smaller systems of equations with fewer unknowns. Moreover, no partial pivoting is necessary, which makes the direct application of profile methods possible. Thanks to these properties, tremendous savings in computing time and computer memory are possible; for practical problems a saving of a factor 10 in computing time is not exceptional. Besides that, the penalty method can directly be applied to 3D problems, which is an advantage compared to the divergence-free elements that will be treated in the next chapter. A disadvantage of the penalty method is that the penalty parameter τ must be chosen carefully, since otherwise loss of accuracy may arise because of the singularity of the resulting matrix.

Chapter 9

Divergence-free elements

Introduction

The numerical solution of the Navier-Stokes equations by the integrated method and the penalty method has been the subject of Chapters 8 and 9. Both methods have some disadvantages: the integrated method is expensive, the penalty function method requires the introduction of a large parameter. The numerical solution of the resulting systems of linear equations by iterative techniques is for both methods impossible. In this Chapter the solenoidal approach is discussed. This method is based on the construction of approximate divergence-free basis functions. The application of these basis functions, uncouples the pressure and velocity without the introduction of a large parameter. The resulting system of linear equations is smaller than that of the penalty method. Moreover the condition of this system is such that iterative techniques may be applied. It will be shown how divergence-free basis functions may be constructed in the case of the Crouzeix-Raviart type elements. It will be made clear that for 3D elements this construction is a heavy one requiring additional software.

9.1 General approach

Once again the system of Stokes equations and continuity equation:

$$- \frac{1}{Re} \Delta \underline{u} + \nabla p = \underline{f} \; , \qquad (9.1.1)$$

$$\text{div } \underline{u} = 0 \; ,$$

will be considered with boundary conditions:

$$\underline{u} = \underline{0} \quad \text{on} \quad \Gamma \text{ boundary of } \Omega. \qquad (9.1.2)$$

In Chapter 7 ((7.1.9), (7.1.10)) it has been shown, that multiplication of the momentum equations by test functions v_i and multiplication of the continuity equation by a test function q results in:

$$\int_\Omega \{\frac{1}{Re} \nabla u_k . \nabla v_k - p \frac{\partial v_k}{\partial x_k}\} d\Omega = \int_\Omega f_k v_k d\Omega, \quad k = 1,2,3, \qquad (9.1.3)$$

$$\int_\Omega q \text{ div } \underline{u} \, d\Omega = 0. \qquad (9.1.4)$$

Summation of (9.1.3) over k, and defining $\underline{v} = (v_1,v_2,v_3)$ yields (see (8.2.2)):

$$\int_\Omega \{\frac{1}{Re} \nabla \underline{u} . \nabla \underline{v} - p \text{ div } \underline{v}\} d\Omega = \int_\Omega \underline{f} . \underline{v} \, d\Omega \; , \qquad (9.1.5)$$

$$\int_\Omega q \text{ div } \underline{u} \, d\Omega = 0 \; .$$

For the construction of an approximate solution, the following expressions were substituted:

$$\tilde{u}_k = \sum_{j=1}^{N} u_{kj} \phi_j(\underline{x}) \; , \qquad (9.1.6)$$

$$\tilde{p} = \sum_{j=1}^{M} p_j \psi_j(\underline{x}) \; ,$$

with N the number of velocity points and M the number of pressure points. The system of Galerkin equations was derived by setting:

$$v_k = \phi_i \; , \qquad k = 1,2,3 \; , \quad i = 1,2,\ldots,N \; , \qquad (9.1.7)$$

and

$$r = \psi_i \ , \qquad i = 1,2,\ldots,M.$$

(9.1.6) can be written in vector notation as:

$$\tilde{\underline{u}} = \sum_{j=1}^{N} \{u_{1j}\underline{\phi}_{1j}(\underline{x}) + u_{2j}\underline{\phi}_{2j}(\underline{x}) + u_{3j}\underline{\phi}_{3j}(\underline{x})\} \ , \qquad (9.1.8)$$

with

$$\underline{\phi}_{1j} = (\phi_j,0,0), \quad \underline{\phi}_{2j} = (0,\phi_j,0), \quad \underline{\phi}_{3j} = (0,0,\phi_j). \qquad (9.1.9)$$

The only requirements to be satisfied by the basis functions are the continuity properties treated in Section 7.2.2. Therefore the approximation does not satisfy the approximate incompressibility constraint (9.1.4) a priori. In the approach with the divergence-free elements, also called the solenoidal approach, the test functions $\underline{v}$ and consequently also the basis functions $\underline{\phi}_i$ are chosen such that the approximate continuity equation (9.1.4) is satisfied.
In principle it is always possible to construct functions $\underline{\phi}_i$ and $\underline{v}$ satisfying (9.1.4), since each velocity vector has a divergence-free and a curl-free part:

$$\underline{v} = \underline{v}_d + \underline{v}_r + \underline{w} \quad \text{with} \quad \text{div}\ \underline{v}_d = \text{div}\ \underline{w} = 0 \ , \qquad (9.1.10)$$

$$\text{curl}\ \underline{v}_r = \text{curl}\ \underline{w} = 0,$$

where $\underline{w}$ is determined by the conditions to be specified at the boundary of the fluid (see Batchelor (1967)). If we limit ourselves to Crouzeix-Raviart elements, where the gradient of the pressure has been eliminated, (9.1.4) can be written as:

$$\int_\Omega \psi_i \ \text{div}\ \tilde{\underline{u}}\ d\Omega = \int_{e_i} \text{div}\ \tilde{\underline{u}}\ d\Omega, \quad i = 1,2,\ldots,M \ , \qquad (9.1.11)$$

with M the number of elements. For the derivation of (9.1.11) we used:

$$\psi_i = 1 \quad \text{within element } i \ ,$$

$$= 0 \quad \text{outside element } i,$$

see Chapters 7 and 8.
When divergence-free basis functions are used, (9.1.5) reduces to

$$\int_\Omega \frac{1}{Re} \nabla\tilde{\underline{u}}.\nabla\underline{v}\ d\Omega = \int_\Omega \underline{f}.\underline{v}\ d\Omega\ , \tag{9.1.12}$$

where $$\tilde{\underline{u}} = \sum_{j=1}^{D} u_{dj}\underline{\phi}_{dj}\ , \tag{9.1.13}$$

$$\underline{v} = \underline{\phi}_{di}\ ,$$

and $\underline{\phi}_{di}$ satisfies:

$$\int_\Omega q\ \mathrm{div}\ \underline{\phi}_{di} d\Omega = 0 \quad \Rightarrow \quad \int_{e_i} \mathrm{div}\ \underline{\phi}_{di} d\Omega = 0\ . \tag{9.1.14}$$

Here D is the number of nodal point degrees of freedom u_{di} for the velocity and $\underline{\phi}_{di}$ are the divergence-free basis functions.

The main advantage of this method is the considerable reduction of the number of degrees of freedom, which makes the solution process more efficient, and the fact that pressure computation and velocity computation have been segregated. A serious problem, however, is the construction of basis functions for the velocity that satisfy the approximate continuity equation. In the next section we shall demonstrate how such basis functions may be constructed in the case of Crouzeix-Raviart elements. For the Taylor-Hood elements no such construction procedure is known at present.

It will be clear that the divergence-free basis functions $\underline{\phi}_{di}$ do not span the complete space of velocity functions, but only the solenoidal part of it. The remaining curl-free part is spanned by the not-divergence-free basis functions, which we shall call $\underline{\phi}_{ri}$. These basis functions can be used to calculate the pressure by solving the system of equations once the velocity has been computed:

$$\int_\Omega \tilde{p} \operatorname{div} \underline{v}_r d\Omega = \int_\Omega \frac{1}{Re} \nabla\tilde{\underline{u}} . \nabla\underline{v}_r d\Omega - \int_\Omega \underline{f} . \underline{v}_r d\Omega , \tag{9.1.15}$$

where for $\underline{v}_r$ the curl-free basis functions must be substituted:

$$\underline{v}_r = \underline{\phi}_{ri} , \qquad i = 1,2,\ldots,R.$$

In the next sections, it will be demonstrated how the basis functions can be separated into a divergence-free and a curl-free part for some Crouzeix-Raviart elements. Furthermore the structure of the system of equations will be considered.

9.2 The construction of divergence-free basis functions for 2D elements

9.2.1 Introduction

Divergence-free elements were first constructed by Crouzeix (1976) and Thomasset (1981). The derivation we follow here is a mixture of the approach of Griffiths (1979, 1981) and the approach of Crouzeix. As mentioned in Section 9.1 we shall say that basis functions $\underline{\phi}_{dj}$ are divergence-free, if they satisfy:

$$\int_\Omega \psi_i \operatorname{div} \underline{\phi}_{dj} d\Omega = 0 , \quad i = 1,2,\ldots,M. \tag{9.2.1}$$

For the Crouzeix-Raviart elements, where the gradient of the pressure has been eliminated, (9.2.1) reduces to:

$$\int_{e_i} \operatorname{div} \underline{\phi}_{dj} d\Omega = 0 , \quad i = 1,2,\ldots,M , \tag{9.2.2}$$

see Section 7.2.3.3.
In fact the basis functions $\underline{\phi}_{dj}$ are not pointwise divergence-free, but divergence-free in the mean. When the velocity approximation $\tilde{\underline{u}}$ is written as:

$$\tilde{\underline{u}} = \sum_{j=1}^{D} u_{dj} \underline{\phi}_{dj} , \tag{9.2.3}$$

then it is clear that $\tilde{\underline{u}}$ satisfies the approximate continuity equation. For an arbitrary element, like for example the element of fig. 9.1, the velocity approximation is defined by general basis functions, for example:

$$\tilde{\underline{u}} = \sum_{j=1}^{N} (u_{1j}\underline{\phi}_{1j}+u_{2j}\underline{\phi}_{2j}). \tag{9.2.4}$$

Substitution of (9.2.4) into the discretized continuity equation for a Crouzeix-Raviart element gives:

$$\int_{e_i} \operatorname{div} \tilde{\underline{u}}\, d\Omega = 0 \ , \tag{9.2.5}$$

and by virtue of the Gauss theorem this can be written as:

$$\int_{\partial e_i} \tilde{\underline{u}}\cdot\underline{n}\, d\Gamma = 0 \ . \tag{9.2.6}$$

It will be clear that one set of divergence-free basis functions could be formed by the tangential components of the general basis functions. On the other hand for the curl of the velocity approximation per element, the Stokes equation gives:

$$\int_{e} \operatorname{curl} \tilde{\underline{u}}\, d\Omega = \int_{\partial e} \underline{u}\cdot\underline{t}\, d\Gamma.$$

This illustrates that the basis functions corresponding to the normal velocity components belong to the curl-free part of the space spanned by the velocity basis functions. Hence to construct divergence-free basis functions, the normal velocity components need to be replaced. To that end the stream function ψ is introduced (see Section 6.2), according to

$$\tilde{\underline{u}} = \left(\frac{\partial\psi}{\partial x_2}, -\frac{\partial\psi}{\partial x_1}\right) \text{ which implies } \int_{e} \operatorname{div} \tilde{\underline{u}}\, d\Omega = 0 \ . \tag{9.2.7}$$

Using the vector relation

$$\tilde{\underline{u}}\cdot\underline{n} = \frac{\partial\psi}{\partial x_2} n_1 - \frac{\partial\psi}{\partial x_1} n_2 = \frac{\partial\psi}{\partial x_2} t_2 + \frac{\partial\psi}{\partial x_1} t_1 = \nabla\psi\cdot\underline{t} \ , \tag{9.2.8}$$

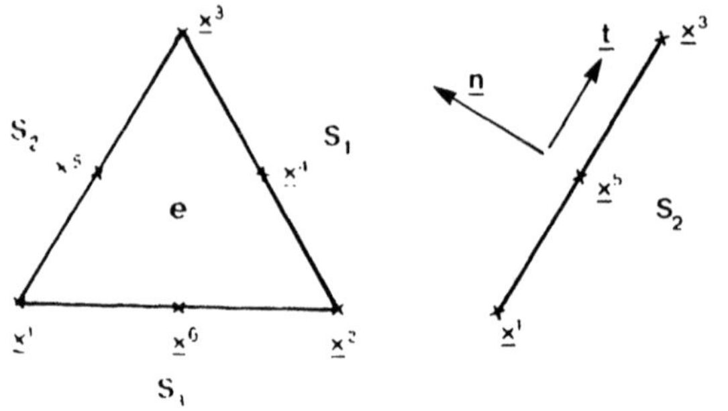

Fig. 9.1 Element e, with boundary S_2 separately plotted. Normal and tangential unity vectors on S_2 are indicated.

where the unit vectors $\underline{n}$ and $\underline{t}$ are related by:

$$\underline{n} = (n_1,n_2), \quad \underline{t} = (t_1,t_2) = (-n_2,n_1) . \tag{9.2.9}$$

For side S_3 of the element e of fig. 9.1 we find:

$$\int_{S_3} \underline{\tilde{u}}\cdot\underline{n} \, d\Gamma = \int_{S_3} \nabla\psi\cdot\underline{t} \, d\Gamma = \psi_2-\psi_1 .$$

Hence: $$\psi_2 = \psi_1 + \int_{S_3} \underline{\tilde{u}}\cdot\underline{n} \, d\Gamma . \tag{9.2.10}$$

For side S_1 we obtain $\psi_3 = \psi_2 + \int_{S_1} \underline{\tilde{u}}\cdot\underline{n} \, d\Gamma$.

By requiring that, when we encircle the element, the initial value ψ_1 of the stream function is equal to the end value ψ_4 defined by:

$$\psi_4 = \psi_3 + \int_{S_2} \underline{\tilde{u}}\cdot\underline{n} \, d\Gamma , \tag{9.2.11}$$

we set:

$$\psi_1 = \psi_4 . \tag{9.2.12}$$

By this construction the divergence-freedom of the velocity $\underline{\tilde{u}}$ and hence of the basis functions $\underline{\phi}_i$ is always satisfied. Notice that the stream function ψ is defined such, that its value in the common nodal point of contiguous elements, is unique. This is caused by the fact that if

$$\int_{\partial e_1} \underline{\tilde{u}}\cdot\underline{n}\, d\Gamma = 0 \quad \text{and} \quad \int_{\partial e_2} \underline{\tilde{u}}\cdot\underline{n}\, d\Gamma = 0 , \tag{9.2.13}$$

then

$\int_{\partial(e_1 \cup e_2)} \underline{\tilde{u}}\cdot\underline{n}\, d\Gamma = 0$, where $\partial(e_1 \cup e_2)$ is the boundary of the elements e_1 and e_2. Due to this property it is possible to introduce the stream function parameters as unknowns in the problem. The divergence-freedom of the element is now enforced by eliminating a number of normal velocities on the boundaries of the elements, in favour of the stream function values in the vertices of the element. In this way the other set of divergence-free basis functions is implicitly defined. In the next sections the actual construction of these basis functions will be treated.

9.2.2 The non-conforming Crouzeix-Raviart element

The reason of treating this element is not its practical applicability, which is very limited (see Section 7.2.3.3), but the fact that the construction of the divergence-free basis functions is demonstrated in a simple way. In fact this construction will be generalized for the more important quadratic elements.

Consider the non-conforming element of fig. 9.2, with midside points as nodal points for the velocity, and nodal point values u_{1i} and u_{2i}. The velocity approximation can be written as:

$$\tilde{u}_1 = \sum_{j=4}^{6} u_{1j}\phi_j , \qquad \tilde{u}_2 = \sum_{j=4}^{6} u_{2j}\phi_j , \tag{9.2.14}$$

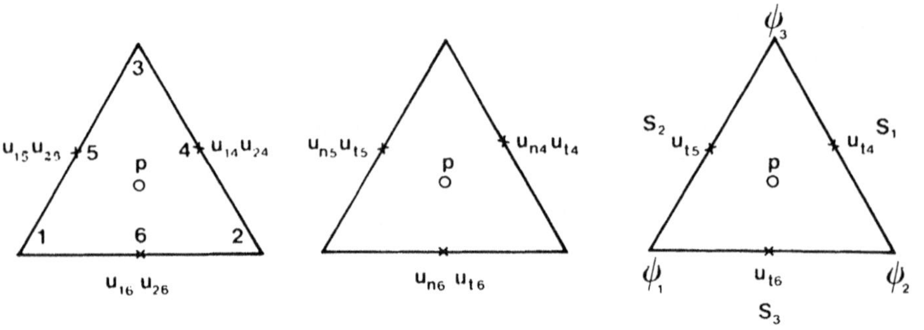

Fig. 9.2 The P_1-P_0 Crouzeix-Raviart element with various possibilities for the velocity nodal point values. The sides are indicated by S_1, S_2 and S_3.

with the linear basis functions for the velocity as treated in Section 4.2.3. In vector notation (9.2.14) becomes:

$$\tilde{\underline{u}} = \sum_{j=4}^{6} (u_{1j}\underline{\phi}_{1j} + u_{2j}\underline{\phi}_{2j}) \ , \tag{9.2.15}$$

with $\underline{\phi}_{1j} = (\phi_j, 0)$, $\underline{\phi}_{2j} = (0, \phi_j)$.

Using the normal and tangential velocity components in the nodal points, (9.2.15) can be expressed as:

$$\tilde{\underline{u}} = \sum_{j=4}^{6} (u_{nj}\underline{\phi}_{nj} + u_{tj}\underline{\phi}_{tj}) \ , \tag{9.2.16}$$

with

$$\underline{\phi}_{nj} = \phi_j \underline{n}_j \ , \quad \underline{\phi}_{tj} = \phi_j \underline{t}_j \ ,$$

and $\underline{n}_j$ and $\underline{t}_j$ the unit vectors in normal and tangential direction corresponding to the sides S_{j-3} (j = 4,5,6).

The approximation (9.2.16) has the following properties:

i) the basis functions ϕ_j are 1 in nodal point j and 0 in the other nodal points, i.e.

$$\phi_j(\underline{x}^i) = \delta_{ij} , \qquad i,j = 4,5,6 .$$

ii) in nodal point j:

$$\underline{\tilde{u}}\cdot\underline{n}_j = u_{nj} , \quad \underline{\tilde{u}}\cdot\underline{t}_j = u_{tj} .$$

The first set of divergence-free basis functions is formed by the basis functions $\underline{\phi}_{tj}$ since clearly $\underline{\phi}_{tj}$ satisfies:

$$\int_{\partial e} \underline{\phi}_{tj}\cdot\underline{n}\, d\Gamma = 0 .$$

For the construction of the second set of basis functions the stream function ψ with values ψ_1, ψ_2 and ψ_3 in the vertices is defined (see Fig. 9.2). From (9.2.10)...(9.2.12) it follows that

$$\psi_j = \psi_i + \int_{S_k} \underline{\tilde{u}}\cdot\underline{n}_k\, d\Gamma , \qquad (9.2.17)$$

with $\underline{n}_k$ the normal on side S_k.

Since $\underline{\tilde{u}}$ is linear in the element:

$$\int_{S_k} \underline{\tilde{u}}\cdot\underline{n}_k\, d\Gamma = L_k u_{nk+3} , \quad k = 1,2,3 , \qquad (9.2.18)$$

with L_k the length of side S_k and u_{nk} the normal component of the velocity in the midside nodal point k+3. From (9.2.17) and (9.2.18) it follows that:

$$\psi_2 = \psi_1 + u_{n6}L_3 ,$$

$$\psi_3 = \psi_2 + u_{n4}L_1 ,$$

$$\psi_1 = \psi_3 + u_{n5}L_2 ,$$

or:

$$u_{n6} = (\psi_2-\psi_1)/L_3 , \qquad (9.2.19)$$

$$u_{n4} = (\psi_3-\psi_2)/L_1 ,$$

$$u_{n5} = (\psi_1-\psi_3)/L_2.$$

Substitution of (9.2.19) into (9.2.16) yields:

$$\tilde{\underline{u}} = \sum_{j=4}^{6} u_{tj}\underline{\phi}_{tj} + \sum_{j=1}^{3} \psi_j\underline{\phi}_{\psi j} , \tag{9.2.20}$$

with

$$\underline{\phi}_{\psi 1} = \frac{\underline{\phi}_{n5}}{L_2} - \frac{\underline{\phi}_{n6}}{L_3}, \ \underline{\phi}_{\psi 2} = \frac{\underline{\phi}_{n6}}{L_3} - \frac{\underline{\phi}_{n4}}{L_1}, \ \underline{\phi}_{\psi 3} = \frac{\underline{\phi}_{n4}}{L_1} - \frac{\underline{\phi}_{n5}}{L_2}. \tag{9.2.21}$$

Thus the second set of divergence-free basis functions is now formed by the basis functions $\underline{\phi}_{\psi j}$, since it is clear from the construction of $\underline{\phi}_{\psi j}$ that

$$\int_{\partial e} \underline{\phi}_{\psi j}\cdot\underline{n}\, d\Gamma = 0. \tag{9.2.22}$$

This completes the construction of 6 basis functions that are divergence-free. With these basis functions the system of equations for the velocity approximation can be solved.

The basis functions $\underline{\phi}_{tj}$ and $\underline{\phi}_{\psi j}$ are characterized by the following properties:

(i) both $\underline{\phi}_{tj}$ and $\underline{\phi}_{\psi j}$ have components that are linear in x_1 and x_2

(ii) $\underline{\phi}_{tj} = 0$ in the midside nodes not equal to j.
$\underline{\phi}_{tj}$ is equal to the unit tangential vector in midside node j

(iii) $\underline{\phi}_{\psi j} = 0$ on the side opposite to vertex j.
$\pm\, \underline{\phi}_{\psi j}$ is equal to the unit normal vector divided by the length of the side, in the two mid-side points of the sides containing the vertex j. The sign must be such that it is opposite for these two points.

In fig. 9.3a the values of the basis function $\underline{\phi}_{tj}$ in the three midside points is sketched. Fig. 9.3b shows the values of the basis function $\underline{\phi}_{\psi j}$ in the midside points of an ensemble of triangles.

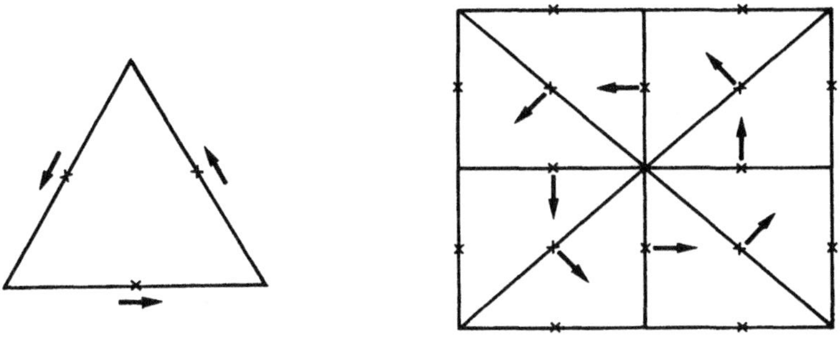

a) b)

Fig. 9.3 Representation of the divergence-free basis functions of the linear non-conforming triangle in the midside points:
a) basis function $\underline{\phi}_{tj}$ for one element
b) basis function $\underline{\phi}_{\psi j}$ for an ensemble of elements. Notice that a counter clockwise direction has been chosen.

In order to construct the divergence-free basis functions the normal velocities have been eliminated. The basis functions corresponding to these normal velocities are not divergence-free and can therefore be used to compute the pressure. Since the pressure is defined up to an additive constant, it may be fixed in one point. We shall fix the pressure in element e_1 (pressure p_1) and calculate for the configuration of fig. 9.4, the value of the pressure in element e_2, using the fact that the pressure is constant in each element.
For the computation of the pressure p_2 in element e_2 we use the curl-free basis function $\underline{\phi}_n$ corresponding to the velocity nodal

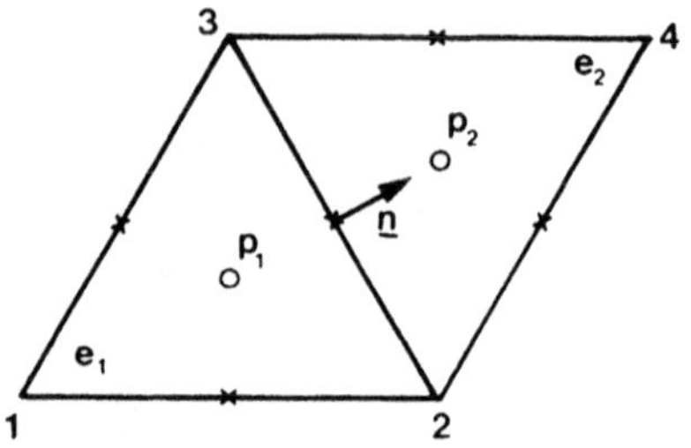

Fig. 9.4 Two adjacent elements e_1 and e_2. The pressure nodal points have been indicated.

point on the common side of the elements e_1 and e_2. Since $\underline{\phi}_n$ is zero outside the elements e_1 and e_2, (9.1.15) reduces to:

$$\int_{e_1 \cup e_2} \tilde{p} \, \text{div} \, \underline{\phi}_n \, d\Omega = \int_{e_1 \cup e_2} \{\frac{1}{Re} \nabla \tilde{\underline{u}} . \nabla \underline{\phi}_n - \underline{f} . \underline{\phi}_n\} d\Omega. \tag{9.2.23}$$

Application of Gauss' theorem to the left-hand side gives

$$\int_{e_1 \cup e_2} \tilde{p} \, \text{div} \, \underline{\phi}_n \, d\Omega = \int_{\partial e_1} \tilde{p} \, \underline{\phi}_n . \underline{n} \, d\Gamma + \int_{\partial e_2} \tilde{p} \, \underline{\phi}_n . \underline{n} \, d\Gamma \tag{9.2.24}$$

$$= (p_1 - p_2) L_{23}.$$

From (9.2.24) en (9.2.23) one can express p_2 in the pressure p_1 in element e_1 and the velocity approximation in both elements. Since the velocity has been computed before, p_2 is determined. Proceeding in the same way for all adjacent elements, the pressure can be computed for the entire region Ω.

9.2.3 The modified P_2^+-P_1 Crouzeix-Raviart element

Consider the P_2^+-P_1 element in fig. 9.5. The velocity approximation in each element is (see Chapter 7):

$$\tilde{u}_k(\underline{x}) = \sum_{j=1}^{7} u_{kj}\phi_j(\underline{x}). \tag{9.2.25}$$

For the pressure approximation we have:

$$\tilde{p}(\underline{x}) = p_7 + (x_1 - x_1^7)\frac{\partial p}{\partial x_1}(\underline{x}_7) + (x_2 - x_2^7)\frac{\partial p}{\partial x_2}(\underline{x}_7). \tag{9.2.26}$$

The first step in the construction of divergence-free basis functions, is the elimination of the velocities $\underline{u}_7$ in the centroid, as well as the gradient of the pressure. This process has been demonstrated in Section 8.3.2.2. The final result of these manipulations is that the velocity approximation can be written as (8.3.14):

$$\tilde{\underline{u}} = \sum_{i=1}^{6} (u_{1i}\underline{\Phi}_{1i} + u_{2i}\underline{\Phi}_{2i}) , \tag{9.2.27}$$

where the components of $\underline{\Phi}_{kj}$ (k=1,2 and j=1,...,6) are linear combinations of the basis functions ϕ_i (i=1,2,...,7).

Analogous to the non-conforming element, the velocity components on the midpoints of the sides are replaced by their normal and tangential components, resulting into:

$$\tilde{\underline{u}} = \sum_{i=1}^{3} \{u_{1i}\underline{\Phi}_{1i} + u_{2i}\underline{\Phi}_{2i}\} + \sum_{i=4}^{6} \{u_{ni}\underline{\Phi}_{ni} + u_{ti}\underline{\Phi}_{ti}\} . \tag{9.2.28}$$

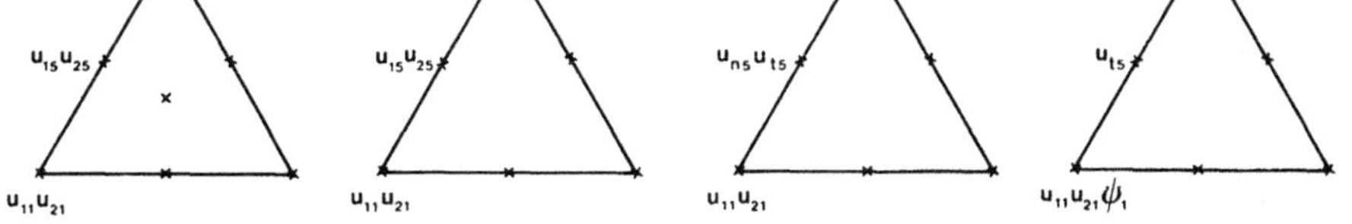

Fig. 9.5 The $P_2^+ - P_1$ Crouzeix-Raviart element with various possibilities for the velocity nodal point values.

The basis functions can be made divergence-free, by the elimination of the normal velocity components in the midside nodal points, in favour of the stream function values in the vertices. This can be done using the relation:

$$\psi_j = \psi_i + \int_{S_k} \tilde{\underline{u}}\cdot\underline{n}_k \, d\Gamma. \tag{9.2.29}$$

Integration of (9.2.29) for a straight side gives:

$$\int_{S_k} \tilde{\underline{u}}\cdot\underline{n}_k \, d\Gamma = \frac{L_k}{6} \{u_{1i}\underline{\phi}_{1i}\cdot\underline{n}_k + u_{2i}\underline{\phi}_{2i}\cdot\underline{n}_k + u_{1j}\underline{\phi}_{1j}\cdot\underline{n}_k + u_{2j}\underline{\phi}_{2j}\cdot\underline{n}_k + 4u_{nk}\}, \tag{9.2.30}$$

with i and j the end points of side S_k.

From (9.2.30) the normal velocity in the midside point k, can be expressed in the velocity components in the vertices of side S_k, the stream function corresponding to these vertices and the length L_k of side S_k:

$$u_{nk} = \frac{3}{2}\frac{\psi_j-\psi_i}{L_k} - \frac{1}{4}\{u_{1i}\underline{\phi}_{1i}\cdot\underline{n}_k + u_{2i}\underline{\phi}_{2i}\cdot\underline{n}_k + u_{1j}\underline{\phi}_{1j}\cdot\underline{n}_k + u_{2j}\underline{\phi}_{2j}\cdot\underline{n}_k\}. \tag{9.2.31}$$

In exactly the same way as in Section 9.2.2, the quantities u_{nk} can be eliminated and divergence-free basis functions have been constructed. We shall not give the explicit expression of these basis functions. The derivation of the system of equations for practical computations will be treated in Section 9.2.6. The computation of the pressure for this element consists of two parts. As is shown in Section 8.3.2.3, two of the three pressure parameters, the so-called pressure derivatives, can be computed from the calculated velocity approximation in each element. The only remaining parameter is the pressure value in the centroid of each element. The computation of this parameter is completely analogous to the non-conforming triangle. The basis function corresponding to the normal component of the velocity in a midside point is used as non-divergence-free basis function. In this way the pressure can be computed from element to element and all pressure parameters can be found. Since the pressure is discontinuous over the element boundaries, the computed pressure values in the vertices must be sampled over the contiguous elements. In this way a "continuous" image of the pressure will be generated, see also Section 7.2.3.3, and fig. 7.3.

9.2.4 The Q_2-P_1 9-node quadrilateral

Consider the Q_2-P_1 quadrilateral of fig. 9.6.

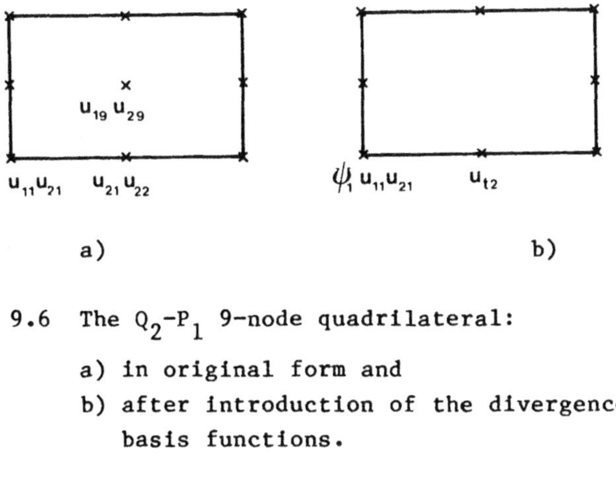

Fig. 9.6 The Q_2-P_1 9-node quadrilateral:

a) in original form and
b) after introduction of the divergence-free basis functions.

For the construction of the divergence-free basis functions exactly the same method as for the modified P_2^+-P_1 element is used. This means that the centroid velocity values are eliminated, using the linear part of the pressure approximation. The normal velocity component is eliminated by the introduction of a stream function in the vertices according to

$$\psi_j = \psi_i + \int_{S_k} \underline{\tilde{u}} \cdot \underline{n} \, d\Gamma. \tag{9.2.32}$$

The basis functions for the original element may be derived on a reference element (see Section 4.5.2). However, the introduction of divergence-free basis functions depends on the computational element, since a normal in the computational element, is not a normal in the reference element. If one takes care of this difference, the derivation of the divergence-free basis functions is straightforward. For a description see Griffiths (1981).

9.2.5 Boundary conditions

For divergence-free elements as derived in 9.2.2, we have introduced new unknowns u_t and ψ. A consequence is that on boundaries we do not only need boundary conditions for u_1 and u_2 alone but also for u_t and ψ. This implies that boundary conditions for u_1 and u_2 must be transformed. The transformation to u_t is straightforward, but for the stream function special care is needed. The problems are caused by the fact that the stream function must be computed from line integrals, since:

$$\psi(\underline{x}^2) = \psi(\underline{x}^1) + \int_{\underline{x}^1}^{\underline{x}^2} \underline{\tilde{u}} \cdot \underline{n} \, d\Gamma. \tag{9.2.33}$$

For the complicated case of the glass oven example of figure 6.2, see Section 6.3.3, the following boundary conditions for ψ are possible:

i) inflow: $u_1 = 0$, $u_2 = u_2(\underline{x})$.

Choose $\psi_0 = 0$ in the left corner, then:

$$\psi(x_1) = \int_0^{x_1} - u_2(x_1) \, dx_1.$$

Let the value in the right corner of boundary i be ψ_1.

ii) free surface: $u_2 = 0$, $u_n = 0$.

Then: $\psi = \psi_1$.

iii) fixed walls: $u_1 = 0$, $u_2 = 0$.

Then: $\psi = \psi_1$.

iv) outflow: ψ unknown.

v) fixed walls: $u_1 = 0$, $u_2 = 0$.

Then: $\psi = \psi_0$.

If there is only one boundary with ψ unknown, then ψ can be computed on all other places of the boundary by following the boundary in clockwise and counter-clockwise direction, starting with the value $\psi = 0$ in one point. When we have for example two outflow parts as in figure 9.7, then there is a boundary condition of the type ψ is unknown constant on the boundary between those parts.

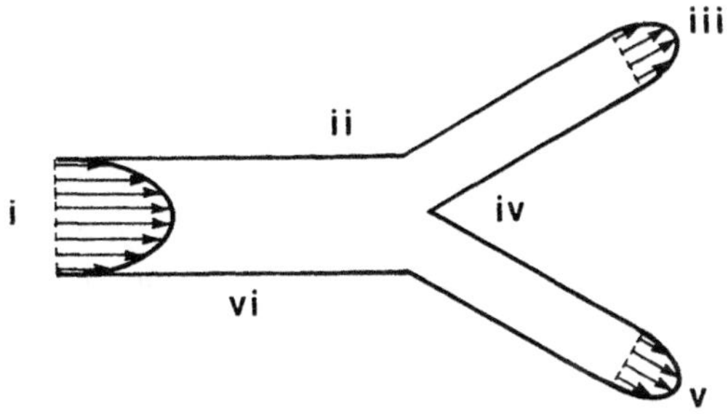

Fig. 9.7 Example of a region with two outflow parts. On the boundary iv we have the boundary condition ψ equals unknown constant.

Boundary conditions of this type, require a special treatment, see Section 9.2.7.

9.2.6 The structure of the system of equations

The Galerkin equations corresponding to the Stokes equations, in the presence of divergence-free basis functions $\underline{\phi}_{di}$:

$$\tilde{\underline{u}} = \sum_{j=1}^{D} u_{dj}\underline{\phi}_{dj} \quad \text{with} \int_e \operatorname{div} \underline{\phi}_{dj}\, d\Omega = 0 , \tag{9.2.34}$$

become:

$$\frac{1}{Re} \int_\Omega \nabla\tilde{\underline{u}} \cdot \nabla\underline{\phi}_{di}\, d\Omega = \int_\Omega \underline{f} \cdot \underline{\phi}_{di}\, d\Omega . \tag{9.2.35}$$

In matrix-vector notation (9.2.35) can be written as:

$$S_d\, \underline{u}_d = \underline{F}_d , \tag{9.2.36}$$

$$\text{with } S_d(i,j) = \frac{1}{Re} \int_\Omega \nabla\underline{\phi}_{dj} \cdot \underline{\nabla\phi}_{di}\, d\Omega , \quad i,j = 1,2,...,D,$$

$$F_d(i) = \int_\Omega \underline{f} \cdot \underline{\phi}_{di}\, d\Omega , \qquad i = 1,2,...,D$$

$$\underline{u}_d^T = [u_{d1},...,u_{dD}].$$

(9.2.36) can directly be computed from the divergence-free basis functions that have been constructed in the preceding sections. An alternative way to derive this system is to start with the original set of equations (7.1.9), (7.1.10). From a programming point of view this is the easiest way of performing. Therefore we shall consider the connection between both systems of equations.

We start with the system of equations (7.4.3) for the integrated method:

$$S\, \underline{u} + L^T \underline{p} = \underline{F} , \tag{9.2.37}$$

$$L\, \underline{u} = \underline{0} .$$

The vector $\underline{u}$ contains the velocity degrees of freedom in the nodal points, and $\underline{p}$ the pressure values in the centroids of the elements. The construction of the divergence-free elements implies the introduction of new unknowns $\underline{u}_d$. The elimination process can be expressed by a matrix R_d according to:

$$\underline{u} = R_d \, \underline{u}_d. \tag{9.2.38}$$

The transformation (9.2.38) is such that the continuity equation is always satisfied:

$$L \, \underline{u} = L \, R_d \, \underline{u}_d = \underline{0} \tag{9.2.39}$$

Since (9.2.39) must be valid for all vectors $\underline{u}_d$, it is necessary that the matrix LR_d is equal to the zero matrix:

$$L \, R_d = 0 \; . \tag{9.2.40}$$

Substitution of (9.2.38) in (9.2.37) and premultiplication by R_d^T yields:

$$R_d^T \, S \, R_d \, \underline{u}_d + R_d^T \, L^T \, \underline{p} = R_d^T \, \underline{F} \tag{9.2.41}$$

and because of (9.2.40):

$$R_d^T \, S \, R_d \, \underline{u}_d = R_d^T \, \underline{F}. \tag{9.2.42}$$

Comparison of (9.2.36) and (9.2.42) shows that:

$$S_d = R_d^T \, S \, R_d \; , \tag{9.2.43}$$

$$\underline{F}_d = R_d^T \, \underline{F} \; .$$

The matrix S is positive definite, the matrix S_d is also positive definite provided the matrix R_d has full rank. That means S_d is positive definite when

$$\underline{u} = \underline{0} \quad \text{if and only if} \quad \underline{u}_d = \underline{0} \; .$$

From the construction it is clear that $\underline{u} = \underline{0}$ whenever $\underline{u}_d = \underline{0}$. Conversely, from (9.2.17) and (9.2.31) it can be seen that

$\underline{u}_d = \underline{0}$ whenever $\underline{u} = \underline{0}$, provided that ψ is set to zero at one vertex. This is in accordance with the fact that the stream function is defined up to an additive constant.

The simplest way to construct the matrix S_d and the vector $\underline{F}_d$, is to build the element matrix S^{e_k} and the element vector $\underline{F}^{e_k}$ for the integrated solution method, and then to compute in each element:

$$(R_d^{e_k})^T S^{e_k} R_d^{e_k} \quad \text{and} \quad (R_d^{e_k})^T \underline{F}^{e_k} , \tag{9.2.44}$$

with $R_d^{e_k}$ the element transformation matrix. $R_d^{e_k}$ follows from the relation:

$$\underline{u}^{e_k} = R_d^{e_k} \underline{u}_d^{e_k} , \tag{9.2.45}$$

with $(\underline{u}^{e_k})^T = [u_{11},u_{12},u_{13},u_{21},u_{22},u_{23},u_{14},u_{15},u_{16},u_{24},u_{25},u_{26}]$,

$$(\underline{u}_d^{e_k})^T = [u_{11},u_{12},u_{13},u_{21},u_{22},u_{23},u_{t4},u_{t5},u_{t6},\psi_1,\psi_2,\psi_3],$$

for the $P_2^+ - P_1$ Crouzeix-Raviart triangle and similar expressions for the other elements.

By adding the element contributions in the usual way, see Section 5.1, the large matrix $R_d^T S R_d$ is constructed. This way of handling saves computer memory since R_d does not have to be stored, and costs relatively little computing time, since the sparseness of the element transformation matrix R_d can be utilized for the multiplications.

Let us consider finally the number of degrees of freedom. If we compare the divergence-free system of equations with the system we found with the integrated method, then it is immediately clear that the system of equations is considerably smaller, resulting in less computing time and memory needed. This is the main advantage of the method. The pressure, however, must be computed afterwards. By the elimination of the normal velocity components in the midside nodes, in favour of the stream function values in the vertices, the number of degrees of freedom is reduced further. This is illustrated in fig. 9.8. A disadvantage is that, once the solution has been computed, it has to be backtransformed to the velocity values in the nodal points using the relation:

$$\underline{u} = R_d \, \underline{u}_d.$$

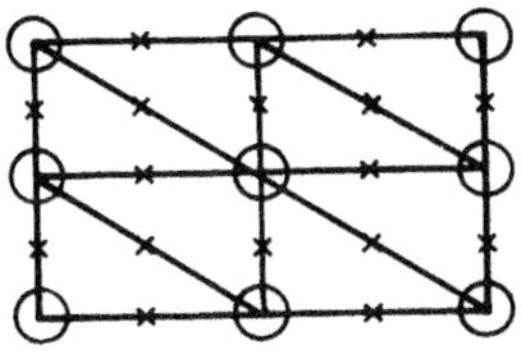

Number of degrees of freedom:

u_{nk} : 16 ψ_i : 9

Fig. 9.8 Region Ω with 8 elements. The nodal points for the normal velocity at the midside points have been indicated by a x, the stream function nodal points by 0.

9.2.7 The implementation of boundary conditions of the type ψ equals unknown constant

In Section 9.2.3 it has been shown that for some types of problems to be solved with divergence-free elements, one is faced with the problem of boundary conditions of the type ψ equals unknown constant. A natural way of dealing with this type of boundary condition, is to introduce one common degree of freedom for all nodes on this boundary. However, such a solution technique would require an essential intervention in existing codes and, moreover, it is very likely that the bandwidth or the profile of the solution matrix will be affected in a negative way. Therefore we shall derive an alternative method, suggested by Segal (1985), to incorporate boundary conditions of this type.

Boundary conditions of the type ψ equals constant on a part of the boundary Γ_p, can also be expressed as:

$$\psi(\underline{x}^i) = \psi(\underline{x}^{i-1}), \qquad (9.2.46)$$

with $\underline{x}^i$ and $\underline{x}^{i-1}$ two neighbouring nodes for ψ on Γ_p, for all points $\underline{x}^i$ on Γ_p.

(9.2.46) can be considered as linear constraints of the form:

$$L^* \underline{\psi} = \underline{0}\ , \qquad (9.2.47)$$

where L^* has the following shape:

$$L^* = \begin{bmatrix} 1 & -1 & & & & 0 \\ & 1 & -1 & & & \\ & & 1 & -1 & & \\ & & & \ddots & & \\ 0 & & & & 1 & -1 \end{bmatrix}$$

When a direct solution technique (see Section 5.6) is used to solve the system of algebraic equations, these constraints can be satisfied by the introduction of a penalty term:

$$\sigma\, L^T L\, \underline{u}_d\ , \qquad (9.2.48)$$

in the same way as is done in the penalty function method, with σ a large constant. In (9.2.48) the matrix L is defined as the extension of L^* with zero rows in order to get the right dimensions.

The system of equations (9.2.42):

$$S_d \underline{u}_d = \underline{F}_d\ , \qquad (9.2.49)$$

under the constraints (9.2.46) then becomes:

$$(S_d + \sigma L^T L) \underline{u}_d = \underline{F}_d \ . \qquad (9.2.50)$$

One easily verifies that the matrix $L^T L$ has only non-zero entries for the rows corresponding to points on the boundary Γ_p. For those rows the matrix $L^T L$ has the following shape:

$$L^{*T} L^{*} = \begin{bmatrix} 1 & -1 & & & 0 \\ -1 & 2 & -1 & & \\ & \cdot & \cdot & \cdot & \\ & & \cdot & \cdot & \cdot \\ & & -1 & 2 & -1 \\ 0 & & & 1 & -1 \end{bmatrix} \qquad (9.2.51)$$

$L^{*T} L^{*}$ is the finite element discretization of the Laplace operator for an equidistant mesh apart from a multiplication factor. This suggests the following way of formulating the boundary condition (9.2.46):

ψ on Γ_p must satisfy:

$$\frac{d^2\psi}{ds^2} = 0 \ , \quad \frac{d\psi}{ds}(\underline{x}^0) = \frac{d\psi}{ds}(\underline{x}^n) = 0 \ , \qquad (9.2.52)$$

where $\underline{x}^0$ denotes the starting point, $\underline{x}^n$ the end point of Γ_p, and s denotes the arclength along Γ_p.
Now the implementation of the matrix $\sigma L^T L$ can be easily carried out by the introduction of line-elements along the boundary Γ_p where the element matrices are equal to the element matrix corresponding to (9.2.52) multiplied by σ.
Hence

$$S^e = \frac{\sigma}{h_e} \begin{bmatrix} 1 & -1 \\ -1 & 1 \end{bmatrix} \qquad (9.2.53)$$

with h_e the length of the line-element.

The preceding method, that has been succesfully applied for some practical problems, Segal (1985), can only be used when the system of equations (9.2.50) is solved by a direct solver. This is caused by the fact that the large penalty parameter σ, influences the condition of the solution matrix badly, and hence iterative solvers converge slowly.

For iterative solvers (see Section 5.6), one can use a so-called projection method to solve (9.2.49) under the constraints (9.2.46). That means that after each iteration for the solution of (9.2.49), the computed solution is projected on the space of functions with ψ constant along Γ_p. Such a projection is for example: make ψ on Γ_p equal to the mean value of the computed ψ. This projection method can easily be implemented in iterative solvers.

9.3 The construction of divergence-free basis functions for 3D elements

9.3.1 Introduction

At this moment only a few papers have been published on the subject of 3D divergence-free elements. The reason for this is that the stream function vector, see Section 6.2, consist of only one non-zero component for 2D flow problems, while for 3D fluid flows three such components exist. In the next section we follow an article of Hecht (1981), who constructs basis functions for the non-conforming linear tetrahedron. This element shows the problem related to the construction of divergence-free elements, as well as a method for solving these problems. The extension to other elements is then straightforward.

9.3.2 A non-conforming Crouzeix-Raviart element in $\mathbb{R}^3$

Consider the tetrahedron in fig. 9.9, with midface points as nodal points for the velocity and the centroid as nodal point for the pressure, see Section 4.3.3. Let the velocity

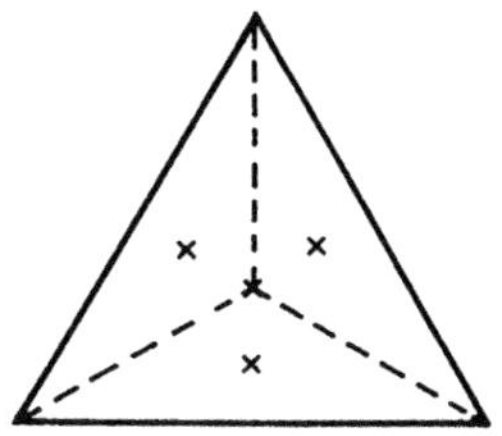

Fig. 9.9 The non-conforming linear tetrahedron

approximation in each element be linear and the pressure approximation be constant. The velocity approximation can be written as:

$$\tilde{u}_k = \sum_{j=1}^{4} u_{kj}\phi_j \ , \tag{9.3.1}$$

with ϕ_j the linear basis functions corresponding to the 4 velocity nodal points. In vector notation (9.3.1) becomes:

$$\underline{\tilde{u}} = \sum_{j=1}^{4} (u_{1j}\underline{\phi}_{1j} + u_{2j}\underline{\phi}_{2j} + u_{3j}\underline{\phi}_{3j}) \ , \tag{9.3.2}$$

with $\underline{\phi}_{1j} = (\phi_j,0,0)$, $\underline{\phi}_{2j} = (0,\phi_j,0)$ and $\underline{\phi}_{3j} = (0,0,\phi_j)$.

(9.3.2) can be written in nodal point values corresponding to normal and tangential velocity components as:

$$\underline{\tilde{u}} = \sum_{j=1}^{4} (u_{nj}\underline{\phi}_{nj} + u_{t_1j}\underline{\phi}_{t_1j} + u_{t_2j}\underline{\phi}_{t_2j}) \ , \tag{9.3.3}$$

with $\underline{\phi}_{nj} = \phi_j\underline{n}_j$, $\underline{\phi}_{t_1j} = \phi_j\underline{t}_1$, $\underline{\phi}_{t_2j} = \phi_j\underline{t}_2$,

where $\underline{t}_1$ and $\underline{t}_2$ are two mutual perpendicular tangential directions. The vectors $\underline{t}_1$ and $\underline{t}_2$ must be defined such that they are identical in common faces of adjacent elements.
The basis functions $\underline{\phi}$ are divergence-free in the mean if they satisfy:

$$\int_e \text{div}\ \underline{\phi}\ d\Omega = 0 \ , \tag{9.3.4}$$

which implies: $\int_{\partial e} \underline{\phi}\cdot\underline{n}\, dS = 0$

The first set of divergence-free basis functions is formed by the basis functions $\underline{\phi}_{t_1 j}$ and $\underline{\phi}_{t_2 j}$ since:

$$\int_{\partial e} \underline{\phi}_{t_1 j}\cdot\underline{n}\, dS = \int_{\partial e} \underline{\phi}_{t_2 j}\cdot\underline{n}\, dS = 0\ . \tag{9.3.5}$$

For the construction of the second set of divergence-free basis functions we use the vector potential $\underline{\psi}$ defined by:

$$\underline{u} = \text{curl}\ \underline{\psi}\ , \tag{9.3.6}$$

for an incompressible fluid.
We also use Stokes' theorem:

$$\int_F \text{curl}(\underline{\alpha})\cdot\underline{n}\, dS = \int_\Gamma \underline{\alpha}\cdot\underline{t}\, d\Gamma\ , \tag{9.3.7}$$

with F a simply connected surface in $\mathbb{R}^3$ and Γ the closed boundary of F, $\underline{\alpha}$ some vector, and $\underline{t}$ the tangential vector on the boundary Γ.
Substitution of (9.3.6) into (9.3.7) yields:

$$\int_F \underline{u}\cdot\underline{n}\, dS = \int_\Gamma \underline{\psi}\cdot\underline{t}\, d\Gamma\ . \tag{9.3.8}$$

(9.3.8) together with the construction of divergence-free basis functions in $\mathbb{R}^2$, suggests that the introduction of the value of the tangential component of the vector potential on an edge as a new degree of freedom, could be used for the derivation of divergence-free basis functions in $\mathbb{R}^3$.
First a set of functions that are divergence-free is introduced. Consider to that end, a face of a tetrahedron as sketched in fig. 9.10. The fluid flow through the face is equal to $\int_F \underline{\tilde{u}}\cdot\underline{n}\, dS$.

Because of the linear non-conforming approximation:

$$\int_F \underline{\tilde{u}}\cdot\underline{n}\, dS = \text{Area}(F)\ \underline{\tilde{u}}\cdot\underline{n}(\underline{x}^{123})\ , \tag{9.3.9}$$

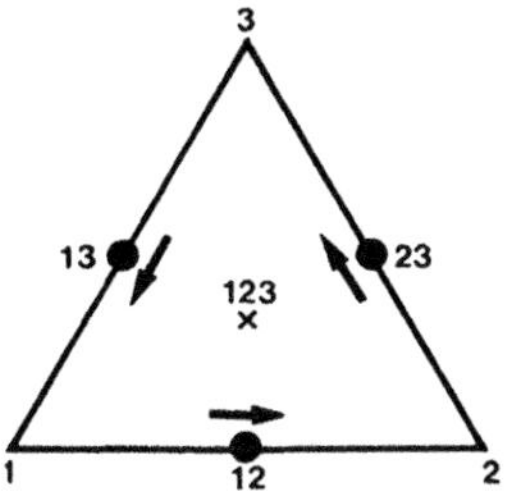

Fig. 9.10 Face F of tetrahedron with velocity nodal point (x) and vector potential nodal points (0). The arrows indicate the positive tangential directions.

and according to (9.3.8) this is equal to:

$$\text{Area}(F)\ \underline{\tilde{u}}\cdot\underline{n}(\underline{x}^{123}) = \int_{\text{side}_{12}} \underline{\psi}\cdot\underline{t}_{12}d\Gamma + \int_{\text{side}_{23}} \underline{\psi}\cdot\underline{t}_{23}d\Gamma + + \int_{\text{side}_{13}} \underline{\psi}\cdot\underline{t}_{13}d\Gamma. \qquad (9.3.10)$$

We now introduce new degrees of freedom α_{ij} by:

$$\alpha_{ij} = \int_{\text{side}_{ij}} \underline{\psi}\cdot\underline{t}_{ij}d\Gamma. \qquad (9.3.11)$$

Notice that $\alpha_{ij} = -\alpha_{ji}$, (9.3.12)

since the tangential directions are taken according to a right threated screw, with the screw in the direction of the normal, see fig. 9.10. Hence according to equation (9.3.8):

$$\int_e \text{div}\ \underline{\tilde{u}}\ d\Omega = \int_{\partial e} \underline{\tilde{u}}\cdot\underline{n}\ dS = \sum_{\text{sides}} (\alpha_{ij}+\alpha_{ji}) = 0. \qquad (9.3.13)$$

So, for a divergence-free set of basis functions, the normal velocity components need to be eliminated in favour of the new degrees of freedom:

$$\underset{\sim}{\tilde{u}} = \sum_{j=1}^{4} (u_{t_1 j} \underline{\phi}_{t_1 j} + u_{t_2 j} \underline{\phi}_{t_2 j}) + \sum_{ij} \alpha_{ij} \underline{\phi}_{\alpha_{ij}} \ . \tag{9.3.14}$$

One easily verifies from (9.3.3), (9.3.10) and (9.3.12) that $\underline{\phi}_{\alpha_{ij}}$ in element e is equal to:

$$\underline{\phi}_{\alpha_{ij}} = \frac{\underline{\phi}_{n_k}}{\text{Area}(F_k)} - \frac{\underline{\phi}_{n_\ell}}{\text{Area}(F_\ell)} , \tag{9.3.15}$$

where F_k and F_ℓ are the faces containing the edge ij. $\underline{n}_k$ and $\underline{n}_\ell$ are the outward directed normals to these faces.

Hence the function $\underline{\phi}_{\alpha_{ij}}$ has the following properties:

i) $\underline{\phi}_{\alpha_{ij}} = \underline{0}$ in the centroid of all faces not containing the edge ij.

ii) $\underline{\phi}_{\alpha_{ij}} = \pm\, \underline{n}/\text{Area}(F)$ in the centroid of F, where F is a face containing the edge ij.

iii) $\underline{\phi}_{\alpha_{ij}}$ is linear.

In fig. 9.11 a representation of $\underline{\phi}_{\alpha_{ij}}$ for a number of tetrahedra is given.
Due to the construction, the functions $\underline{\phi}_{\alpha_{ij}}$ are approximately divergence-free in the sense of (9.3.4), i.e.

$$\int_e \text{div}\, \underline{\phi}_{\alpha_{ij}}\, d\Omega = 0 , \tag{9.3.16}$$

since

$$\int_e \text{div}\, \underline{\phi}_{\alpha_{ij}}\, d\Omega = \int_{\partial e} \underline{\phi}_{\alpha_{ij}} \cdot \underline{n}\, dS =$$

$$\int_{F_k \cup F_\ell} \left(\frac{\underline{\phi}_{n_k}}{\text{Area}(F_k)} - \frac{\underline{\phi}_{n_\ell}}{\text{Area}(F_\ell)}\right) dS = \int_{F_k} \frac{dS}{\text{Area}(F_k)} - \int_{F_\ell} \frac{dS}{\text{Area}(F_\ell)} = 0.$$

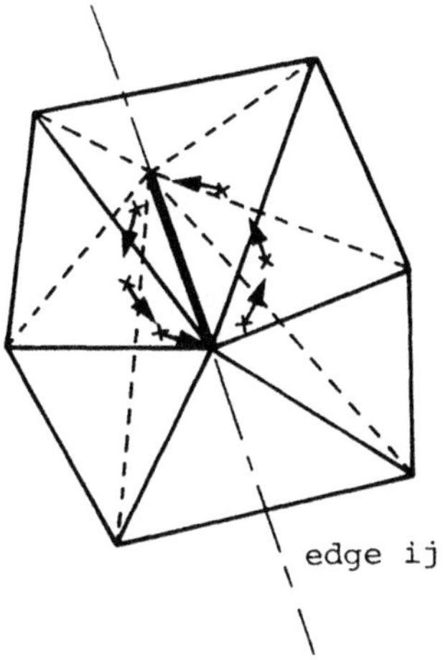

Fig. 9.11 Representation of the function $\underline{\phi}_{\alpha_{ij}}$ for the tetrahedra containing edge ij. The values in the nodal points are indicated by arrows.

However, the functions $\underline{\phi}_{\alpha_{ij}}$ do not form a basis, since they are not linearly independent. The fact that the functions $\underline{\phi}_{\alpha_{ij}}$ are dependent can easily be demonstrated if we consider one tetrehedron (see fig. 9.9). For one tetrahedron there are four faces, twelve velocity unknowns and one pressure degree of freedom. There is one equation for the divergence-freedom, hence after elimination of this equation, eleven degrees of freedom for the velocity remain. Next, from the tangential velocity components eight divergence-free basis functions have been constructed, so that another three independent basis functions are needed. Since the number of edges is six, three of the functions $\underline{\phi}_{\alpha_{ij}}$ must be linearly dependent.

From these observations it is clear that a number of functions $\underline{\phi}_{\alpha_{ij}}$ must be removed in order to get a linear independent set, thus forming a basis. Unfortunately the process of constructing a linear independent set is, from a practical point of view, a complicated one.

9.3.3 The construction of a divergence-free basis

For the construction we consider so-called paths of edges defined in the following way:

> A path of edges consists of one or more edges, each provided with a direction, where the last point of one edge is the initial point of the next edge, and where each edge is used only once. (9.3.17)

Furthermore:

> A closed path is a path of edges with common initial and end point. (9.3.18)

First we shall show, that for the construction of a basis, it is necessary that of each possible path of edges, at least one edge must be used together with its corresponding basis function $\underline{\phi}_{\alpha_{ij}}$.

Let γ be a closed path consisting of edges s_{ij} and corresponding functions $\underline{\phi}_{\alpha_{ij}}$. Each function $\tilde{\underline{v}}$ in the space spanned by the functions $\underline{\phi}_{\alpha_{ij}}$ can be written as:

$$\tilde{\underline{v}} = \sum_{ij} \alpha_{ij}\underline{\phi}_{\alpha_{ij}} . \qquad (9.3.19)$$

Let F be a surface spanned by γ and consisting of faces of the triangulation only. Then the flux of the function $\underline{\phi}_{\alpha_{ij}}$ through this surface is equal to zero when the side s_{ij} does not belong to γ. This is clear when the edge s_{ij} is not contained in F. When the edge s_{ij} is contained in F, but does not belong to γ then:

$$\int_F \underline{\phi}_{\alpha_{ij}} \cdot \underline{n}\, dS = \int_{F_k} \frac{\underline{\phi}_{n_k}}{\text{Area}(F_k)} \cdot \underline{n}_k dS - \int_{F_\ell} \frac{\underline{\phi}_{n_\ell}}{\text{Area}(F_\ell)} \cdot \underline{n}_\ell dS \qquad (9.3.20)$$

$$= 1 - 1 = 0 \ ,$$

with F_k and F_ℓ the faces containing the edge α_{ij}, that do belong to F, see fig. 9.12.

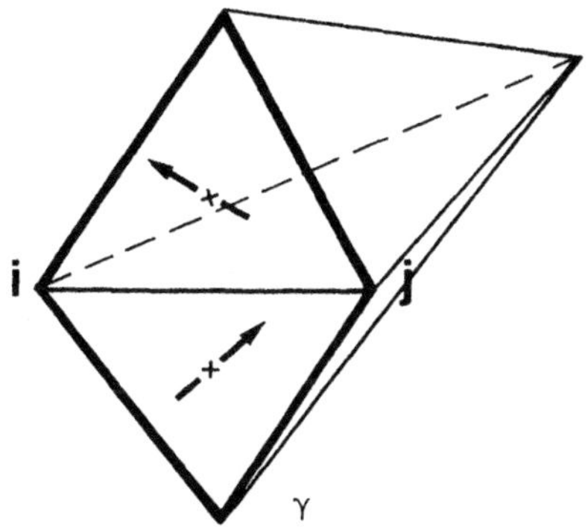

Fig. 9.12 Closed path γ in region F consisting of 2 tetrahedra. Flux of a function $\underline{\phi}_{\alpha_{ij}}$ is equal to:

$$\int_F \underline{\phi}_{\alpha_{ij}} \cdot \underline{n}\, dS.$$

The values of the function $\underline{\phi}_{\alpha_{ij}}$ corresponding to the edge ij have been plotted in the nodal points.

Only when s_{ij} belongs to γ, the flux through F is nonzero and equal to plus or minus one. This statement can be easily checked from equation (9.3.15); for a general proof see Hecht (1981). Hence the flux of a function $\tilde{\underline{v}}$ through F is only dependent on the flux of the functions $\underline{\phi}_{\alpha_{ij}}$ that correspond to γ. When none of the basis functions belong to γ, then the flux through F is always zero, which is in general not true for all functions $\tilde{\underline{v}}$ spanned by functions $\underline{\phi}_{\alpha_{ij}}$.

Next we shall show that it is sufficient to add functions $\underline{\phi}_{\alpha_{ij}}$, corresponding to one of the edges of any closed path, to the basis, in order that these functions $\underline{\phi}_{\alpha_{ij}}$ span the complete space. Hence the basis consists of the minimum number of functions $\underline{\phi}_{\alpha_{ij}}$ that are necessary in order that each closed path contains an edge corresponding to a function $\underline{\phi}_{\alpha_{ij}}$.
Each function $\tilde{\underline{v}}$ in the space spanned by the functions $\underline{\phi}_{\alpha_{ij}}$ is defined by its normal decomposition on all the faces, hence $\tilde{\underline{v}}$ is also defined by the flux through each face F_k (relation 9.3.15). Each face F_k corresponds to a closed path of edges spanned by its three edges. If the function $\underline{\phi}_{\alpha_{ij}}$ corresponding to one of the edges, belongs to the basis, then the flux of $\tilde{\underline{v}}$ through F_k can be expressed with the aid of this basis function only, since:

$$\int_{F_k} \tilde{\underline{v}}\cdot\underline{n}_k \, dS = \sum_{ij} \alpha_{ij} \int_{F_k} \underline{\phi}_{\alpha_{ij}}\cdot\underline{n}_k dS \qquad (9.3.21)$$

$$= \sum_{ij} \alpha_{ij} \int_{F_k} \frac{\phi_{\underline{n}_k}}{\text{Area}(F_k)}\cdot\underline{n}_k dS = \sum_{ij} \alpha_{ij}$$

$$= (\sum_{ij} \alpha_{ij}) \int_{F_k} \underline{\phi}_{\alpha_{pq}}\cdot\underline{n}_k dS,$$

where $\underline{\phi}_{\alpha_{pq}}$ is the function $\underline{\phi}_{\alpha_{ij}}$ from the basis.
Next consider a surface F spanned by a closed path γ, and consisting of faces F_k of the triangulation only. Let at least one function $\underline{\phi}_{\alpha_{ij}}$ correspond to an edge s_{ij} belonging to γ. Then the flux through F is completely defined by the sum of the fluxes through each of the faces F_k.
Since each face F_k contains at least one edge s_{ij} with

corresponding basis function $\Phi_{\alpha_{ij}}$, the fluxes through each face F_k and hence through the surface F are completely determined (see fig. 9.13). This proves that it is indeed sufficient to

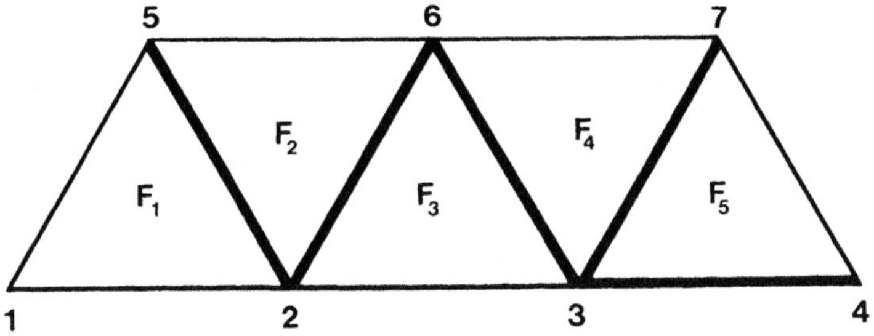

Fig. 9.13 5 faces F_k spanning the surface F.
The flux through F can completely be determined in terms of the basis functions $\Phi_{\alpha_{25}}$, $\Phi_{\alpha_{26}}$, $\Phi_{\alpha_{36}}$, $\Phi_{\alpha_{37}}$ and $\Phi_{\alpha_{34}}$.

take at least one edge of each closed path. For a thorough proof, using graph theory, the reader is referred to Hecht (1981).

Combining the preceding results, suggests the following algorithm for removing functions $\Phi_{\alpha_{ij}}$ in order to construct a basis.

(i) Start: Make H equal to the empty set $\emptyset$; set k = 1.

(ii) When $H \cup \{edge_{ij}\}$ does not contain a closed path then put $edge_{ij}$ in the set H.

(iii) k = k+1; return to step (ii) until all edges have been considered.

Finally the set H contains the maximal number of edges, that does not contain closed paths. All functions $\Phi_{\alpha_{ij}}$ corresponding

to edges in the set H must be removed in order to get a linear independent set of basis functions.
When Dirichlet boundary conditions are given on the face, the same construction must be used, however, the process must start at the faces where Dirichlet boundary conditions are given. See Hecht (1981).
The practical elaboration of this method falls beyond the scope of this book. However, the technique described here illustrates that also in 3D the construction of divergence-free elements is possible.

9.4 Evaluation of the solenoidal method

The main advantage of the solenoidal method, is that less computing time and computer memory is required than for both other methods. On one hand this is caused by the fact that velocity computation and pressure computation have been decoupled, on the other hand because the system of equations has a structure that makes it possible to use iterative solution techniques to solve the linearized equations, see Chapter 6. Another important advantage compared to the penalty function method, is that no large parameter τ is necessary. This parameter can cause inaccuracies for computers with relatively short word length.
A disadvantage of the method is that it can be applied only for Crouzeix-Raviart type elements. Another, more severe disadvantage, is that the method, compared to for example the penalty function method, requires more complicated software especially in 3D problems. This extra software is required since the boundary conditions must be transformed and the solution must be transformed back to the original degrees of freedom. In 3D problems also extra software is needed for the elimination of the redundant parameters.

Chapter 10

The instationary Navier-Stokes equations

Introduction

This Chapter is devoted to the solution of the instationary Navier-Stokes equations and the continuity equation. The general approach will be the application of the Galerkin method in the space variables. In that way the PDEs are reduced to a system of ordinary differential equations. This system can be solved with classical techniques, one of which will be considered. The problems related to the continuity equation are the same as for the stationary Navier-Stokes equations. The solution of these problems will be discussed. Besides the methods introduced in Chapters 8 and 9 a new method, the pressure-correction method, will be treated. This method is only applicable for time-dependent equations. In Section 10.4 some attention is paid to a new upwinding technique for finite element methods: streamline upwinding.

10.1 General approach

In this chapter we are dealing with the instationary Navier-Stokes equations (see 6.1.3):

$$\rho \frac{\partial u_i}{\partial t} - \mu \Delta u_i + \rho\left(u_1 \frac{\partial u_i}{\partial x_1} + u_2 \frac{\partial u_i}{\partial x_2} + u_3 \frac{\partial u_i}{\partial x_3}\right) + \frac{\partial p}{\partial x_i} = \rho f_i \;,$$

$$i = 1,2,3, \qquad (10.1.1)$$

together with the continuity equation:

$$\frac{\partial u_1}{\partial x_1} + \frac{\partial u_2}{\partial x_2} + \frac{\partial u_3}{\partial x_3} = 0 \ . \qquad (10.1.2)$$

The dimensionless form of (10.1.1) is given by (6.1.6):

$$\frac{\partial u_i}{\partial t} - \frac{1}{Re} \Delta u_i + u_1 \frac{\partial u_i}{\partial x_1} + u_2 \frac{\partial u_i}{\partial x_2} + u_3 \frac{\partial u_i}{\partial x_3} + \frac{\partial p}{\partial x_i} = f_i \ ,$$

$$i = 1,2,3 \ . \qquad (10.1.3)$$

For the unique solvability of (10.1.2), (10.1.3) it is necessary to give both initial conditions and boundary conditions (see Section 6.3). The boundary conditions must be of the same type as for the stationary Navier-Stokes equations (Section 6.3), except that they may depend on time. The initial conditions consist of a given velocity at the initial time $t = 0$, i.e.:

$$u_i(\underline{x},t) = \overset{0}{u_i}(\underline{x}) \ , \qquad i = 1,2,3. \qquad (10.1.4)$$

Of course the initial velocity field $\overset{0}{u_i}(\underline{x})$ must satisfy the incompressibility constraint (10.1.2). Throughout this chapter Dirichlet boundary conditions are assumed:

$$u_i(\underline{x},t) = g_i(\underline{x},t) \ , \quad \underline{x} \in \Gamma \ , \ (i=1,2,3). \qquad (10.1.5)$$

One easily verifies that the pressure in that case is fixed up to an additive function of time (not of space), i.e. $p(x,t) = p_c(x,t) + c(t)$.
In order to prescribe the function $c(t)$ we demand:

$$\int_\Omega p \, d\Omega = 0 \ , \qquad \text{for all } t. \qquad (10.1.6)$$

The treatment of other types of boundary conditions is exactly the same as for the stationary case, see Section 7.2.1.
For the numerical solution of (10.1.2), (10.1.3) with boundary conditions (10.1.5) and initial conditions (10.1.4), Galerkins method will be applied (see Section 7.1). Thereto the equations

(10.1.3) are multiplied by arbitrary, time-independent, test functions v_i and the continuity equation (10.1.2) by test functions q. Integration over the domain results in:

$$\int_\Omega \frac{\partial u_i}{\partial t} v_i d\Omega + \int_\Omega \left\{ -\frac{1}{Re} \Delta u_i + u_1 \frac{\partial u_i}{\partial x_1} + u_2 \frac{\partial u_i}{\partial x_2} + u_3 \frac{\partial u_i}{\partial x_3} + \frac{\partial p}{\partial x_i} \right\} v_i d\Omega$$

$$= \int_\Omega f\, v_i \, d\Omega \,, \qquad i = 1,2,3 \qquad (10.1.7)$$

$$\int_\Omega q\left(\frac{\partial u_1}{\partial x_1} + \frac{\partial u_2}{\partial x_2} + \frac{\partial u_3}{\partial x_3}\right) d\Omega = 0 \,. \qquad (10.1.8)$$

Application of the Gauss theorem (compare with Section 7.1) gives:

$$\int_\Omega \frac{\partial u_i}{\partial t} v_i d\Omega + \int_\Omega \left\{ -\frac{1}{Re} (\nabla u_i . \nabla v_i) + \left(u_1 \frac{\partial u_i}{\partial x_1} + u_2 \frac{\partial u_i}{\partial x_2} + \right.\right.$$

$$\left.\left. + u_3 \frac{\partial u_i}{\partial x_3}\right) v_i \right\} d\Omega - \int_\Omega p \frac{\partial v_i}{\partial x_i} d\Omega =$$

$$= \int_\Omega f_i v_i d\Omega + \int_\Gamma (\underline{\nabla u}_i . \underline{n})\, v_i d\Gamma - \int_\Gamma p\, v_i n_i d\Gamma, \; i=1,2,3, \qquad (10.1.9)$$

$$\int_\Omega q \operatorname{div} \underline{u}\, d\Omega = 0 \,. \qquad (10.1.10)$$

The boundary integrals vanish due to the boundary conditions:

$$v_i(\underline{x}) = 0 \,, \quad \underline{x} \in \Gamma. \qquad (10.1.11)$$

For the weak formulation of problem (10.1.2)-(10.1.5) the same function spaces $\underline{V}_0$, $\underline{V}_g$ and Q as in Section 7.1 are introduced: $\underline{V}_0$ is the space of vector functions $\underline{v} = (v_1,v_2,v_3)$ satisfying (10.1.11) and $\underline{V}_g$ the space of vector functions $\underline{v}$ satisfying (10.1.5). Q is the space of functions satisfying (10.1.6). Then the weak formulation of problem (10.1.2)-(10.1.5) reads:

Find $\underline{u}(t) \in \underline{V}_g$ and $p(t) \in Q$ such that

$$\int_\Omega \frac{\partial u_i}{\partial t} v_i \, d\Omega + \int_\Omega \left\{ -\frac{1}{Re} (\underline{\nabla} u_i \cdot \underline{\nabla} v_i) + (u_1 \frac{\partial u_i}{\partial x_1} + u_2 \frac{\partial u_i}{\partial x_2}\right.$$

$$\left. + u_3 \frac{\partial u_i}{\partial x_3}) v_i \right\} d\Omega - \int_\Omega p \frac{\partial v_i}{\partial x_i} \, d\Omega = \int_\Omega f_i v_i d\Omega, \; i=1,2,3, \tag{10.1.12}$$

$$\int_\Omega q \, \mathrm{div} \, \underline{u} \, d\Omega = 0 \; . \tag{10.1.13}$$

for all $\underline{v} \in \underline{V}_0$ and all $q \in Q$.

For the construction of the approximation of the solution, u_1, u_2, u_3 and p are written as linear combinations of time-independent basis functions with time-dependent coefficients:

$$u_i(\underline{x},t) = u_{i0}(\underline{x},t) + \sum_{j=1}^{\infty} u_{ij}(t) \; \phi_j(\underline{x}) \; , \tag{10.1.14}$$

$$p(\underline{x},t) = \sum_{j=1}^{\infty} p_j(t) \; \psi_j(\underline{x}) \; , \tag{10.1.15}$$

$$\text{with } u_{i0}(\underline{x},t) = g_i(\underline{x},t) \; , \quad \underline{x} \in \Gamma \; . \tag{10.1.16}$$

The basis functions ϕ_i and ψ_i are exactly the same as in the stationary case of chapter 7.
The basis functions ϕ_i are substituted subsequently for the test functions v_i in (10.1.12) and the test function q is replaced by ψ_i. If we limit ourselves to approximate solutions, constructed by a finite number of basis functions i.e.

$$\tilde{u}_i(\underline{x},t) = \tilde{u}_{i,0}(\underline{x},t) + \sum_{j=1}^{N} u_{ij}(t) \; \phi_j(\underline{x}) \; , \tag{10.1.17}$$

$$\tilde{p}(\underline{x},t) = \sum_{j=1}^{M} p_j(t) \; \psi_j(\underline{x}) \; , \tag{10.1.18}$$

then the following system of Galerkin equations remains:

$$\int_\Omega \frac{\partial \tilde{u}_1}{\partial t} \phi_i d\Omega + \int_\Omega \left\{ -\frac{1}{Re} (\nabla \tilde{u}_1 \cdot \nabla \phi_i) + (\tilde{u}_1 \frac{\partial \tilde{u}_1}{\partial x_1} + \tilde{u}_2 \frac{\partial \tilde{u}_1}{\partial x_2}\right.$$

$$+ \tilde{u}_3 \frac{\partial \tilde{u}_1}{\partial x_3}) \phi_i \} d\Omega - \int_\Omega \tilde{p} \frac{\partial \phi_i}{\partial x_1} d\Omega = \int_\Omega f_1 \phi_i d\Omega, \quad i=1,2,\ldots,N.$$

$$\int_\Omega \frac{\partial \tilde{u}_2}{\partial t} \phi_i \, d\Omega + \int_\Omega \{- \frac{1}{Re} (\nabla \tilde{u}_2 . \nabla \phi_i) + (\tilde{u}_1 \frac{\partial \tilde{u}_2}{\partial x_1} + \tilde{u}_2 \frac{\partial \tilde{u}_2}{\partial x_2}$$

$$+ \tilde{u}_3 \frac{\partial \tilde{u}_3}{\partial x_3}) \phi_i \} d\Omega - \int_\Omega \tilde{p} \frac{\partial \phi_i}{\partial x_2} d\Omega = \int_\Omega f_2 \phi_i d\Omega \ , \quad i=1,2,\ldots,N.$$

$$\int_\Omega \frac{\partial \tilde{u}_3}{\partial t} \phi_i d\Omega + \int_\Omega \{- \frac{1}{Re} (\nabla \tilde{u}_3 . \nabla \phi_i) + (\tilde{u}_1 \frac{\partial \tilde{u}_3}{\partial x_1} + \tilde{u}_2 \frac{\partial \tilde{u}_3}{\partial x_2}$$

$$+ \tilde{u}_3 \frac{\partial \tilde{u}_3}{\partial x_3}) \phi_i \} d\Omega - \int_\Omega \tilde{p} \frac{\partial \phi_i}{\partial x_3} d\Omega = \int_\Omega f_3 \phi_i d\Omega \ , \quad i=1,2,\ldots,N.$$

$$\int_\Omega \psi_i (\frac{\partial \tilde{u}_1}{\partial x_1} + \frac{\partial \tilde{u}_2}{\partial x_2} + \frac{\partial \tilde{u}_3}{\partial x_3}) d\Omega = 0 \ , \quad i=1,2,\ldots,M. \qquad (10.1.19)$$

In matrix-vector notation:

$$M \dot{\underline{u}} + S \underline{u} + N(\underline{u})\underline{u} + L^T \underline{p} = \underline{F} \ , \qquad (10.1.20)$$

$$L \underline{u} = 0 \ , \qquad (10.1.21)$$

where S, N, L, $\underline{F}$ $\underline{u}$ and $\underline{p}$ are defined as in Section 7.4, and M is the so-called mass matrix:

$$M = \begin{bmatrix} M_1 & 0 & 0 \\ 0 & M_1 & 0 \\ 0 & 0 & M_1 \end{bmatrix} \qquad (10.1.22)$$

$$M_1(i,j) = \int_\Omega \phi_i \phi_j d\Omega \quad \text{and} \quad \dot{\underline{u}} = \frac{\partial \underline{u}}{\partial t} \ .$$

The vectors $\underline{u}$ and $\underline{p}$ depend on time.

(10.1.20), (10.1.21) show that the Galerkin method reduces the instationary Navier-Stokes equations, to a system of ordinary differential equations. Such systems can be solved using standard techniques from the theory of numerical solution of ordinary differential equations. The problems with respect to the continuity equation, mentioned in Chapter 7, still remain for the instationary Navier-Stokes equations, since the continuity equation does not contain the pressure nor time derivatives. Therefore the penalty method or the divergence-free approach will be used to eliminate the pressure from the equations (10.1.20), (10.1.21). In the next section some classical methods for the solution of systems of ordinary differential equations will be considered. The solution of the system of equations (10.1.20), (10.1.21) will be the subject of Section 10.3.

10.2 The numerical solution of systems of ordinary differential equations

10.2.1 Introduction

In this section we shall consider some classical methods for the solution of a system of ordinary differential equations:

$$\dot{\underline{u}} = \underline{F}(\underline{u},t) \,, \tag{10.2.1}$$

with initial condition $\underline{u}(0) = \underline{u}_0$.

There is a large amount of literature for this type of problems. See for example Stoer and Bulirsch (1973), Lambert (1973), Gear (1971).
Methods to solve ordinary differential equations are distinguished in one-step and multi-step methods, and in explicit and implicit methods. A one-step method is characterized by the fact that for the computation of the solution at time-level t^n only information of the preceding time-level t^{n-1} is necessary, whereas multi-step methods

require information at time-levels t^{n-1}, t^{n-2},...,t^{n-k}. Multi-step methods therefore need a starting procedure to find the solution at time-levels $t^1, t^2, \ldots, t^{n-1}$ when the initial condition at t^0 is given. Although multi-step methods may have advantages for complicated right-hand-side functions $\underline{F}(\underline{u},t)$ (Lambert (1973)), the treatment of these methods is beyond the scope of this text. Explicit methods are methods in which the solution at time-level t^n can be computed by simple algebraic manipulations from the solution at preceding time-levels. Implicit methods require the solution of a set of simultaneous equations. Explicit methods are therefore conceptually simpler, but in general smaller time-steps are needed. The choice of which type of method must be used, is determined by accuracy and stability requirements.
In this section we limit ourselves to one class of simple one-step methods, that can be used to solve the system of ordinary differential equations (10.1.20), (10.1.21): The so-called θ-methods. θ-methods are frequently used in the literature for the solution of time-dependent PDEs, which does not imply that they always form the best scheme.

The θ-method applied to (10.2.1) reads:

$$\frac{\underline{u}^{n+1}-\underline{u}^{n}}{\Delta t} = \theta\ \underline{F}(\underline{u}^{n+1},t^{n+1}) + (1-\theta)\underline{F}(\underline{u}^{n},t^{n}), \qquad 0 \leq \theta \leq 1. \tag{10.2.2}$$

t^n denotes the nth time-level, and $\underline{u}^n$ the solution of (10.2.1) at $t = t^n$. Δt denotes the time-step and Δt may be variable.

Special values of θ are:

$$\theta = 0 \text{ (explicit Euler scheme)}: \frac{\underline{u}^{n+1}-\underline{u}^{n}}{\Delta t} = \underline{F}(\underline{u}^{n},t^{n}).$$

$\theta = \frac{1}{2}$ (Crank-Nicolson scheme):

$$\frac{\underline{u}^{n+1}-\underline{u}^{n}}{\Delta t} = \tfrac{1}{2}(\underline{F}(\underline{u}^{n},t^{n}) + \underline{F}(\underline{u}^{n+1},t^{n+1})).$$

$\theta = 1$ (implicit Euler scheme): $\dfrac{\underline{u}^{n+1}-\underline{u}^{n}}{\Delta t} = \underline{F}(\underline{u}^{n+1},t^{n+1}).$

The accuracy of the θ-scheme can easily be found by Taylor-series expansion, see for example Lambert (1973). It is found that the global error of the scheme is $O(\Delta t)$, except for the case that $\theta = \frac{1}{2}$ where the global error is of $O(\Delta t^2)$. In the next section the stability of the θ-method is investigated.

10.2.2 Stability of the θ-method

The general approach to investigate the stability of the system of ordinary differential equations (10.2.1) is to consider the linearized equations:

$$\dot{\underline{u}}(t) = \underline{F}(\underline{u}(t),t) \simeq \underline{F}(\underline{u}(t_0),t_0) + \sum_k (u_k(t) - u_k(t_0))$$

$$\frac{\partial \underline{F}}{\partial u_k}(\underline{u}(t_0),t_0) + (t-t_0)\frac{\partial \underline{F}}{\partial t}(\underline{u}(t_0),t_0), \qquad (10.2.3)$$

where $(\underline{u}(t_0),t_0)$ is a point in the neighbourhood of $(\underline{u}(t),t)$.

(10.2.3) can be written as:

$$\dot{\underline{u}} = A\underline{u} + \underline{g}\ , \qquad (10.2.4)$$

where A is a matrix with elements $A_{ij} = \dfrac{\partial F_i}{\partial u_j}(\underline{u}(t_0),t_0)$ and $\underline{g}$ does not depend on $\underline{u}(t)$. In the sequel we shall only investigate the stability of the linearized equations (10.2.4) with initial condition

$$\underline{u}(t_0) = \underline{u}_0. \qquad (10.2.5)$$

However, it must be remarked that stability of (10.2.4), (10.2.5) is not sufficient for the stability of (10.2.1). For

practical purposes this linear theory usually suffices.
The system of equations (10.2.4) with initial condition (10.2.5) will be called stable if a small perturbation of the initial condition results in a small perturbation of the solution for all t.
Let $\underline{v}$ be the solution of the perturbed system:

$$\dot{\underline{v}} = A\underline{v} + \underline{g} \,, \tag{10.2.6}$$

$$\underline{v}(t_0) = \underline{u}_0 + \underline{\varepsilon}_0 \,.$$

Then the difference $\underline{\varepsilon} = \underline{v}-\underline{u}$ satisfies:

$$\dot{\underline{\varepsilon}} = A\underline{\varepsilon} \,, \tag{10.2.7}$$

$$\underline{\varepsilon}(t_0) = \underline{\varepsilon}_0 .$$

The solution of (10.2.7) is given by:

$$\underline{\varepsilon}(t) = \exp(A(t-t_0))\underline{\varepsilon}_0 \,. \tag{10.2.8}$$

In order that the error $\underline{\varepsilon}(t)$ remains finite for all t, it is necessary that the real part of all eigenvalues λ_A of the matrix A is negative: $\mathrm{Re}(\lambda_A) < 0.$ (10.2.9)

In the remaining of this section we suppose that (10.2.9) is satisfied. When the system of equations (10.2.4) is stable then it is natural to demand that the discretization is also stable.
Application of the θ-method to (10.2.4), (10.2.5) gives:

$$\frac{\underline{u}^{n+1}-\underline{u}^{n}}{\Delta t} = \theta A\underline{u}^{n+1} + (1-\theta)A\underline{u}^{n} + \theta\underline{g}^{n+1} + (1-\theta)\underline{g}^{n}, \tag{10.2.10}$$

or:

$$(I-\theta\Delta tA)\underline{u}^{n+1} = (I+(1-\theta)\Delta tA)\underline{u}^{n}+\Delta t(\theta\underline{g}^{n+1}+(1-\theta)\underline{g}^{n}). \tag{10.2.11}$$

Since for stability only the homogeneous equations are relevant (compare with equation 10.2.7), we consider:

$$\underline{\varepsilon}^{n+1} = (I-\theta\Delta tA)^{-1}(I+(1-\theta)\Delta tA)\underline{\varepsilon}^{n} = G\,\underline{\varepsilon}^{n}. \tag{10.2.12}$$

The matrix G is called amplification matrix. The method is stable if $|\lambda_G| < 1$, where λ_G denotes the eigenvalues of G and unstable if there is at least one λ_G such that $|\lambda_G| > 1$. We shall prove this statement for the case that the matrix A, and hence the matrix G, is not-defect, i.e. the number of eigenvectors of A is equal to the order of the matrix A.

When A is not-defect then the eigenvectors $\underline{w}_i$ $(i=1,2,\ldots,m)$ of A span the complete space $\mathbb{R}^m$, where m is the order of the matrix. Each vector $\underline{\varepsilon} \in \mathbb{R}^m$ can be written as:

$$\underline{\varepsilon} = \sum_{i=1}^{m} \alpha_i \underline{w}_i \; . \tag{10.2.13}$$

Consider the homogeneous equation:

$$\dot{\underline{\varepsilon}} = A\underline{\varepsilon} \; . \tag{10.2.14}$$

Substitution of (10.2.13) in (10.2.14) gives:

$$\dot{\alpha}_i = \alpha_i \lambda_i \; , \tag{10.2.15}$$

where λ_i are the eigenvalues of A corresponding to $\underline{w}_i$.
The general solution of (10.2.15) is:

$$\alpha_i(t) = \alpha_{i0} e^{\lambda_i t} \; , \tag{10.2.16}$$

which shows that $\alpha_i(t)$ is decreasing when $\mathrm{Re}(\lambda_i) < 0$.
One easily verifies that the eigenvectors $\underline{w}_i$ of A are also eigenvectors of the matrix G with corresponding eigenvalues:

$$\mu_i = \frac{1 + (1-\theta)\Delta t \; \lambda_i}{1 - \theta \; \Delta t \; \lambda_i} \; . \tag{10.2.17}$$

The numerical solution of (10.2.14) by the θ-method is given by:

$$\underline{\varepsilon}^{n+1} = G \; \underline{\varepsilon}^{n} = G^{n+1} \underline{\varepsilon}^{0} . \tag{10.2.18}$$

Let $\underline{\varepsilon}^0$ be written as:

$$\underline{\varepsilon}^0 = \sum_{i=1}^{m} \alpha_{i0}\underline{w}_i ,$$

then $\underline{\varepsilon}^{n+1} = \sum_{i=1}^{m} \alpha_{i0}\mu_i^{n+1}\underline{w}_i$.

and the method is stable if $|\mu_i| < 1$. When there is at least one μ_i with $|\mu_i| > 1$ then the solution increases for increasing n and the method is unstable.
It is easily verified that $|\mu_i| < 1$ for $0.5 \leq \theta \leq 1$ (Re $\lambda_i < 0$). For $\theta < 0.5$ the condition $|\mu_i| < 1$ limits the allowed stepsize Δt. For example when $\theta = 0$:

$$|\mu_i| = |1 + \Delta t\ \lambda_i| < 1. \qquad (10.2.19)$$

(10.2.19) implies that Δt must be chosen such that $\Delta t\lambda_i$ is within a circle with radius 1 and midpoint (-1,0) in the complex plane. In fig. 10.1, the stability regions (i.e. the regions in which the method is stable) are sketched in the complex plane for θ = 0, 0.5, and 1.

For the solution of time-dependent PDEs, not only the stability, but also the behaviour of a method for $\mathrm{Re}(\lambda_i\Delta t) \to -\infty$ can be important. This is because frequently the corresponding matrices have eigenvalues with large negative real part. The solution (10.2.16) for $-\mathrm{Re}(\lambda_i)$ large damps very fast since

$$e^{\lambda_i t} \to 0 \quad \text{when} \quad \mathrm{Re}(\lambda_i) \to -\infty , \qquad t > 0. \qquad (10.2.20)$$

So the effect of coefficients α_{i0} in the solution decreases rapidly when $-\mathrm{Re}(\lambda_i)$ is large.
For the implicit θ methods, however, there is an essential difference in the behaviour of the schemes with $\theta \approx \frac{1}{2}$ and the schemes with $\theta \approx 1$. For example when $\theta = \frac{1}{2}$ and λ_i is real negative we have:

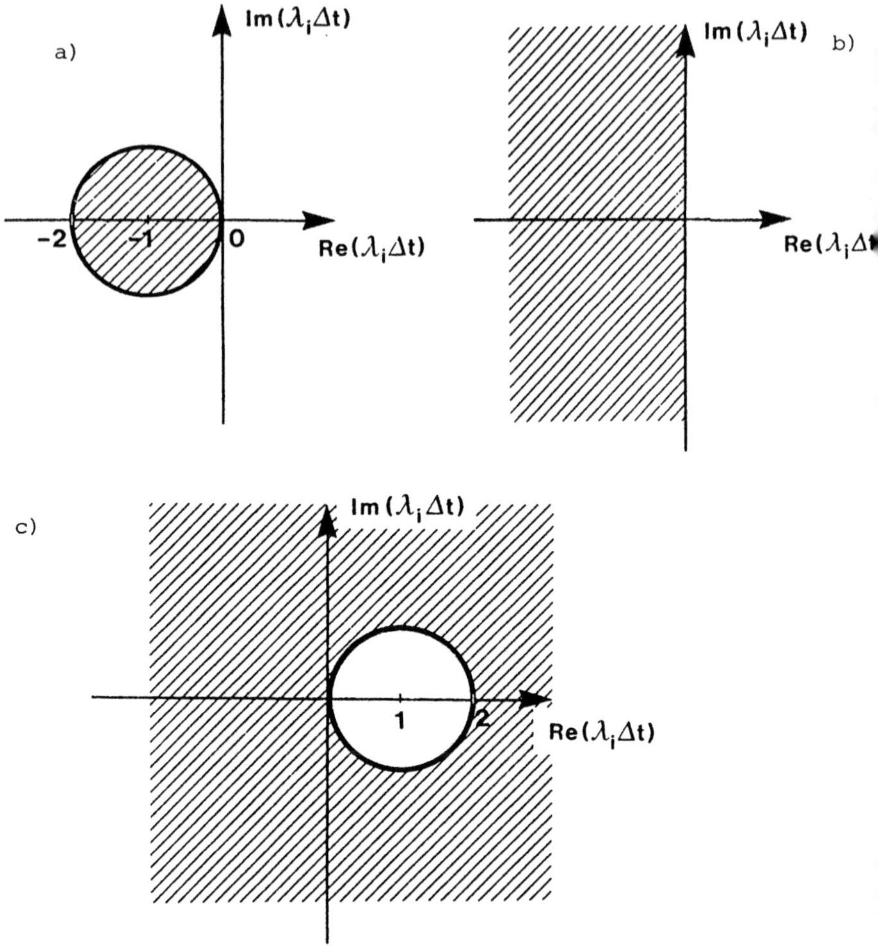

Fig. 10.1 Stability regions for the θ-method,
a) θ = 0 b) θ = 0.5 c) θ = 1.
The method is stable in the shaded region.

$$\mu_i = \frac{1 + \frac{1}{2}\Delta t \; \lambda_i}{1 - \frac{1}{2}\Delta t \; \lambda_i} \text{, and } \mu_i \to -1 \text{ when } \lambda_i \to -\infty. \qquad (10.2.21)$$

When $\theta = 1$: $\mu_i = \dfrac{1}{1 - \Delta t \; \lambda_i}$ so $\mu_i \to 0$ when $\lambda_i \to -\infty$.

In other words for large negative eigenvalues λ_i we may expect an oscillating behaviour of the components α_i corresponding to λ_i when Crank-Nicolson is used, whereas the solution is damped for the Euler implicit scheme. Whether this oscillating effect is important or not depends largely on the size of the coefficients α_{i0}. When the coefficients α_{i0} corresponding to eigenvalues with large negative real part are small compared to the other components, then there is no necessity for damping. However, when these components are important for the initial condition (which is usually the case for non-smooth solutions), then the solution for large t may be unacceptable perturbed. Generally stated: when the solution of a PDE is smooth, then Crank-Nicolson is an accurate method allowing for large time-steps, when the solution is non-smooth (steep gradients), then the Euler implicit scheme should be preferred. In fig. 10.2 the amplification factors of Euler implicit and Crank-Nicolson for real negative values of λ_i are plotted.

10.3 The solution of the systems of ordinary differential equations resulting from the Galerkin method applied to the Navier-Stokes equations

Consider the equations (10.1.20), (10.1.21):

$$M\,\dot{\underline{u}} + S\,\underline{u} + N(\underline{u})\underline{u} + L^T\underline{p} = \underline{F}\,, \qquad (10.3.1)$$

$$L\,\underline{u} = \underline{0}\,, \qquad (10.3.2)$$

with initial condition $\underline{u}(0) = \underline{u}_0$.

The first observation is that the constraint $L\underline{u} = \underline{0}$ must be satisfied at any time. So in general it is not possible to use explicit time integration schemes, unless we are able to satisfy the constraint in some way at the new time level. Application of the θ-method to system (10.3.1), (10.3.2) results in:

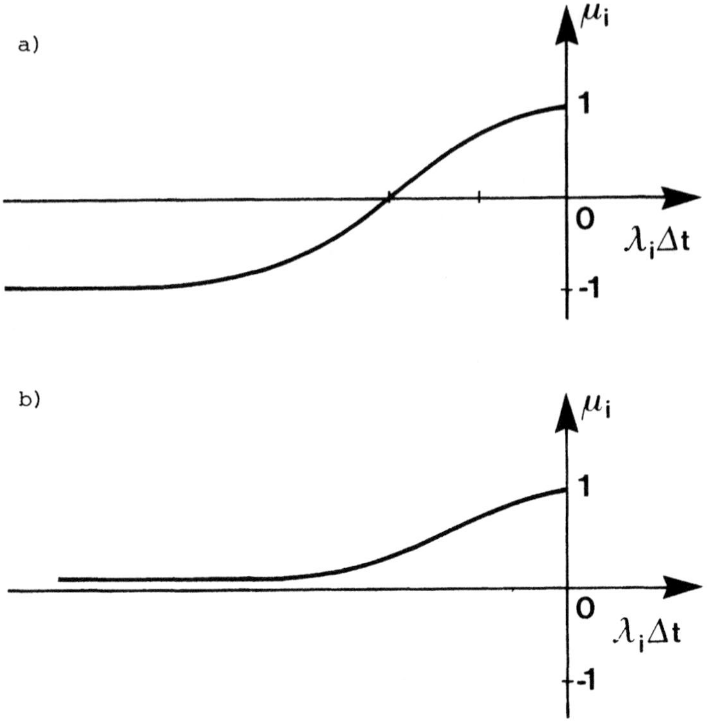

Fig. 10.2 Amplification factors for real negative values of λ_i: a) Crank Nicolson

b) Euler implicit scheme.

$$M\frac{\underline{u}^{n+1}-\underline{u}^{n}}{\Delta t} + S(\theta\underline{u}^{n+1}+(1-\theta)\underline{u}^{n}) + \theta N(\underline{u}^{n+1})\underline{u}^{n+1} + (1-\theta)N(\underline{u}^{n})\underline{u}^{n}$$

$$+ L^{T}(\theta\underline{p}^{n+1}+(1-\theta)\underline{p}^{n}) = \theta\underline{F}^{n+1}+(1-\theta)\underline{F}^{n} \qquad (10.3.3)$$

$$L\,\underline{u}^{n+1} = \underline{0}\ . \qquad (10.3.4)$$

with Δt : the time step,

$\underline{u}^{n+1}$, $\underline{p}^{n+1}$: the velocity and pressure at the new time level $n+1(t+\Delta t)$,

$\underline{u}^n$, $\underline{p}^n$: the velocity and pressure at the preceding time level n(t),

θ : parameter in the θ-method $(0 \leq \theta \leq 1)$.

The system of equations (10.3.3), (10.3.4) can be reformulated as:

$$(M+\theta\Delta t(S+N(\underline{u}^{n+1})))\underline{u}^{n+1} + \theta\Delta tL^T\underline{p}^{n+1} = (M-(1-\theta)\Delta t(S+N(\underline{u}^n)))\underline{u}^n$$

$$- (1-\theta)\Delta tL^T\underline{p}^n + \theta\Delta t\underline{F}^{n+1} + (1-\theta)\Delta t\underline{F}^n \, , \qquad (10.3.5)$$

$$L\,\underline{u}^{n+1} = \underline{0} \, . \qquad (10.3.6)$$

(10.3.5), (10.3.6) have exactly the same structure as the system of equations (7.4.3) for the Navier-Stokes equations. Therefore it is very likely that the same methods as for the stationary Navier-Stokes may be used.
So the non-linearity of equation (10.3.5) may be solved by a linearization technique (Picard, Newton, Quasi-Newton). In contradistinction to the stationary case, usually no iteration is used to solve the non-linear equations (10.3.5). For Δt small enough the difference between two succeeding time-steps is small and the preceding time-step gives a sufficient initial guess for the linearization method.
The following schemes are commonly used to solve (10.3.5):

$$(M+\theta\Delta tS)\underline{u}^{n+1} + \theta\Delta tN(\underline{u}^{n+1})\underline{u}^n + \theta\Delta tL^T\underline{p}^{n+1} = \qquad (10.3.7)$$

$$(M-(1-\theta)\Delta t(S+N(\underline{u}^n)))\underline{u}^n - (1-\theta)\Delta tL^T\underline{p}^n + \theta\Delta t\underline{F}^{n+1} + (1-\theta)\Delta t\underline{F}^n,$$

and

$$(M+\theta\Delta tS)\underline{u}^{n+1} + \theta\Delta tN(\underline{u}^{n+1})\underline{u}^n + \theta\Delta tN(\underline{u}^n)\underline{u}^{n+1} + \theta\Delta tL^T\underline{p}^{n+1} =$$

$$(M-(1-\theta)\Delta t(S+N(\underline{u}^n)))\underline{u}^n - (1-\theta)\Delta tL^T\underline{p}^n + \theta\Delta tN(\underline{u}^n)\underline{u}^n$$

$$+ \theta\Delta t\underline{F}^{n+1} + (1-\theta)\Delta t\underline{F}^n \, . \qquad (10.3.8)$$

For $\theta > \frac{1}{2}$, the θ-method is unconditionally stable for the

linear Stokes equations (see Section 10.2.2). Temam (1979) proves the stability of scheme (10.3.7) for $\theta=1$ and for $\theta=\frac{1}{2}$. One may expect that both the schemes (10.3.7) and (10.3.8) are unconditionally stable for all $\theta > \frac{1}{2}$, however a simple proof is not available.
The Crank-Nicolson scheme ($\theta=\frac{1}{2}$) is the most accurate one for solutions that are smooth in time ($O(\Delta t^2)$). For non-smooth solutions in space, Crank-Nicolson may cause numerical oscillations when large time steps are used (see Section 10.2.2). In that case $\theta > 0.5$ should be used, for example $\theta = 0.75$ or $\theta = 1$.

For some problems no accurate initial condition is available. In that case one is faced with a transient, that is of no interest for the solution. For such problems it is advisible to start with a scheme with strong damping properties (for example the Euler implicit scheme). When the transient has no visible effect on the solution, the Crank-Nicolson scheme could be used for an accurate solution of the problem. Such an approach has been proven to be succesfully for a pulsatile blood flow. See for example v.d. Vosse et al (1985).

Concerning the problems related to the continuity equation, we have seen in Chapters 7 to 9 that one could use the following 3 methods:

(i) the direct solution method,
(ii) the penalty function method,
(iii) the method with divergence-free elements.

Since the direct method requires too much computing time, (see Chapter 7) only the latter two are serious alternatives. In the case of instationary equations there are two additional methods known in the literature:

(iv) an alternative artificial compressibility method,
(v) the so-called pressure-correction method or projection method.

In the next sections we shall treat the methods (ii), (iii), (iv) and (v) and give some estimation of the amount of work related to these methods.

10.3.1 The penalty function method and artificial compressibility methods

The penalty method as treated in Chapter 8 is the most commonly used method to solve the system of equations (10.3.5), (10.3.6). The method can be derived in exactly the same way as is done in Section 8.1. Introduction of a perturbation εp of the continuity equation according to equation 8.1.3, yields in discretized form:

$$\underline{p}^{n+1} = -\tau D^{-1} L \underline{u}^{n+1} , \qquad (10.3.9)$$

with D the pressure mass matrix (8.1.20) and $\tau = 1/\varepsilon$. Substituting (10.3.9) in (10.3.5) at levels n and n+1 and after some rearranging we get:

$$(M + \theta\Delta t(S+N(\underline{u}^{n+1})+\tau L^T D^{-1} L))\underline{u}^{n+1} =$$

$$(M - (1-\theta)\Delta t(S+N(\underline{u}^n) + \tau L^T D^{-1} L))\underline{u}^n + \theta\Delta t\underline{F}^{n+1} + (1-\theta)\Delta t\underline{F}^n. \qquad (10.3.10)$$

Equation (10.3.10) closely resembles the system of equations (8.1.23) and therefore one can expect the same behaviour as for the penalty method in the stationary case.

The penalty function method as treated here is based on the discretization of the perturbed continuity equation:

$$\varepsilon\, p + \text{div}\, \underline{u} = 0 . \qquad (10.3.11)$$

Another perturbation of the continuity equation is known as the artificial compressibility method (Peyret, Taylor (1983), Temam (1979), Cuvelier (1976)):

$$\frac{\partial p}{\partial t} + c^2 \mathrm{div}\, \underline{u} = 0 \ , \qquad (10.3.12)$$

or in discretized form:

$$D\, \dot{\underline{p}} - c^2\, L\, \underline{u} = 0 \ . \qquad (10.3.13)$$

If one replaces (10.3.13) by the discretization:

$$D\, \frac{\underline{p}^{n+1} - \underline{p}^{n}}{\Delta t} - c^2\, L\, \underline{u}^{n+1} = 0 \ , \qquad (10.3.14)$$

then a new method has been derived, that for large values of c^2 very much resembles the penalty function method.

If one considers equation (10.3.10) or its linearized form, it might seem possible to use an explicit time stepping scheme (i.e. $\theta = 0$). Equation (10.3.10) reduces in that case to:

$$M\, \underline{u}^{n+1} = (M - \Delta t(S + N(\underline{u}^n) + \tau L^T D^{-1} L))\underline{u}^n + \Delta t\, \underline{F}^n \ . \qquad (10.3.15)$$

For stability it is necessary that the modulus of the eigenvalues of the matrix:

$$A = I - \Delta t\, M^{-1}(S + N(\underline{u}^n) + \tau L^T D^{-1} L) \ ,$$

are smaller than one.
It is very likely that for large values of τ, the eigenvalues of A are proportional to $\Delta t \tau$, and therefore it is necessary that $\Delta t = O(\varepsilon)$, which is a very restrictive demand for the time-step.
From these observations we conclude that equation (10.3.10) must be solved with an unconditionally scheme, i.e. $\theta \geq \frac{1}{2}$. Numerical experiments confirm this observation.
From Section 10.2 it is clear that a Crank-Nicolson scheme gives an oscillatory behaviour for non-smooth solutions. Practical computations (see v.d. Vosse et al 1985) have shown that even for smooth velocity fields, the computed pressures were always oscillating in time. Therefore in practice a penalty method must always be combined with a method with good damping properties for large eigenvalues. The Euler implicit

scheme ($\theta = 1$) is the simplest scheme with such behaviour.
Equation (10.3.10) reduces for $\theta = 1$ to:

$$(M+\Delta t(S+N(\underline{u}^{n+1})+\tau L^T D^{-1} L))\underline{u}^{n+1} = M\underline{u}^n + \Delta t\underline{F}^{n+1}. \qquad (10.3.16)$$

10.3.2 Divergence-free elements

The construction of divergence-free basis functions has been the subject of Chapter 9. In that chapter it has been shown that the introduction of divergence-free basis functions can be considered as the construction of new unknowns $\underline{u}_d$ according to the linear transformation:

$$\underline{u}^{n+1} = R_d\underline{u}_d \ , \quad \text{such that} \quad LR_d = 0. \qquad (10.3.17)$$

Application of the transformation (10.3.17) to (10.3.5), (10.3.6) and premultiplication by R_d^T yields:

$$R_d^T(M+\theta\Delta t(S+N(R_d\underline{u}_d^{n+1})))R_d\underline{u}_d^{n+1} = R_d^T(M-(1-\theta)\Delta t(S+N(\underline{u}^n)))R_d\underline{u}_d^n +$$

$$+ \theta\Delta t R_d^T\underline{F}^{n+1} + (1-\theta)\Delta t R_d^T\underline{F}^n \ . \qquad (10.3.18)$$

(10.3.18) can be solved both by explicit and implicit methods. However, even for the Euler explicit method ($\theta = 0$), in each step a (constant) system of equations must be solved, since (10.3.18) reduces to:

$$R_d^T M R_d\underline{u}_d^{n+1} = R_d^T(M-\Delta t(S+N(\underline{u}^n)))R_d\underline{u}_d^n + \Delta t R_d^T\underline{F}^n \ , \qquad (10.3.19)$$

for $\theta = 0$. In general the matrix $R_d^T M R_d$ is not a diagonal matrix, so in each time-step the system of equations (10.3.19) must be solved. Furthermore, generally explicit methods require small time steps, especially when small space steps are used. Therefore for practical applications unconditionally stable methods ($\theta \geq \frac{1}{2}$) are favourable.
Since the large matrix does not contain a large parameter as in

the case of the penalty function method, one may expect that the system of equations (10.3.19) can be solved both by direct and by iterative solvers. Concluding one may state that the method of divergence-free elements, seems to be a very promising one for time-dependent problems.

10.3.3 The pressure-correction method

The pressure-correction method, also called projection method (Peyret and Taylor (1983)) has been introduced by Chorin (1968) and Temam (1969) for finite-difference formulations. Application of this method to finite-element formulations has been reported by Donea et al (1982). Following van Kan (1985), it can be considered as a solution method for the system of equation (10.3.5), (10.3.6):

$$(M+\theta\Delta t(S+N(\underline{u}^{n+1})))\underline{u}^{n+1} + \theta\Delta t L^T\underline{p}^{n+1} =$$

$$(M-(1-\theta)\Delta t(S+N(\underline{u}^n)))\underline{u}^n - (1-\theta)\Delta t L^T\underline{p}^n + \theta\Delta t\underline{F}^{n+1} + (1-\theta)\Delta t\underline{F}^n \tag{10.3.20}$$

$$L\,\underline{u}^{n+1} = \underline{0}\ . \tag{10.3.21}$$

It is convenient to introduce a new variable $\underline{q}$ by

$$M\,L^T\underline{q} = L^T\underline{p}\ . \tag{10.3.22}$$

Then (10.3.20) is replaced by:

$$(M+\theta\Delta t(S+N(\underline{u}^{n+1})))\underline{u}^{n+1} + \theta\Delta t ML^T\underline{q}^{n+1} = (M-(1-\theta)\Delta t(S+N(\underline{u}^n)))\underline{u}^n$$

$$- (1-\theta)\Delta t ML^T\underline{q}^n + \theta\Delta t\underline{F}^{n+1} + (1-\theta)\Delta t\underline{F}^n\ . \tag{10.3.23}$$

In order to solve (10.3.22), (10.3.23) a new unknown $\underline{u}^*$ is introduced such that:

$$(M+\theta\Delta t(S+N(\underline{u}^*)))\underline{u}^* = (M-(1-\theta)\Delta t(S+N(\underline{u}^n)))\underline{u}^n - ML^T\underline{q}^n$$

$$+ \ \theta\Delta t\underline{F}^{n+1} + (1-\theta)\Delta t\underline{F}^{n} \ , \tag{10.3.24}$$

and $\underline{u}^{n+1}$ is computed from

$$M(\underline{u}^{n+1}-\underline{u}^{*}) = - \ \theta\Delta t M L^{T}(\underline{q}^{n+1}-\underline{q}^{n}) \ . \tag{10.3.25}$$

Comparing (10.3.24) and (10.3.25) with (10.3.23) shows that the term

$$\theta\Delta t(S+N(\underline{u}^{n+1}))\underline{u}^{n+1} - \theta\Delta t(S+N(\underline{u}^{*}))\underline{u}^{*} \ , \tag{10.3.26}$$

has been neglected. van Kan (1985) shows that the error due to this neglection is equal to $O(\Delta t^2)$ for Crank-Nicolson ($\theta = \frac{1}{2}$) and equal to $O(\Delta t)$ for the other θ-methods, which is in accordance with these methods.

(10.3.21) together with (10.3.25) gives:

$$- \ L \ \underline{u}^{*} = - \ \theta \ \Delta t \ LL^{T}(\underline{q}^{n+1}-\underline{q}^{n}) \ , \tag{10.3.27}$$

and therefore:

$$\underline{q}^{n+1} = \underline{q}^{n} + \frac{1}{\theta\Delta t} (LL^{T})^{-1} L \ \underline{u}^{*} \ . \tag{10.3.28}$$

So the pressure-correction method can be formulated as:

$$(M+\theta\Delta t(S+N(\underline{u}^{*})))\underline{u}^{*} = (M-(1-\theta)\Delta t(S+N(\underline{u}^{n})))\underline{u}^{n} - \Delta t M L^{T}\underline{q}^{n}$$

$$+ \ \theta\Delta t\underline{F}^{n+1} + (1-\theta)\Delta t\underline{F}^{n} \ , \tag{10.3.29}$$

$$\underline{q}^{n+1} = \underline{q}^{n} + \frac{1}{\theta\Delta t}(LL^{T})^{-1}L\underline{u}^{*} \ , \tag{10.3.30}$$

$$M\underline{u}^{n+1} = M\underline{u}^{*} - \theta\Delta t M L^{T}(\underline{q}^{n+1}-\underline{q}^{n}) \ . \tag{10.3.31}$$

The steps (10.3.30) and (10.3.31) are such that the velocities are projected on the space of divergence-free vector functions (see van Kan (1985)).

The pressure $\underline{p}$ can be computed from equation (10.3.22) by premultiplication by L: $LL^T\underline{p} = LML^T\underline{q}$. (10.3.32)

So when one is interested in the pressure at a certain time, the same type of equation as for $\underline{q}$ must be solved (compare with 10.3.30).

For the computation of the initial value $\underline{q}^0$ we start with equations (10.3.1) and (10.3.2) and use relation (10.3.22):

$$M\dot{\underline{u}} + S\underline{u} + N(\underline{u})\underline{u} + ML^T\underline{q} = \underline{F} , \tag{10.3.33}$$

$$L\underline{u} = \underline{0} . \tag{10.3.34}$$

From (10.3.34) it follows that $L\dot{\underline{u}} = \underline{0}$.

Premultiplication of (10.3.33) by LM^{-1} results in:

$$L\dot{\underline{u}} + LM^{-1}(S\underline{u}+N(\underline{u})\underline{u}) + LL^T\underline{q} = LM^{-1}\underline{F} , \tag{10.3.35}$$

and therefore $\underline{q}^0$ can be computed from:

$$LL^T\underline{q}^0 = LM^{-1}(\underline{F}-S\underline{u}^0-N(\underline{u}^0)\underline{u}^0). \tag{10.3.36}$$

It is finally remarked that for the quantity $\underline{u}^*$ the same boundary conditions as for $\underline{u}^{n+1}$ can be used. Since for smooth pressures $\underline{q}^{n+1}-\underline{q}^n = O(\Delta t)$, it can be seen from (10.3.31) that the error in the boundary conditions is of $O(\Delta t^2)$.

For the pressure correction method (10.3.29)...(10.3.31) we have to solve three systems of equations for each time step, two of which are linear ((10.3.30) and (10.3.31)). When the mass matrix M is lumped, i.e. transferred into a diagonal matrix by numerical integration, the number of systems of equations to be solved reduces to two. Of course one can solve these equations by a direct solver. However, in that case the methods treated in 10.3.1 and 10.3.2 are in general more economic. For a fast solution of (10.3.29)...(10.3.31) we expect that an iterative solver will save a lot of computing time and computer memory compared to a direct solution method. A severe practical problem is that the matrix LL^T has a structure that differs from the other finite element matrices. For a fast implementation of the pressure-correction method a complicated intervention in existing finite element packages is necessary. However, for finite difference methods, the pressure-correction method has shown to be one of the few successful techniques to solve the instationary Navier-Stokes equations and therefore also for finite element methods one may expect good results.

10.4 Streamline upwinding

In Chapter 2 it has been shown that incorrect boundary conditions or too coarse grid spacings may cause oscillations in the computed velocities. Such oscillations may also be the result of an obstacle in the fluid. In the finite difference literature this behaviour is often suppressed by upwinding techniques. Extensions of upwinding techniques to finite element methods have been introduced by Christie et al (1976) and Heinrich et al (1977) and Hughes et al (1979). These methods are based on the so-called Petrov-Galerkin method, in which basis functions and test functions have a different shape. Since these methods are subject to the same disadvantages as the upwind finite differences, i.e. introduction of artificial viscosity and inaccurate solutions, we shall not pay attention to it. Recently several authors have proposed finite element upwind techniques where the upwinding takes place in the direction of the streamlines. The advantage of such a method is that no cross-wind artificial viscosity is introduced. Gresho et al (1984) for example, introduce an explicit artificial viscosity of the form $\mu_{ij} = c\ \Delta t\ u_i u_j$, where μ_{ij} is the viscosity in ij direction, Δt denotes the time-step, u_i is a component of the velocity and c a constant depending on the type of time discretization. A more natural procedure is proposed by Douglas et al (1982) and Pironneau (1982) which combine the method of characteristics with the FEM. In this section we shall treat this "characteristic" method following Pironneau's article.

First we consider the transport equation:

$$\frac{\partial \rho}{\partial t} + \underline{u}.\nabla\rho = f, \quad \text{in} \quad \Omega \times (0,T), \tag{10.4.1}$$

$$\rho(t=0) = \rho^0 \qquad \text{in} \quad \Omega \ .$$

Equation (10.4.1) is of pure hyperbolic character and can be written as: $\frac{D\rho}{Dt} = f$ (10.4.2)

where $\frac{D\rho}{Dt}$ denotes the total (or material) derivative of ρ in the flow ($D/Dt = \frac{\partial}{\partial t} + \underline{u}.\nabla$).

The solution of (10.4.2) is:

$$\rho(x,t) = \rho^0(X,0) + \int_0^t f(X,\tau)\, d\tau \ , \tag{10.4.3}$$

where X(x,t) is the solution of the ordinary differential equation $\frac{dX}{d\tau} = u(X,\tau)$. (10.4.4)

$X(x,\tau)$ is the position of a fluid particle at time τ which will be in x at time t. Hence $X(x,t) = x$. In fact (10.4.4) defines the characteristics of equation (10.4.1) and (10.4.3) gives the solution of (10.4.1) along a characteristic. The methods of characteristics is based on the relations (10.4.3), (10.4.4). In this particular example it is possible to integrate (10.4.4) accurately by some numerical scheme and then to solve (10.4.3).

In practical applications, however, such an approach is not possible since u is unknown and one proceeds as follows. Suppose we have computed ρ^m in a finite element mesh, where m denotes the number of time-steps. In order to compute ρ^{m+1} it is necessary to derive the characteristics from (10.4.4). To that end u is approximated by a piecewise constant or piecewise linear function $\tilde{u}$ of time and (10.4.4) is integrated for each nodal point of the mesh. At $t = t^{m+1}$ we then have:

$$\frac{dX}{d\tau} = \tilde{u}(x,\tau) \ , \text{ with } X(x,t^{m+1}) = x \tag{10.4.5}$$

When $\tilde{u}$ is piecewise constant in time, the solution of (10.4.5) is: $X(x,t^m) = x - \Delta t\tilde{u}(x,t^m)$. (10.4.6)
If f is approximated by $\tilde{f}$ in the same way as u by $\tilde{u}$ then the solution of (10.4.2) is:

$$\rho^{m+1}(x) = \rho^m(X(x,t^m)) + \Delta t f(x,t^{m+1}) \ , \tag{10.4.7}$$

where $\rho^m(X(x,t^m))$ is the value of ρ at time level t^m and position $X(x,t^m)$.
In general $X(x,t^m)$ is not a nodal point, even if x is a nodal point. Therefore the value of $\rho^m(X(x,t^m))$ must be found by interpolation in the finite element mesh. The interpolation procedure as well as the approximation by piecewise functions in time may be the source of artificial diffusion. However, by using higher order interpolations and approximations one may expect a small amount of artificial diffusion. Moreover the approach is such that all information is transported approximately along characteristics, which is the natural thing to do in hyperbolic equations.

The procedure followed for the pure convective equation (10.4.1) can be extended to the Navier-Stokes equations. Although these equations are parabolic rather than hyperbolic, for large Reynolds numbers (dominant first derivatives) they act as if they were hyperbolic, except in the neighbourhood of fixed walls.
Consider the time-dependent Navier-Stokes equations (10.1.2), (10.1.3):

$$\frac{\partial \underline{u}}{\partial t} + \underline{u}.\nabla\underline{u} - \frac{1}{Re}\Delta\underline{u} + \nabla p = \underline{f} \ , \tag{10.4.8}$$

$$\text{div } \underline{u} = 0 \ . \tag{10.4.9}$$

(10.4.8), (10.4.9) can be written as:

$$\frac{D\underline{u}}{Dt} - \frac{1}{Re} \Delta\underline{u} + \nabla p = \underline{f} \ , \tag{10.4.10}$$

$$\text{div } \underline{u} = 0 \ ,$$

where $\frac{D\underline{u}}{Dt}$ is the total derivative of $\underline{u}$.

(10.4.10), (10.4.9) can be solved by some implicit scheme, for example Crank-Nicolson. For $t = t^{m+1}$ we then find:

$$\frac{\underline{u}^{m+1} - \underline{u}^{m}(X(x,t^{m}))}{\Delta t} - \frac{1}{2Re} \Delta\underline{u}^{m+1} + \nabla p^{m+1} = \tfrac{1}{2}(\underline{f}^{m} + \underline{f}^{m+1}) + \frac{1}{2Re} \Delta\underline{u}^{m},$$

$$\text{div } \underline{u}^{m+1} = 0 \ , \tag{10.4.11}$$

where $X(x,t^{m})$ is the solution of

$$\frac{dX}{d\tau} = \underline{u}(X,\tau) \ , \text{ with } \ X(x,t^{m+1}) = x \ . \tag{10.4.12}$$

Equation (10.4.12) can be solved by assuming an implicit Euler scheme:

$$X(x,t^{m}) = x - \Delta t \ \underline{u}(x,t^{m}) \ . \tag{10.4.13}$$

Pironneau shows that system (10.4.11), (10.4.13) is unconditionally stable. A disadvantage is that (10.4.13) is only first order accurate, whereas Crank-Nicolson is of second order. A more accurate approximation of (10.4.12) is:

$$X(x,t^{m}) = x - \frac{\Delta t}{2} (\underline{u}(x,t^{m+1}) + \underline{u}(x - \Delta t\underline{u}(x,t^{m+1}))). \tag{10.4.14}$$

(10.4.14) requires the knowledge of the unknown velocity $\underline{u}$ at the new time level. In the finite difference literature, see for example Fox (1962), Colella (1985), this problem is solved by a predictor-corrector method:

i) estimate $\underline{u}(x,t^{m+1})$ by solving (10.4.11) and (10.4.13)

ii) use the result of i) to solve (10.4.11) and (10.4.14) where the velocity computed in i) is used to calculate the right-hand side of (10.4.14).

An important advantage of this "characteristic" approach is that in each time-step we have only to solve the same type of linear Stokes equations. Furthermore the method is unconditionally stable, so there are no restrictions on the time-step. The experience with this approach is rather limited.

PART III

THEORETICAL ASPECTS OF THE FINITE ELEMENT METHOD

Introduction

In this section we shall be concerned with the more mathematical aspects of linear elliptic PDEs of the second order and their finite element approximation. In Chapter 11 the continuous problem will be studied which covers, for instance, the Dirichlet and Neumann problem for a second order PDE with variable coefficients and the Stokes equations. Chapter 12 is dedicated to the finite element approximation of the continuous problem. An error analysis of the FEM, that is to say an estimate of the difference between the solution of the continuous problem and the solution of the approximate problem, will be the subject of Chapter 13. A special FEM, called the mixed FEM, which is very useful in the study of the (Navier-) Stokes equations will be treated in Chapter 14.
The major results are stated, but proofs are not included, unless they are basic for the understanding of the FEM. For the proofs we refer to the cited literature.

Chapter 11

Second Order Elliptic PDEs

Introduction

In this Chapter we will study linear second order PDEs of elliptic type with various kinds of boundary conditions. First we shall give a weak (or variational or generalized) formulation of the Dirichlet, Neumann and mixed Dirichlet-Neumann problem for the Laplace operator. These formulations will be generalized in order to give a "general" (weak) formulation of elliptic PDEs. We shall prove that within certain conditions the general weak formulation has precisely one solution. The proof of this result will be given, since it explains the idea of Galerkin's method which has already been used in Chapter 3 for the derivation of the Galerkin equations. Examples of weak formulations will be given for a second order elliptic operator with variable coefficients and non-homogeneous Dirichlet, Neumann, mixed Dirichlet-Neumann and Robbins boundary conditions. The Stokes equations will be studied and results for existence and uniqueness of a solution of the Navier-Stokes equations will be stated. Finally we shall make some remarks concerning the regularity (i.e. smoothness properties) of the solution of the weak formulation.
For a thorough treatment of the theory of weak formulations of (non-) linear PDEs we refer to Brézis (1983), Temam (1977), Ladyzhenskaya (1969), Trèves (1975), Lions (1962, 1969, 1973), Necas (1967), Adams (1975), Agmon (1965).

11.1 Dirichlet problem for the Laplace operator

Let us consider the following problem. We look for a function u, defined in a bounded region $\bar{\Omega} \subset \mathbb{R}^n$ with smooth boundary Γ, satisfying the following Dirichlet problem for the Laplace operator

$$-\Delta u + \lambda u = f \quad \text{in } \Omega \tag{11.1.1}$$

$$u = 0 \quad \text{on } \Gamma \tag{11.1.2}$$

where f is a given funtion and λ a non-negative constant. In order to derive a variational formulation for this problem, we multiply (11.1.1) by a test function v which vanishes on Γ and integrate the result over Ω

$$\int_\Omega -\Delta u \; v \; d\Omega + \lambda \int_\Omega u \; v \; d\Omega = \int_\Omega f \; v \; d\Omega \tag{11.1.3}$$

Introducing the space

$$L_2(\Omega) = \{v : \Omega \to \mathbb{R} \mid \int_\Omega |v(\underline{x})|^2 d\Omega < \infty\}$$

which is a Hilbert space for the following scalar product and norm:

$$(u,v)_{L_2} = \int_\Omega u(\underline{x}) \; v(\underline{x}) \; d\Omega \; , \qquad ||v||_{L_2} = (v,v)_{L_2}^{\frac{1}{2}}$$

we can write (11.1.3) as

$$-(\Delta u,v)_{L_2} + \lambda(u,v)_{L_2} = (f,v)_{L_2} \tag{11.1.4}$$

An application of Greens' formula to the left hand side of (11.1.4) leads to

$$\sum_{i=1}^{n} \left(\frac{\partial u}{\partial x_i},\frac{\partial v}{\partial x_i}\right)_{L_2} - \int_\Gamma \frac{\partial u}{\partial n} \; v \; d\Gamma + \lambda(u,v)_{L_2} = (f,v)_{L_2} \tag{11.1.5}$$

Because v is chosen such that v = 0 on Γ, it follows from (11.1.5) that

$$a(u,v) = (f,v)_{L_2} \tag{11.1.6}$$

with

$$a(u,v) = \sum_{i=1}^{n} \left(\frac{\partial u}{\partial x_i},\frac{\partial v}{\partial x_i}\right)_{L_2} + \lambda(u,v)_{L_2} \tag{11.1.7}$$

For the boundedness of the integrals $a(u,v)$ and $(f,v)_{L_2}$ it is sufficient to take $u,v \in H^1(\Omega)$ and $f \in L_2(\Omega)$, where $H^1(\Omega)$ is the Sobolev space of order 1 defined by

$$H^1(\Omega) = \left\{v : \Omega \to \mathbb{R} \mid v \in L_2(\Omega),\ \frac{\partial v}{\partial x_i} \in L_2(\Omega),\ i=1,2,\ldots,n\right\}$$

in which the following Hilbert-structure is given

$$(u,v)_{H^1} = (u,v)_{L_2} + \sum_{i=1}^{n} \left(\frac{\partial u}{\partial x_i},\frac{\partial v}{\partial x_i}\right)_{L_2} \qquad ||v||_{H^1} = (v,v)^{\frac{1}{2}}_{H^1}$$

Moreover u is zero on Γ (i.e. condition (11.1.2)) and so does v. This implies that we have to take $u,v \in H_0^1(\Omega)$ which is defined as

$$H_0^1(\Omega) = \left\{v \in H^1(\Omega) \mid v = 0 \text{ on } \Gamma\right\}$$

endowed with the Hilbert structure

$$(u,v)_{H_0^1} = \sum_{i=1}^{n} \left(\frac{\partial u}{\partial x_i},\frac{\partial v}{\partial x_i}\right)_{L_2} \qquad ||v||_{H_0^1} = (v,v)^{\frac{1}{2}}_{H_0^1}$$

The Hilbert structure in $H_0^1(\Omega)$ is equivalent to the one induced by $H^1(\Omega)$, by virtue of the Poincaré inequality:

$$||v||_{L_2} \leq c\,||v||_{H_0^1} \qquad \text{for all} \quad v \in H_0^1(\Omega)$$

The variational (or weak) formulation of problem (11.1.1), (11.1.2) is the following

$$\left|\begin{array}{l} \text{Find } u \in H_0^1(\Omega) \text{ such that} \\[1ex] a(u,v) = (f,v)_{L_2} \qquad \text{for all} \quad v \in H_0^1(\Omega) \end{array}\right. \tag{11.1.8}$$

It is clear that if u is a classical solution of (11.1.1), (11.1.2), i.e. $u \in C^2(\Omega) \cap C(\bar{\Omega})$, it is also a solution of (11.1.8). On the other hand we have that if u is a solution of (11.1.8) which belongs to the space $C^2(\Omega) \cap C(\bar{\Omega})$, it is a solution of (11.1.1), (11.1.2). Indeed, the boundary condition (11.1.2) is fulfilled, since for $u \in C(\bar{\Omega}) \cap H^1(\Omega)$ the following two properties are equivalent (i) $u = 0$ on $\partial\Omega$, (ii) $u \in H_0^1(\Omega)$. When we write (11.1.8) with $v = \phi \in C^1(\Omega) \cap H_0^1(\Omega)$, we obtain

$$0 = \sum_{i=1}^{n} \left(\frac{\partial u}{\partial x_i}, \frac{\partial \phi}{\partial x_i}\right)_{L_2} + \lambda(u,\phi) - (f,\phi) =$$

$$= (-\Delta u + \lambda u - f, \phi)_{L_2} = \int_{\Omega} (-\Delta u + \lambda u - f)\phi \, d\Omega \qquad (11.1.9)$$

for all $\phi \in C^1(\Omega) \cap H_0^1(\Omega)$, which implies, by virtue of DU BOIS REYMOND's lemma, that

$$-\Delta u + \lambda u = f \quad \text{in} \quad \Omega$$

11.2 Neumann problem for the Laplace operator

We are looking for a function u, defined on $\bar{\Omega} \subset \mathbb{R}^n$, which satisfies

$$-\Delta u + \lambda u = f \quad \text{in} \quad \Omega, \quad \lambda > 0 \qquad (11.2.1)$$

$$\frac{\partial u}{\partial n} = 0 \quad \text{on} \quad \Gamma \qquad (11.2.2)$$

We take the $L_2(\Omega)$ scalar product of (11.2.1) and a test function v which is arbitrary on the boundary Γ. We apply Green's formula and we substitute the boundary condition (11.2.2) to obtain

$$a(u,v) = (f,v)_{L_2} \qquad (11.2.3)$$

with

$$a(u,v) = \sum_{i=1}^{n} \left(\frac{\partial u}{\partial x_i}, \frac{\partial v}{\partial x_i}\right)_{L_2} + \lambda(u,v)_{L_2}$$

In fact we have

$$(-\Delta u,v)_{L_2} + \lambda(u,v)_{L_2} = -\int_\Gamma \frac{\partial u}{\partial n} v \, d\Gamma + \sum_{i=1}^{n} \left(\frac{\partial u}{\partial x_i},\frac{\partial v}{\partial x_i}\right)_{L_2} + \lambda(u,v)_{L_2} = a(u,v)$$

For the boundedness of the terms $a(u,v)$ and $(f,v)_{L_2}$ it is sufficient to choose $u,v \in H^1(\Omega)$ and $f \in L_2(\Omega)$. The variational (or weak) formulation of problem (11.2.1), (11.2.2) is now the following

$$\left|\begin{array}{ll} \text{Find } u \in H^1(\Omega) \text{ such that} & \\ a(u,v) = (f,v)_{L_2} & \text{for all } v \in H^1(\Omega) \end{array}\right. \tag{11.2.4}$$

With the assumption that $u \in C^2(\Omega) \cap C^1(\bar{\Omega})$, (11.2.4) is equivalent to (11.2.1), (11.2.2). If $u \in C^2(\Omega) \cap C^1(\bar{\Omega})$ is a solution of (11.2.1), (11.2.2) it is trivially a solution of (11.2.4). If $u \in C^2(\Omega) \cap C^1(\bar{\Omega})$ is a solution of (11.2.4), we write (11.2.4) with $v = \phi \in C^1(\bar{\Omega}) \cap H_0^1(\Omega)$ and we find (11.2.1) in the same way as in section 11.1. With the smoothness assumption for u, an application of Green's formula shows that, for all $v \in H^1(\Omega)$:

$$a(u,v) = (-\Delta u+\lambda u,v) + \int_\Gamma \frac{\partial u}{\partial n} v \, d\Gamma = (f,v)_{L_2}$$

which implies

$$\int_\Gamma \frac{\partial u}{\partial n} v \, d\Gamma = 0 \qquad \text{for all } v \in H^1(\Omega)$$

and from this relation we can deduce that

$$\frac{\partial u}{\partial n} = 0 \quad \text{on } \Gamma$$

For our aim it is sufficient to remark that the variational formulation is a generalization of the original (classical) problem and that for smooth solutions the two formulations are equivalent.

Mixed Dirichlet-Neumann boundary conditions

Consider the following problem. Let $\Omega \subset \mathbb{R}^n$ and let Γ_0 be a smooth non-empty subset of Γ and let $\Gamma_1 = \Gamma \setminus \Gamma_0$.
The variational formulation of the following problem

$$\left| \begin{array}{lll} -\Delta u = f & \text{in } \Omega, & f \in L_2(\Omega) \\ u = 0 & \text{on } \Gamma_0 & \\ \dfrac{\partial u}{\partial n} = 0 & \text{on } \Gamma_1 & \end{array} \right. \qquad (11.2.5)$$

is given by

$$\left| \begin{array}{ll} \text{Find } u \in W \text{ such that} & \\ a(u,v) = (f,v)_{L_2} & \text{for all } v \in W \end{array} \right. \qquad (11.2.6)$$

with $W = \{v \in H^1(\Omega) \mid v|_{\Gamma_0} = 0\}$ equiped with the Hilbert structure induced by $H^1(\Omega)$ or $H_0^1(\Omega)$

$$a(u,v) = \sum_{i=1}^{n} \left(\frac{\partial u}{\partial x_i}, \frac{\partial v}{\partial x_i}\right)_{L_2}$$

11.3 General variational formulation; existence, uniqueness

In the examples of the Dirichlet, the Neumann and the mixed Dirichlet-Neumann problem for the Laplace operator we can distinguish the following data:

(i) a Hilbert space V: $V = H_0^1(\Omega)$ in (11.1.8)
$V = H^1(\Omega)$ in (11.2.4)
$V = W$ in (11.2.6)

(ii) a continuous bilinear form a(.,.) on V × V:

$$a(u,v) = \sum_{i=1}^{n} (\frac{\partial u}{\partial x_i}, \frac{\partial v}{\partial x_i})_{L_2} + \lambda(u,v) \text{ in (11.1.8), (11.2.4)}$$

$$a(u,v) = \sum_{i=1}^{n} (\frac{\partial u}{\partial x_i}, \frac{\partial v}{\partial x_i})_{L_2} \quad \text{in} \quad (11.2.6)$$

(iii) a continuous linear form L(.) on V:

$$L(v) = (f,v)_{L_2} \quad \text{in (11.1.8), (11.2.4), (11.2.6)}$$

The general variational (or weak) formulation of linear problems of elliptic type is the following:

$$\left|\begin{array}{l} \text{Find } u \in V \text{ such that} \\ a(u,v) = L(v) \quad \text{for all } v \in V \end{array}\right. \qquad (11.3.1)$$

For the well-posedness of this problem, we have to impose some extra condition on a(.,.), because if $a(u,v) \equiv 0$, the problem would be meaningless. For our aim it is sufficient to demand that the form a(.,.) is V-elliptic (or V-coercive), which means that there exists a positive constant α such that

$$a(v,v) \geq \alpha||v||_V^2 \qquad \text{for all } v \in V \qquad (11.3.2)$$

One easily shows that this condition is satisfied for the examples of sections 11.1, 11.2.

In the following theorem we give a result on the unique solvability of problem (11.3.1)

Theorem (11.3.3)

Let V be a Hilbert space.
Let a(.,.) be a bilinear form which satisfies:
(i) a(.,.) is V × V continuous, i.e. there exists $M > 0$ such that

$$|a(u,v)| \leq M||u||_V||v||_V \quad \text{for all } u,v \in V$$

(ii) $a(.,.)$ is V-elliptic, i.e. there exists $\alpha > 0$ such that

$$a(v,v) \geq \alpha||v||_V^2 \quad \text{for all } v \in V$$

Let $L(.)$ be a continuous linear form on V, i.e. there exists $\ell > 0$ such that

$$|L(v)| \leq \ell||v||_V \quad \text{for all } v \in V.$$

Then there exists one and only one solution of problem (11.3.1).

Proof

We first prove the uniqueness of the solution. Suppose there are two solutions u_1 and u_2. Then

$$a(u_i,v) = L(v) \quad \text{for all } v \in V \quad i = 1,2$$

Subtraction of the equality for $i = 2$ from the equality for $i = 1$ gives:

$$a(u_1-u_2,v) = 0 \qquad \text{for all } v \in V$$

Take $v = u_1-u_2 \in V$, then by the V-ellipticity of $a(.,.)$ we find

$$\alpha||u_1-u_2||_V^2 \leq a(u_1-u_2,u_1-u_2) = 0$$

and hence $u_1 = u_2$.
For the proof of existence of a solution, which is based on Galerkin's method, we assume that V has a countable basis $w_1,w_2,\ldots,w_N,\ldots$. Let V_N be the space spanned by $w_1,\ldots,w_N$. For each fixed integer N we define an approximate solution of (11.3.1) by

$$\left|\begin{array}{ll} \text{Find } u_N \in V_N \text{ such that} & \\ a(u_N,v) = L(v) \qquad \text{for all } v \in V_N & \end{array}\right. \qquad (11.3.4)$$

Notice that (11.3.4) is a system of N linear equations with N unknowns. If we set

$$u_N = \sum_{j=1}^{N} g_{jN} w_j \qquad (g_{jN},\ j=1,2,\ldots,N,\ \text{unknown})$$

then (11.3.4) is equivalent to

$$a(u_N, w_i) = L(w_i) \qquad i = 1,2,\ldots,N$$

i.e.

$$\sum_{j=1}^{N} g_{jN}\, a(w_j, w_i) = L(w_i) \qquad i = 1,2,\ldots,N$$

The existence and uniqueness of u_N is proved if we show that this linear system is non-singular. To show this, it is sufficient to prove that the homogeneous linear system has only the zero-vector as solution. We multiply each equation

$$\sum_{j=1}^{N} g_{jN}\, a(w_j, w_i) = 0$$

by the corresponding g_{iN} and we add these equations to obtain

$$\sum_{i=1}^{N} \sum_{j=1}^{N} g_{iN} g_{jN} a(w_j, w_i) = a\Big(\sum_{i=1}^{N} g_{iN} w_i, \sum_{j=1}^{N} g_{jN} w_j\Big) = 0$$

By the V-ellipticity of $a(.,.)$ and the linear independence of the $w_1,\ldots,w_N$, we find that

$$g_{iN} = 0 \qquad i = 1,2,\ldots,$$

The conclusion is that problem (11.3.4) has exactly one solution. The mapping $v \to L(v)$ satisfies

$$|L(v)| \leq \ell ||v||_V \qquad \text{for all } v \in V$$

We take $v = u_N$ in (11.3.4) and we deduce

$$\alpha ||u_N||_V^2 \leq a(u_N, u_N) = L(u_N) \leq \ell ||u_N||_V$$

which implies

$$||u_N||_V \leq \frac{\ell}{\alpha} \tag{11.3.5}$$

The sequence $\{u_N\}$ belongs to a bounded subset of V, so it contains a subsequence $\{u_\mu\}$ which converges weakly in V to some $\phi \in V$, i.e.

$$(u_\mu,v)_V \to (\phi,v)_V \quad \text{as} \quad \mu \to \infty \quad \text{for all } v \in V.$$

Let $v \in V_{N_0}$, N_0 fixed. For each $\mu \geq N_0$, we have

$$a(u_\mu,v) = L(v) \tag{11.3.6}$$

because $V_{N_0} \subset V_\mu$.

It is now possible to take the limit on the left hand side of (11.3.6):

$$a(u_\mu,v) \to a(\phi,v) \quad \text{as} \quad \mu \to \infty \tag{11.3.7}$$

Combining (11.3.6) and (11.3.7), we deduce that

$$a(\phi,v) = L(v) \quad \text{for all } v \in V_{N_0}, \quad N_0 \text{ fixed.}$$

But since N_0 is arbitrarily chosen and due to the property that $V_N \to V$ as $N \to \infty$ (i.e. $\bigcup_{j=1}^{\infty} V_j$ is dense in V), we find that

$$a(\phi,v) = L(v) \quad \text{for all } v \in V$$

This proves that ϕ is a solution of (11.3.1) and by the uniqueness of a solution of (11.3.1) we have $\phi = u$.

In the preceding theorem we have proved the weak convergence of u_μ (subsequence of u_N) to u. In fact, it is possible to prove that the whole sequence satisfies

$$||u_N-u||_V \leq \frac{M}{\alpha} \inf_{v\in V_N} ||v-u||_V$$

which implies the strong convergence of u_N to u as $N \to \infty$:

$$||u_N-u||_V \to 0 \quad \text{as} \quad N \to \infty.$$

Let us emphasize that there are infinitely many bases in the space V. In order to make the calculation of u_N as easy as possible, it is of great importance to choose the basis of V in such a way that the number of coefficients of the matrix $a(w_j,w_i)$ that are equal to zero is optimal. The choice of an appropriate basis for V is one of the fundamental problems in the numerical analysis of PDEs.

11.4 Examples

To illustrate the preceding results we will treat in this section some examples of variational problems of elliptic type and the corresponding BVP.

Let $\Omega \subset IR^n$ be an open bounded set with smooth boundary Γ. Consider the following bilinear form

$$a(u,v) = \sum_{i,j=1}^{n} \left(a_{ij} \frac{\partial u}{\partial x_i}, \frac{\partial v}{\partial x_j}\right)_{L_2} + (a_0 u,v)_{L_2} \tag{11.4.1}$$

where the coefficients $a_{ij} = a_{ij}(\underline{x})$ and $a_0 = a_0(\underline{x})$ satisfy

$$\left| \begin{array}{l} |a_{ij}| < \infty, \quad i,j = 1,\dots,n \qquad |a_0| < \infty \\ \sum_{i,j=1}^{n} a_{ij}(\underline{x})\xi_i\xi_j \geq \alpha \sum_{i=1}^{n} \xi_i^2 \quad \text{for} \quad \alpha > 0, \text{ for all } \underline{\xi}\in IR^n, \ \underline{x}\in\Omega. \end{array} \right. \tag{11.4.2}$$

The following lemma can be proved:

Lemma (11.4.3)

(i) if (11.4.2) holds, then the form $a(.,.)$ is continuous on $H^1(\Omega) \times H^1(\Omega)$

(ii) if, apart from (11.4.2),

$$a_0(\underline{x}) \geq \alpha_0 > 0, \tag{11.4.4}$$

then $a(.,.)$ is $H^1(\Omega)$-elliptic

(iii) if, apart from (11.4.2),

$$a_0(\underline{x}) \geq 0 \qquad (11.4.5)$$

then $a(.,.)$ is $H_0^1(\Omega)$- elliptic.

Using theorem (11.3.3) and lemma (11.4.3) we can prove the following propositions:

Proposition (non-homogeneous Dirichlet problem) (11.4.6)

Let $a(.,.)$ be defined by (11.4.1) and satisfy (11.4.2), (11.4.5).
Let $f \in L_2(\Omega)$ and let $g_0 : \Gamma \to \mathbb{R}$ be such that there exists a function $G_0 \in H^1(\Omega)$ with $G_0|_\Gamma = g_0$. Then the following variational problem

$$\left| \begin{array}{l} \text{Find } \; u \in H^1_{g_0} = \{v \in H^1(\Omega) | \; v|_\Gamma = g_0\} \; \text{ such that} \\ \\ a(u,v) = (f,v)_{L_2} \qquad \text{for all } v \in H_0^1(\Omega) \end{array} \right. \qquad (11.4.7)$$

has a unique solution. Problem (11.4.7) is the weak formulation of the following BVP

$$\left| \begin{array}{l} -\sum\limits_{i,j=1}^{n} \dfrac{\partial}{\partial x_j} (a_{ij} \dfrac{\partial u}{\partial x_i}) + a_0 u = f \quad \text{in } \Omega \\ \\ u|_\Gamma = g_0 \qquad \text{on } \Gamma \end{array} \right. \qquad (11.4.8)$$

The preceding problem is covered by the general formulation (11.3.1) when the problem is written in terms of $w = u - G_0$. In fact, the variational formulation for w is:

$$\left| \begin{array}{l} \text{Find } \; w \in H_0^1(\Omega) \; \text{ such that} \\ \\ a(w,v) = L(v) \qquad \text{for all } v \in H_0^1(\Omega) \end{array} \right. \qquad (11.4.9)$$

with $L(v) = (f,v)_{L_2} - a(G_0,v)$

Proposition (non-homogeneous Neumann problem) (11.4.10)

Let $a(\cdot,\cdot)$ be defined by (11.4.1) and satisfy (11.4.2), (11.4.4). Let $f \in L_2(\Omega)$ and $g_1 \in L_2(\Gamma)$. Then the following variational problem is uniquely solvable:

$$\left| \begin{array}{l} \text{Find} \quad u \in H^1(\Omega) \quad \text{such that} \\ \\ a(u,v) = L(v) \qquad \text{for all } v \in H^1(\Omega) \end{array} \right. \tag{11.4.11}$$

with $L(v) = (f,v)_{L_2} + (g_1, v|_\Gamma)_{L_2(\Gamma)}$

Problem (11.4.11) is the weak formulation of the following BVP

$$\left| \begin{array}{ll} -\sum\limits_{i,j=1}^{n} \frac{\partial}{\partial x_j} (a_{ij} \frac{\partial u}{\partial x_i}) + a_0 u = f & \text{in } \Omega \\ \\ \frac{\partial u}{\partial n_a} = g_1 & \text{on } \Gamma \end{array} \right. \tag{11.4.12}$$

where $\frac{\partial u}{\partial n_a}$ denotes the normal derivative of u on Γ with respect to $a(\cdot,\cdot)$

$$\frac{\partial u}{\partial n_a} = \sum_{i,j=1}^{n} a_{ij}(x) \frac{\partial u}{\partial x_i} \cos(\underline{n}, x_j)$$

where $\cos(\underline{n}, x_j)$ is the directional cosine.

Proposition (mixed Dirichlet-Neumann problem) (11.4.13)

Suppose that Γ_0 is a smooth subset of Γ with positive measure. Let $a(\cdot,\cdot)$ be given by (11.4.1) and satisfy (11.4.2), (11.4.5). Let $f \in L_2(\Omega)$, $g_0 : \Gamma_0 \to \mathbb{R}$ be such that there exists a function $G_0 \in H^1(\Omega)$ with $G_0|_{\Gamma_0} = g_0$ and let $g_1 \in L_2(\Gamma_1)$. Then the weak problem

$$\left| \begin{array}{l} \text{Find} \quad u \in V_{g_0} = \{v \in H^1(\Omega) \mid v|_{\Gamma_0} = g_0\} \quad \text{such that} \\ \\ a(u,v) = L(v) \qquad \text{for all } v \in V \end{array} \right. \tag{11.4.14}$$

with

$$L(v) = (f,v)_{L_2} + (g_1, v|_{\Gamma_1})_{L_2(\Gamma_1)}$$

$$V = \{v \in H^1(\Omega) \mid v|_{\Gamma_0} = 0\}$$

is equivalent to the following BVP

$$\left| \begin{array}{ll} -\sum_{i,j=1}^{n} \frac{\partial}{\partial x_j}\left(a_{ij} \frac{\partial u}{\partial x_i}\right) + a_0 u = f & \text{in } \Omega \\ u|_{\Gamma_0} = g_0 & \text{on } \Gamma_0 \\ \frac{\partial u}{\partial n_a} = g_1 & \text{on } \Gamma_1 \end{array} \right. \tag{11.4.15}$$

Proposition (Robbins problem) (11.4.16)

Let $a(.,.)$ be given by (11.4.1) satisfying (11.4.2), (11.4.5). Let $f \in L_2(\Omega)$, $g_2 \in L_2(\Gamma)$, $\alpha = \alpha(\underline{x})$, $\underline{x} \in \Gamma$ with $0 < \alpha(\underline{x}) < \infty$. Then the BVP

$$\left| \begin{array}{ll} -\sum_{i,j=1}^{n} \frac{\partial}{\partial x_j}\left(a_{ij} \frac{\partial u}{\partial x_i}\right) + a_0 u = f & \text{in } \Omega \\ \frac{\partial u}{\partial n_a} + \alpha\, u = g_2 & \text{on } \Gamma \end{array} \right. \tag{11.4.17}$$

has the following weak formulation

$$\left| \begin{array}{l} \text{Find} \quad u \in H^1(\Omega) \quad \text{such that} \\ \\ \tilde{a}(u,v) = L(v) \qquad \text{for all } v \in H^1(\Omega) \end{array} \right. \tag{11.4.18}$$

where

$$\tilde{a}(u,v) = a(u,v) + (\alpha u|_\Gamma, v|_\Gamma)_{L_2(\Gamma)}$$

$$L(v) = (f,v)_{L_2} + (g_2, v|_\Gamma)_{L_2(\Gamma)}$$

11.5 (Navier-) Stokes equations

In this section we shall study first the existence and uniqueness of a solution of the Stokes equations for the no-slip boundary conditions. Secondly we shall study the Stokes problem for Neumann-type boundary conditions, i.e. conditions for the normal and tangential stress (cf. (1.6.23), (1.6.24), see also Chapter 6). Since the study of existence and uniqueness of the full Navier-Stokes equations requires an advanced knowledge of functional analysis, we shall only give the results. In fact, the Navier-Stokes equations contain non-linear terms (the convective terms), so that the general theorem (11.3.3) for linear equations is not applicable.

The Stokes problem consists in finding a vector function $\underline{u} = \{u_1, u_2, \ldots, u_n\}$ representing the velocity and a scalar function p, the pressure, related by the following BVP (See Section 6.1)

$$\begin{array}{ll} \text{(i)} & \left| \; -\frac{1}{Re}\Delta\underline{u} + \nabla p = \underline{f} \quad \text{in } \Omega \right. \\ \text{(ii)} & \left| \; \nabla\cdot\underline{u} = 0 \quad \text{in } \Omega \right. \\ \text{(iii)} & \left| \; \underline{u} = \underline{0} \quad \text{on } \Gamma \right. \end{array} \qquad (11.5.1)$$

where $\underline{f} = \{f_1, \ldots, f_n\}$ is a given volumic force per unit mass. In order to give a weak formulation of (11.5.1), we introduce the following Hilbert space

$$\underline{V} = \{\underline{v} \in \underline{H}_0^1(\Omega) \mid \nabla\cdot\underline{v} = 0 \text{ in } \Omega\}, \; \underline{H}_0^1(\Omega) \equiv (H_0^1(\Omega))^n \quad (11.5.2)$$

with scalar product and norm induced by $\underline{H}_0^1(\Omega)$:

$$(\underline{u},\underline{v})_{\underline{V}} = \sum_{i,j=1}^{n} \left(\frac{\partial u_i}{\partial x_j},\frac{\partial v_i}{\partial x_j}\right)_{L_2}, \qquad ||\underline{v}||_{\underline{V}} = (\underline{v},\underline{v})_{\underline{V}}^{\frac{1}{2}}$$

Taking the scalar product of (11.5.1) (i) with a function $\underline{v} \in \underline{V}$ we obtain

$$\int_\Omega \left(-\frac{1}{Re}\,\Delta\underline{u} + \nabla p\right)\cdot\underline{v}\; d\Omega = \int_\Omega \underline{f}\cdot\underline{v}\; d\Omega$$

We apply Green's formula to the left hand side of this equality and use the relation

$$\int_\Omega \nabla p\cdot\underline{v}\; d\Omega = -\int_\Omega p(\nabla\cdot\underline{v})\; d\Omega + \int_\Gamma p(\underline{v}\cdot\underline{n})\; d\Gamma \qquad (11.5.3)$$

$$\text{for all } \underline{v} \in \underline{H}^1(\Omega) \equiv (H^1(\Omega))^n$$

with $\underline{n}$ the unit outer normal on Γ, we find for $\underline{v} \in \underline{V}$:

$$\frac{1}{Re}\sum_{i,j=1}^{n} \left(\frac{\partial u_i}{\partial x_j},\frac{\partial v_i}{\partial x_j}\right)_{L_2} = \sum_{i=1}^{n} (f_i,v_i)_{L_2}$$

which leads to the following weak formulation:

$$\left|\begin{array}{l} \text{Find } \underline{u} \in \underline{V} \text{ such that} \\ \\ a(\underline{u},\underline{v}) = L(\underline{v}) \quad \text{for all } \underline{v} \in \underline{V} \end{array}\right. \qquad (11.5.4)$$

where

$$a(\underline{u},\underline{v}) = \frac{1}{Re}\sum_{i,j=1}^{n} \left(\frac{\partial u_i}{\partial x_j},\frac{\partial v_i}{\partial x_j}\right)_{L_2}$$

$$L(\underline{v}) = \sum_{i=1}^{n} (f_i,v_i)_{L_2} \qquad \left(\equiv (\underline{f},\underline{v})_{\underline{L}_2} = \int_\Omega \underline{f}\cdot\underline{v}\; d\Omega\right)$$

and we see that the pressure has been eliminated from the problem.

Proposition (11.5.5)

For Re > 0 and $\underline{f} \in \underline{L}_2(\Omega) \equiv (L_2(\Omega))^n$, problem (11.5.4) has a unique solution $\underline{u}$.

Proof

One easily verifies that the conditions of theorem (11.3.3) are satisfied:

$$|a(\underline{u},\underline{v})| \leqslant \frac{1}{Re} ||\underline{u}||_{\underline{V}} \; ||\underline{v}||_{\underline{V}}$$

$$a(\underline{v},\underline{v}) \geqslant \frac{1}{Re} ||v||^2_{\underline{V}}$$

$$|L(\underline{v})| \leqslant c||\underline{f}||_{\underline{L}_2} \; ||v||_{\underline{V}}$$

Once we have solved problem (11.5.4), we can introduce the pressure p in the following way. Consider the linear form

$$\mathcal{L} : \underline{H}^1_0(\Omega) \to \mathbb{R} ,$$

defined by

$$\mathcal{L} : \underline{v} \to a(\underline{u},\underline{v}) - (\underline{f},\underline{v})_{\underline{L}_2}$$

This form is continuous on $\underline{H}^1_0(\Omega)$ and vanishes on $\underline{V} \subset \underline{H}^1_0(\Omega)$. According to a theorem of de Rham (see Temam (1977)) there exists a unique function $p \in \tilde{L}_2(\Omega) = \{q \in L_2(\Omega) | \int_\Omega q(\underline{x}) \, d\Omega = 0\}$ such that

$$\mathcal{L}(\underline{v}) = (p,\nabla\cdot\underline{v})_{L_2} \qquad \text{for all} \quad \underline{v} \in \underline{H}^1_0(\Omega)$$

and this function p can be interpreted as the pressure.

The result of existence and uniqueness can be extended to the case of non-homogeneous boundary conditions:

$$\underline{u} = \underline{g}_0 \quad \text{on} \quad \Gamma \tag{11.5.6}$$

provided the following conditions are satisfied:

(i) there exists a function $\underline{G}_0 \in \underline{H}^1(\Omega)$ such that $\underline{G}_0|_\Gamma = \underline{g}_0$

(ii) $\int_\Omega \underline{g}_0 \cdot \underline{n} \, d\Gamma = 0$ (compatibility condition, cf. (1.8.3))

In this non-homogeneous case $\underline{u}$ is sought in the set of solenoidal vector fields in $\underline{G}_0 + \underline{H}_0^1(\Omega)$.

When instead of the no-slip boundary conditions (11.5.1) (iii) or (11.5.6), the stresses are prescribed on the boundary we proceed as follows.
The Stokes equations are written in terms of the stress tensor:

$$\begin{cases} -\nabla \cdot \underline{\underline{\sigma}} = \underline{f} & \text{in} \quad \Omega \\ \nabla \cdot \underline{u} = 0 & \text{in} \quad \Omega \end{cases} \tag{11.5.7}$$

where $\underline{\underline{\sigma}}$, for an incompressible fluid, is defined by (cf. Section 1.6)

$$\underline{\underline{\sigma}} = -p\,\underline{\underline{I}} + \frac{2}{Re}\,\underline{\underline{e}}\,, \quad e_{ij}(\underline{u}) = \tfrac{1}{2}\left(\frac{\partial u_i}{\partial x_j} + \frac{\partial u_j}{\partial x_i}\right)$$

Suppose, for instance, that the boundary Γ consists of two parts $\Gamma_0 \neq \emptyset$ and $\Gamma_1 = \Gamma \setminus \Gamma_0$. On Γ_0 we impose homogeneous no-slip boundary conditions, while on Γ_1 we prescribe the stresses:

$$\underline{u} = \underline{0} \quad \text{on } \Gamma_0$$

$$\underline{\underline{\sigma}} \cdot \underline{n} = \underline{g} \quad \text{on } \Gamma_1$$

The condition on Γ_1 can be written equivalently as (see Section 1.6)

$$\sigma_n \equiv (\underline{\underline{\sigma}} \cdot \underline{n}) \cdot \underline{n} = \underline{g} \cdot \underline{n} \equiv g_n$$

$$\underline{\sigma}_t \equiv \underline{\underline{\sigma}} \cdot \underline{n} - \sigma_n \underline{n} = \underline{g} - g_n \underline{n} \equiv \underline{g}_t$$

Using now the following Green's formula:

$$- \int_\Omega (\nabla \cdot \underline{\underline{\sigma}}) \cdot \underline{v} \, d\Omega = \int_\Omega \underline{\underline{\sigma}} \cdot \nabla \underline{v} \, d\Omega - \int_\Gamma (\underline{\underline{\sigma}} \cdot \underline{n}) \cdot \underline{v} \, d\Gamma$$

$$= \int_\Omega \underline{\underline{\sigma}} \cdot \underline{\underline{e}}(\underline{v}) \, d\Omega - \int_\Gamma (\underline{\underline{\sigma}} \cdot \underline{n}) \cdot \underline{v} \, d\Gamma \qquad (11.5.8)$$

$$= \frac{2}{Re} \int_\Omega \underline{\underline{e}}(\underline{u}) \cdot \underline{\underline{e}}(\underline{v}) \, d\Omega - (p, \nabla \cdot \underline{v})_{L_2} - \int_\Gamma (\underline{\underline{\sigma}} \cdot \underline{n}) \cdot \underline{v} \, d\Gamma$$

then problem (11.5.7) admits the following weak formulation:

$$\left| \begin{array}{l} \text{Find } \underline{u} \in \underline{V} = \{\underline{v} \in \underline{H}^1(\Omega) \mid \nabla \cdot \underline{v} = 0 \text{ in } \Omega, \ \underline{v}|_{\Gamma_0} = \underline{0}\} \text{ such that} \\ a(\underline{u}, \underline{v}) = L(\underline{v}) \qquad \text{for all } \underline{v} \in \underline{V} \end{array} \right. \qquad (11.5.9)$$

with $\underline{V}$ endowed with the Hilbert structure induced by $\underline{H}^1(\Omega)$

$$a(\underline{u}, \underline{v}) = \frac{2}{Re} \int_\Omega \underline{\underline{e}}(\underline{u}) \cdot \underline{\underline{e}}(\underline{v}) \, d\Omega$$

$$L(\underline{v}) = \int_\Omega \underline{f} \cdot \underline{v} \, d\Omega + \int_\Gamma \underline{g} \cdot \underline{v} \, d\Gamma$$

Proposition (11.5.10)

We assume that $\underline{f} \in \underline{L}_2(\Omega)$ and $\underline{g} \in \underline{L}_2(\Gamma)$. Then, for Re > 0, problem (11.5.9) has a unique solution.

Proof

It is easily verified that the form a(.,.) is bilinear and continuous on $\underline{V} \times \underline{V}$, and that the form L(.) is linear and continuous on $\underline{V}$. To prove the $\underline{V}$-ellipticity of the form a(.,.), we need the following inequality which is known as Korn's inequality (see Duvaut, Lions (1971)):

$$\left| \begin{array}{l} \text{There exists a positive constant c such that} \\ \int_\Omega \underline{\underline{e}}(\underline{v}) \cdot \underline{\underline{e}}(\underline{v}) \, d\Omega + ||\underline{v}||^2_{\underline{L}_2} \geq c||\underline{v}||^2_{\underline{H}^1} \quad \text{for all } \underline{v} \in \underline{H}^1(\Omega) \end{array} \right. \qquad (11.5.11)$$

The V-ellipticity is proved by first showing that the mapping

$$\underline{v} \in \underline{V} \rightarrow ||\underline{v}|| = \int_{\Omega} \underline{\underline{e}}(\underline{v}) \cdot \underline{\underline{e}}(\underline{v}) \, d\Omega$$

defines a norm on $\underline{V}$. Next Korn's inequality is used to show that $||\cdot||$ is equivalent to $||\cdot||_{\underline{V}}$.

The extension to non-homogeneous Dirichlet boundary conditions and to other combinations of boundary conditions is immediate. A practical example has been given in Section 6.3.2.

Concerning the existence and uniqueness of a solution of the non-linear Navier-Stokes equations we only give the results, since the non-linearity seriously complicates the theory. We consider here only the Navier-Stokes equations with homogeneous Dirichlet boundary conditions in the two and three dimensional case. The problem is formulated as follows (cf. Section 6.1)

Find $\underline{u}$ and p such that

$$\begin{array}{ll} \text{(i)} & -\frac{1}{Re} \Delta \underline{u} + (\underline{u} \cdot \nabla)\underline{u} + \nabla p = \underline{f} \quad \text{in} \quad \Omega \\ \text{(ii)} & \nabla \cdot \underline{u} = 0 \quad \text{in} \quad \Omega \\ \text{(iii)} & \underline{u} = \underline{0} \quad \text{on} \quad \Gamma \end{array} \qquad (11.5.12)$$

For the derivation of a weak formulation of problem (11.5.12), we proceed as for the Stokes equations (11.5.1). We take the scalar product of (11.5.12) (i) with $\underline{v} \in \underline{V}$, where $\underline{V}$ is defined by (11.5.2). Application of Green's formula and (11.5.3) leads to the following formulation

Find $\underline{u} \in \underline{V}$ such that

$$a(\underline{u},\underline{v}) + b(\underline{u},\underline{u},\underline{v}) = L(\underline{v}) \quad \text{for all } \underline{v} \in \underline{V} \qquad (11.5.13)$$

with

$$a(\underline{u},\underline{v}) = \frac{1}{Re} \sum_{i,j=1}^{n} \left(\frac{\partial u_i}{\partial x_j},\frac{\partial v_i}{\partial x_j}\right)_{L_2}$$

$$b(\underline{u},\underline{u},\underline{v}) = \int_\Omega (\underline{u}.\nabla)\underline{u}.\underline{v}\ d\Omega$$

$$L(\underline{v}) = \int_\Omega \underline{f}.\underline{v}\ d\Omega$$

The next proposition is proved in Ladyshenskaya (1969) and Temam (1977).

Proposition (11.5.14)

Problem (11.5.13) has at least one solution $\underline{u} \in \underline{V}$ and there exists $p \in L_2(\Omega)$ such that (11.5.12) is satisfied.
If, for $n = 2,3$, Re, $\underline{f}$ and Ω satisfy the condition

$$Re^2 ||\underline{f}||_{\underline{L}_2} < c \qquad c = c(n,\Omega) = \text{constant} \qquad (11.5.15)$$

then the solution $\underline{u}$ is unique.

This proposition shows the existence of a solution $\underline{u}$ in both the 2D and 3D situation. The uniqueness of $\underline{u}$ is guarenteed for "sufficient small" external forces or for "sufficiently small" values of the Reynolds number.
For non-homogeneous boundary conditions $\underline{u} = g_0$ on Γ, existence can be proved when (cf. Temam (1977)):

(i) there exists a function $\underline{G}_0 \in \underline{H}^1(\Omega)$ such that $\underline{G}_0|_\Gamma = g_0$

(ii) $\int_\Gamma g_0 . \underline{n}\ d\Gamma = 0$

If moreover the following condition is satisfied
(iii) g_0 sufficiently small and

$$Re^2 ||\tilde{\underline{f}}||_{\underline{L}_2} < c$$

with $\tilde{\underline{f}} = \underline{f} + \frac{1}{Re} \Delta \underline{G}_0 + (\underline{G}_0 . \nabla)\underline{G}_0$

then the solution $\underline{u}$ is unique.
Non-uniqueness results have been reported when condition (iii) is not satisfied (cf. Temam (1977), (1983)).

11.6 Regularity of the solution of the variational problem

In this section we shall give some results on regularity (or smoothness) of the solution u of particular cases of the variational problem (11.3.1). In other words the question is: Suppose we strengthen the regularity properties of the data (for instance: right hand side, boundary conditions), what can be said about the regularity of the solution u?
We introduce the space $C^\infty(\bar{\Omega})$ of infinitely differentiable functions on $\bar{\Omega}$ and the Sobolev space of order m, denoted by $H^m(\Omega)$, which consists of functions belonging to $L_2(\Omega)$ as well as all their partial derivatives of order less than or equal to m.
We consider the operator A (cf. Section 11.4)

$$Au = -\sum_{i,j=1}^{n} \frac{\partial}{\partial x_j}\left(a_{ij}\frac{\partial u}{\partial x_i}\right) + a_0 u \tag{11.6.1}$$

where the coefficients satisfy now (cf. (11.4.2), (11.4.4), (11.4.5))

$$\left|\begin{array}{ll} a_{ij}\ ,\ a_0 \in C^\infty(\bar{\Omega}) & \\ \sum_{i,j=1}^{n} a_{ij}(\underline{x})\xi_i\xi_j \geq \alpha \sum_{i=1}^{n} \xi_i^2 \quad \text{for } \alpha > 0, & \\ \qquad\qquad \text{for all } \underline{\xi} \in IR^n,\ \underline{x} \in \Omega & \\ \alpha_0(\underline{x}) \geq \alpha_0 > 0 \quad \text{(Neumann-problem)} & \\ \alpha_0(\underline{x}) \geq 0 \quad \text{(Dirichlet-problem)} & \end{array}\right. \tag{11.6.2}$$

We assume also that Ω is bounded with C^∞ smooth boundary Γ.

Proposition (regularity of non-homogeneous Dirichlet problem). (11.6.3)

Consider the following non-homogenous Dirichlet problem for the operator A (cf. proposition (11.4.6)):

$$\begin{cases} A\,u = f & \text{in } \Omega \\ u|_{\Gamma} = g_0 & \text{on } \Gamma \end{cases} \tag{11.6.4}$$

(i) If, for $m \geq 0$, $f \in H^m(\Omega)$ and $g_0 : \Gamma \to \mathbb{R}$ is such that there exists a function $G_0 \in H^{m+2}(\Omega)$, then the weak formulation of (11.6.4) is uniquely solvable and $u \in H^{m+2}(\Omega)$

(ii) In particular $f \in C^\infty(\bar{\Omega})$ and $g_0 \in C^\infty(\Gamma)$ imply $u \in C^\infty(\bar{\Omega})$.

Proposition (regularity of non-homogeneous Neumann problem) (11.6.5)

Consider the following non-homogeneous Neumann problem for the operator A (cf. proposition (11.4.10):

$$\begin{cases} A\,u = f & \text{in } \Omega \\ \dfrac{\partial u}{\partial \underline{n}_a} = g_1 & \text{on } \Gamma \end{cases} \tag{11.6.6}$$

(i) If, for $m \geq 0$, $f \in H^m(\Omega)$ and $g_1 : \Gamma \to \mathbb{R}$ is such that there exists a function $G_1 \in H^{m+1}(\Omega)$, then the weak formulation of (11.6.6) is uniquely solvable and $u \in H^{m+2}(\Omega)$

(ii) In particular $f \in C^\infty(\bar{\Omega})$ and $g_1 \in C^\infty(\Gamma)$ imply $u \in C^\infty(\bar{\Omega})$.

Proposition (regularity of Stokes problem) (11.6.7)

Let $\{\underline{u},p\} \in \underline{H}^1(\Omega) \times L_2(\Omega)/\mathbb{R}$ be the weak solution of the Stokes problem:

$$\left|\begin{array}{l} -\dfrac{1}{Re}\,\Delta \underline{u} + \nabla p = \underline{f} \quad \text{in} \quad \Omega \\ \nabla.\underline{u} = 0 \quad \text{in} \quad \Omega \\ \underline{u}|_{\Gamma} = g_0 \quad \text{on} \quad \Gamma \qquad (\int_{\Gamma} g_0.\underline{n}\, d\Gamma = 0) \end{array}\right. \tag{11.6.8}$$

(i) If, for $m \geq 0$, $\underline{f} \in \underline{H}^m(\Omega)$ and $g_0 : \Gamma \to \mathbb{R}$ is such that there exists a function $\underline{G}_0 \in \underline{H}^{m+2}$, then $\underline{u} \in \underline{H}^{m+2}(\Omega)$, $p \in H^{m+1}(\Omega)$

(ii) $\underline{f} \in C^{\infty}(\bar{\Omega})$ and $g_0 \in C^{\infty}(\Gamma)$ imply $\underline{u} \in C^{\infty}(\bar{\Omega})$, $p \in C^{\infty}(\bar{\Omega})$.

Proposition (regularity of Navier-Stokes problem) (11.6.9)

Let $\underline{f}$ be given in $C^{\infty}(\bar{\Omega})$, then any solution $\{\underline{u},p\}$ of the Navier-Stokes problem (11.5.13) belongs to $\underline{C}^{\infty}(\bar{\Omega}) \times C^{\infty}(\bar{\Omega})$.

Using the regularity results for weak solutions, it is possible to give conditions for which a weak solution is in fact a classical solution, which means for second order PDEs that $u \in C^2(\Omega)$. The relation between the Sobolev spaces $H^{m+2}(\Omega)$ and the spaces $C^k(\bar{\Omega})$ is given by Sobolev's inequality (see Ciarlet (1978), Brézis (1983)):

$$H^{m+2}(\Omega) \subset C^k(\bar{\Omega}) \text{ with } k = \text{entier } (m+2-\frac{n}{2}). \tag{11.6.10}$$

One easily deduces now that the solution u of the propositions (11.6.3) and (11.6.5) belongs to $C^2(\bar{\Omega})$ when $m > \frac{n}{2}$.
A solution $\{\underline{u},p\}$ of the Stokes equations is classical if $\underline{u} \in C^2(\bar{\Omega})$ and $p \in C^1(\bar{\Omega})$, which is satisfied for $m > \frac{n}{2}$.

Chapter 12

Finite Element Approximations of Variational Problems

Introduction

In this chapter we shall approximate the weak formulation of elliptic PDEs using a FEM. We shall indicate in which sense the solution of the approximate problem converges to the solution of the original weak formulation. The concept of internal approximation of a Hilbert space will be introduced and applied to the approximation of the spaces $L_2(\Omega)$, $H_0^1(\Omega)$ and $H^1(\Omega)$. With the discretization of the Stokes equations in mind, we shall introduce the notion of external approximation. Applications will be given to the Dirichlet-Neumann problem for the Laplace operator and to the Stokes equations. In this section we shall restrict ourselves to simplicial (i.e. triangular, tetrahedral, ...) finite elements. For a general theory on finite elements applied to elliptic problems (including quadrilateral and hexahedral finite elements), we refer to Ciarlet (1978); see also Temam (1973), Aubin (1972). Theoretical aspects of finite element approximations to the (Navier-) Stokes equations can be found in Temam (1977) and Girault, Raviart (1979).

12.1 Internal approximation of Hilbert spaces

A Hilbert space V can be approximated in two different ways: we can choose the approximation V_h of V such that (i) $V_h \subset V$ or such that (ii) $V_h \not\subset V$. First we shall be concerned with the first type of approximation, the second type will be discussed in Section 12.3.6.

Let us give a simple example. Let $V = H_0^1(\Omega)$ and $\Omega = (0,1)$. We subdivide the interval $[0,1]$ into $N = \frac{1}{h}$ subintervals of equal length h and we define the space V_h as the space spanned by the following set of functions $\{w_1,\dots w_{N-1}\}$, where w_j is piece-wise linear with $w_j(ih) = \delta_{ij}$, $j = 1,2,\dots,N-1$, $i = 0,1,\dots,N$ (see fig. 12.1).

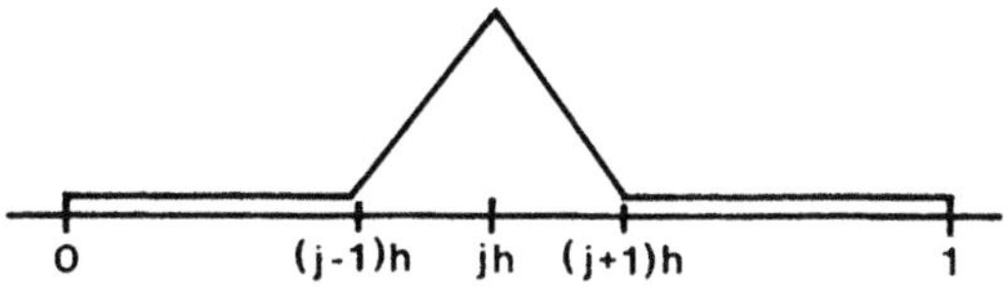

Fig. 12.1 Function w_j.

Notice that $w_j \in L_2(\Omega)$, $\frac{d}{dx} w_j \in L_2(\Omega)$, so that $w_j \in H_0^1(\Omega)$, which implies $V_h \subset V$.

In order to translate the variational problem (11.3.1) into an equation where the space V_h is concerned, we have to define a correspondence between the functions $v \in V$ and the functions $v_h \in V_h$. This could be done in the following way. Let $v \in V = H_0^1(\Omega)$, then

$$v_h(x) = \sum_{j=1}^{N-1} v(jh)\, w_j(x)$$

belongs to the space V_h. This mapping from V to V_h is called the restriction operator and is denoted by ρ_h. It is clear that a mapping from V_h into V can be defined in many ways. The simplest mapping is the identity. A mapping from V_h into V is called the prolongation operator and is denoted π_h.

We now give the general definition:

Definition (12.1.1)

An internal approximation of a Hilbert space V is a family of triples $\{V_{h_1}, \pi_h, \rho_h\}_{h \in H}$, where

(i) H is a parameter set of triangulations E_h of Ω;
$h = \max_{e \in E_h} \text{diam}(e)$

(ii) V_h is a Hilbert space, with norm $||\cdot||_h$

(iii) π_h is a continuous linear operator from V_h into V

(iv) ρ_h is an operator from V into V_h.

The notions of stability and convergence are of major importance in the theory of approximation of Hilbert spaces:

Definition (12.1.2)

An internal approximation $\{V_h, \pi_h, \rho_h\}_{h \in H}$ of a Hilbert space V is called stable if

$$||\pi_h|| \equiv \sup_{\substack{v_h \in V_h \\ v_h \neq 0}} \frac{||\pi_h v_h||_V}{||v_h||_h} \qquad (12.1.3)$$

can be majorized independently of h.
An internal approximation is called convergent if

$$\lim_{h \to 0} ||\pi_h \rho_h v - v||_V = 0 \quad \text{for all } v \in \mathcal{V},\ \mathcal{V} \text{ dense subspace of V.} \qquad (12.1.4)$$

A consequence of the definition of convergence is that it is sufficient to define the restriction operator ρ_h on a dense $\mathcal{V}$ subspace of V. It can be proved that the definition of ρ_h on can be extended to the whole space V.

12.2 Discretized variational problem

Let us consider the following problem (cf. (11.3.1)):

$$\left| \begin{array}{l} \text{Find } u \in V \text{ such that} \\ a(u,v) = L(v) \quad \text{for all } v \in V \end{array} \right. \qquad (12.2.1)$$

where V, a(.,.) and L(.) satisfy the conditions of theorem (11.3.3). Our purpose is to approximate problem (12.2.1) by a family of problems depending on the parameter h, the solution of which lies in the approximation V_h of V.
Let there be given a stable convergent internal approximation $\{V_h, \pi_h, \rho_h\}_{h\in H}$ of V. For each h $\in H$ we associate to problem (12.2.1) the following approximate problem:

$$\left| \begin{array}{l} \text{Find } u_h \in V_h \text{ such that} \\ a(u_h,v_h) = L(v_h) \quad \text{for all } v_h \in V_h \end{array} \right. \qquad (12.2.2)$$

The convergence of u_h to u is formulated in the following theorem:

Theorem (12.2.3)

Let $\{V_h, \pi_h, \rho_h\}_{h\in H}$ be a stable convergent internal approximation of V. Then the unique solution u_h of (12.2.2) converges, for $h \to 0$, to the solution u of (12.2.1) in the following sense

$$\lim_{h\to 0} ||\pi_h u_h - u||_V = 0 \qquad (12.2.4)$$

Proof

Existence and uniqueness of a solution u_h of (12.2.2) follow immediately from theorem (11.3.3). The convergence result is proved using similar arguments as employed in the proof of theorem (11.3.3).

12.3 Finite element approximations of Sobolev spaces

In this section we shall indicate how to construct finite element approximations of the spaces $L_2(\Omega)$, $H_0^1(\Omega)$, $H^1(\Omega)$ and $\underline{V}$.

12.3.1 Definition of finite element

Let there be given n+1 points $\underline{x}^1, \underline{x}^2, \ldots, \underline{x}^{n+1}$ in $\mathbb{R}^n$ with coordinates $(x_1^i, x_2^i, \ldots, x_n^i)$, $i = 1,2,\ldots,n+1$. We suppose that these points are not contained in the same hyperplane, which amounts to saying that the matrix

$$X = \left| \begin{matrix} x_1^1 & x_1^2 & \cdots\cdots & x_1^{n+1} \\ x_2^1 & x_2^2 & & x_2^{n+1} \\ \cdot & \cdot & & \cdot \\ \cdot & \cdot & & \cdot \\ x_n^1 & x_n^2 & \cdots\cdots & x_n^{n+1} \\ 1 & 1 & \cdots\cdots & 1 \end{matrix} \right|$$

is not singular: $\det(X) = 0$. Consequently for any point $\underline{x} = (x_1, \ldots, x_n) \in \mathbb{R}^n$ there exist n+1 uniquely determined numbers $\lambda_i = \lambda_i(\underline{x})$, $i = 1,2,\ldots,n+1$ such that

$$X \left| \begin{matrix} \lambda_1 \\ \lambda_2 \\ \cdot \\ \cdot \\ \cdot \\ \lambda_{n+1} \end{matrix} \right| = \left| \begin{matrix} x_1 \\ x_2 \\ \cdot \\ \cdot \\ x_n \\ 1 \end{matrix} \right| \qquad (12.3.1)$$

The numbers $\lambda_1, \ldots, \lambda_{n+1}$ are called the barycentric coordinates of $\underline{x}$ with respect to the n+1 points $\underline{x}^1, \underline{x}^2, \ldots \underline{x}^{n+1}$. For examples in 1D, 2D, 3D we refer to the Sections 4.1, 4.2.1 and 4.3.1 respectively. One easily verifies the following properties of barycentric coordinates:

(i) the barycentric coordinates are independent of the choice of the basis in $\mathbb{R}^n$

(ii) the convex hull e of the n+1 points $\underline{x}^i$, $i = 1,2,\dots,n+1$ is the set of points in $\mathbb{R}^n$ with barycentric coordinates satisfying $0 \leq \lambda_i \leq 1$, $i = 1,2,\dots,n+1$. Such a convex hull is called n-simplex

(iii) the barycentric coordinates of the barycentre with respect to $\underline{x}^1,\dots,\underline{x}^{n+1}$ are $\{\frac{1}{n+1},\frac{1}{n+1},\dots,\frac{1}{n+1}\}$

(iv) $\lambda_i(\underline{x}^j) = \delta_{ij}$

(v) the mid-point $\underline{x}^{k\ell}$ of the edge $[\underline{x}^k,\underline{x}^\ell]$, $1 \leq k < \ell \leq n+1$ satisfies

$$\lambda_i(\underline{x}^{k\ell}) = \tfrac{1}{2}(\delta_{ik}+\delta_{i\ell})$$

(vi) $$\int_e \lambda_1^{m_1}\lambda_2^{m_2}\dots\lambda_{n+1}^{m_{n+1}}\,de = \frac{m_1!m_2!\dots m_{n+1}!}{(m_1+m_2+\dots+m_{n+1}+n)!}\,n!\,|e|$$

where m_i, $i = 1,2,\dots,n+1$, are non-negative integers and $|e|$ denotes the volume of the n-simplex e. For a proof of this integration result we refer to Stroud (1971).

We now give the general definition of a finite element:

Definition (12.3.2)

A simplicial finite element in $\mathbb{R}^n$ is a triple $\{e,P(e),\Sigma_e\}$ defined by

(i) e is a closed n-simplex in $\mathbb{R}^n$ with non-empty interior

(ii) P(e) is a space of functions defined on e

(iii) Σ_e is a P(e)-unisolvent set of N points $\underline{x}^1,\underline{x}^2,\dots,\underline{x}^N$ which means that to any collection of real numbers α_i, $i = 1,2,\dots,N$, there exists a unique function $p \in P(e)$ such that $p(\underline{x}^i) = \alpha_i$, $i = 1,2,\dots,N$.

In the sequal we shall proceed as follows for the construction of finite element approximations V_h of the space V. The domain Ω will be subdivided into a finite number of n-simplices e with a corresponding set of points Σ_e and a corresponding function space P(e). The space V_h will then be defined as the function space consisting of all functions defined on Ω whose restriction to an n-simplex e belongs to the space P(e). Before we shall construct finite element approximations of the spaces $L_2(\Omega)$, $H_0^1(\Omega)$, $H^1(\Omega)$ and $\underline{V}$, we give some examples of finite elements.

The following proposition enables us to define the linear (or affine) finite element in $\mathbb{R}^n$.

Proposition (linear finite element) (12.3.3)

Let $\underline{x}^1, \underline{x}^2, \ldots, \underline{x}^{n+1}$ be n+1 points in $\mathbb{R}^n$ not included in a hyperplane. Let P(e) be the space of polynomials of degree ≤ 1 (in n variables) defined on e, where e is the n-simplex with vertices $\underline{x}^1, \ldots, \underline{x}^{n+1}$ and let $\Sigma_e = \{\underline{x}^1, \ldots, \underline{x}^{n+1}\}$. Then Σ_e is P(e)-unisolvent

Notice that on the one hand the set $\Sigma_e = \{\underline{x}^1, \ldots, \underline{x}^{n+1}\}$ is P(e)-unisolvent and since on the other hand each function $u \in P(e)$ can be written as

$$u(\underline{x}) = \sum_{i=1}^{n+1} u(\underline{x}^i)\lambda_i(\underline{x}) \qquad (12.3.4)$$

the functions λ_i, $i = 1,2,\ldots,n+1$ form a basis for the function space P(e), which is denoted by $P_1(e)$. Examples of this linear finite element are given in Chapter 3 and 4; we refer to Section 3.5.1 for an example in one dimension (see also the beginning of Section 4.1), for the two dimensional linear finite element we refer to Section 4.2.2 and for the three dimensional linear case to Section 4.3.2.

The following proposition introduces the quadratic finite element in IR^n.

Proposition (quadratic finite element) (12.3.5)

Let $\underline{x}^1,\underline{x}^2,\ldots,\underline{x}^{n+1}$ be n+1 points in IR^n which are not contained in a hyperplane, and let $\underline{x}^{ij}$, $1 \leq i < j \leq n+1$, denote the mid-point of the edge $[\underline{x}^i,\underline{x}^j]$. The space P(e) is the space of polynomials of degree ≤ 2 defined on e, with e the n-simplex generated by $\underline{x}^1,\underline{x}^2,\ldots,\underline{x}^{n+1}$. Let $\Sigma_e = \{\underline{x}^i \mid i=1,\ldots,n+1\} \cup \{\underline{x}^{ij} \mid 1 \leq i < j \leq n+1\}$. Then Σ_e is P(e)-unisolvent and a function in P(e) can be written uniquely as

$$u(\underline{x}) = \sum_{i=1}^{n+1} \lambda_i(\underline{x})(2\lambda_i(\underline{x})-1)u(\underline{x}^i) + 4 \sum_{\substack{i,j=1\\ i<j}}^{n+1} \lambda_i(\underline{x})\lambda_j(\underline{x})u(\underline{x}^{ij}) \qquad (12.3.6)$$

The set of functions $\phi_i(\underline{x}) = \lambda_i(\underline{x})(2\lambda_i(\underline{x})-1)$, $i=1,2,\ldots,n+1$, and $\phi_{ij}(\underline{x}) = 4\lambda_i(\underline{x})\lambda_j(\underline{x})$, $1 \leq i < j \leq n+1$ forms a basis in P(e), which is denoted by $P_2(e)$

For examples in 1D and 2D, the reader is referred to the Sections 4.2 and 4.4.4 respectively.

Other elements have been given in Chapter 4:

(i) 1D cubic finite element: (12.3.7)
P(e) = $P_3(e)$ space spanned by $\{\phi_1,\phi_2,\phi_{12},\phi_{21}\}$,
ϕ given by (4.1.10)

$$\Sigma_e = \{x^1,x^2,x^{12},x^{21}\}, \quad x^{\bullet} \text{ given by (4.1.11)}$$

(ii) 2D linear finite element with reduced continuity: (12.3.8)
P(e) = $P_1(e)$ space spanned by $\{\phi_{12},\phi_{13},\phi_{23}\}$,
ϕ given by (4.2.4)

$$\Sigma_e = \{\underline{x}^{12},\underline{x}^{13},\underline{x}^{23}\}$$

(iii) 2D extended quadratic finite element: (12.3.9)
$P(e) = P_2^+(e)$ space spanned by
$\{\phi_1, \phi_2, \phi_3, \phi_{12}, \phi_{13}, \phi_{23}, \phi_{123}\}$, ϕ given by (4.2.11)

$$\Sigma_e = \{\underline{x}^1, \underline{x}^2, \underline{x}^3, \underline{x}^{12}, \underline{x}^{13}, \underline{x}^{23}, \underline{x}^{123}\}, \ \underline{x}^{123} \text{ barycentre of } e$$

(iv) 3D linear element with reduced continuity: (12.3.10)
$P(e)$ = space spanned by $\{\phi_{123}, \phi_{124}, \phi_{134}, \phi_{234}\}$,
ϕ given by (4.3.3)

$$\Sigma_e = \{\underline{x}^{123}, \underline{x}^{124}, \underline{x}^{134}, \underline{x}^{234}\}, \ \underline{x} \text{ barycentre of face.}$$

Before we pass on to the construction of finite element spaces V_h, we establish a triangulation E_h of the set $\bar{\Omega} \subset \mathbb{R}^n$, i.e. the set $\bar{\Omega}$ is subdivided into a finite number of non-degenerate closed n-simplices having a diameter less than h.
We assume that the following properties are satisfied:

(i) $\bigcup_{e \in E_h} e = \Omega$

(ii) If e_k and e_ℓ belong to E_h, then either $e_k = e_\ell$, or $e_k \cap e_\ell = \emptyset$, or $e_k \cap e_\ell$ = m-face of e_k and e_ℓ $(0 \leq m \leq n-1)$. An m-face of an n-simplex $(0 \leq m \leq n-1)$ is any m-simplex generated by m+1 vertices of the n-simplex. A 0-face is a vertex.

Finally we introduce the following quantities:

$$h_e = \text{diameter of } e \ , \quad h = \sup_{e \in E_h} h_e$$

h'_e = diameter of greatest ball contained in e

$$\sigma(h) = \sup_{e \in E_h} \frac{h_e}{h'_e} \quad \text{which we assume bounded for } h \to 0.$$

12.3.2 Linear finite element approximation of $L_2(\Omega)$

In this section we will define an internal approximation of the space $L_2(\Omega)$ by means of the linear finite element introduced in proposition (5.2.11).

In order to define an internal approximation $\{V_h,\pi_h,\rho_h\}_{h\in H}$, we define the space V_h as the space of continuous functions on $\bar{\Omega}$ whose restriction to an n-simplex $e \in E_h$ belongs to $P_1(e)$ and which vanish on Γ:

$$V_h = \{v_h \in C(\bar{\Omega}) \mid v_h|_e \in P_1(e)\ ,\ v_h|_\Gamma = 0\} \qquad (12.3.11)$$

In V_h we define the Hilbert structure induced by $L_2(\Omega)$.

The prolongation operator $\pi_h : V_h \to L_2(\Omega)$ is defined by

$$\pi_h v_h = v_h \qquad (12.3.12)$$

It is sufficient to define the restriction operator ρ_h on a dense subspace of $L_2(\Omega)$, for instance the space $C_0^m(\bar{\Omega})$, $m > 0$, of $C^m(\bar{\Omega})$-functions on $\bar{\Omega}$ having vanishing derivatives of order $< m$ on Γ:

$$\rho_h : C_0^m(\bar{\Omega}) \to V_h$$

$$\rho_h v(\underline{x}) = \sum_{i=1}^{n+1} v(\underline{x}^i)\lambda_i(\underline{x})\ ,\quad \underline{x} \in e, \text{ for all } e \in E_h \qquad (12.3.13)$$

with $\{\lambda_1(\underline{x}),\ldots,\lambda_{n+1}(\underline{x})\}$ the barycentric coordinates of $\underline{x}$ with respect to the vertices of e.

The following theorem can be proved:

Theorem (12.3.14)

The family $\{V_h,\pi_h,\rho_h\}_{h\in H}$ is a stable, convergent, internal approximation of $L_2(\Omega)$. In fact we can prove

$$||v-\rho_h v||_{L_2} \le c_2(v)h^2 \quad \text{for all } v \in V = C_0^2(\bar{\Omega}) \qquad (12.3.15)$$

where $c_2(v)$ is a constant depending on the second derivatives of v.

A second example of an internal approximation of $L_2(\Omega)$ can be constructed using the finite elements with reduced continuity (12.3.8), (12.3.10). We define

$$V_h = \{v_h : \bar{\Omega} \to \mathbb{R} \mid v_h|_e \in P_1(e) , \quad v_h \text{ is continuous in the barycentres of the } (n-1)\text{-faces of } e,\ e \in \mathcal{E}_h\} \subset L_2(\Omega) \tag{12.3.16}$$

The prolongation operator $\pi_h : V_h \to L_2(\Omega)$ is given by

$$\pi_h v_h = v_h \tag{12.3.17}$$

The restriction operator ρ_h is defined on $C^2(\bar{\Omega})$ and given by

$$\rho_h v(\underline{x}) = \sum_g v(\underline{x}^g)\ \phi_g(\underline{x}), \quad \underline{x} \in e, \text{ for all } e \in \mathcal{E}_h \tag{12.3.18}$$

where the summation is taken over all barycentres $\underline{x}^g$ of (n-1) faces of e and where ϕ_g denotes the corresponding basis function.

Theorem (12.3.19)

The family $\{V_h, \pi_h, \rho_h\}_{h \in H}$ is a stable, convergent, internal approximation of $L_2(\Omega)$ and

$$||v-\rho_h v||_{L_2} \leq c_2(v)\ h^2 \qquad \text{for all } v \in C^2(\bar{\Omega}) \tag{12.3.20}$$

12.3.3 Linear finite element approximation of $H_0^1(\Omega)$

The space V_h, the prolongation operator ρ_h and the restriction operator π_h are defined by (12.3.11), (12.3.12) and (12.3.13)

respectively. We endow the space V_h with the Hilbert structure induced by $H_0^1(\Omega)$.
We can prove the following theorem:

Theorem (12.3.21)

The family $\{V_h, \pi_h, \rho_h\}_{h \in H}$ is a stable, convergent, internal approximation of $H_0^1(\Omega)$.
In fact we have

$$||v-\rho_h v||_{H_0^1} \leq c_2(v)\ \sigma(h)\ h \quad \text{for all } v \in C_0^2(\bar{\Omega}) \tag{12.3.22}$$

where $c_2(v)$ depends on second derivatives of v.

12.3.4 Quadratic finite element approximation of $H_0^1(\Omega)$

The space V_h consists of all continuous functions on $\bar{\Omega}$ which are polynomials of degree 2 on each n-simplex $e \in E_h$ and which vanish on Γ:

$$V_h = \{v_h \in C_0(\bar{\Omega})\,|\, v_h|_e \in P_2(e)\} \tag{12.3.23}$$

The topology in V_h is induced by $H_0^1(\Omega)$. The prolongation operator $\pi_h : V_h \to H_0^1(\Omega)$ is given by

$$\pi_h v_h = v_h \tag{12.3.24}$$

and the restriction operator ρ_h, which is defined on the dense subspace $C_0^3(\bar{\Omega})$ of $H_0^1(\Omega)$, is given by

$$\rho_h v(\underline{x}) = \sum_{i=1}^{n+1} v(\underline{x}^i)\lambda_i(\underline{x})(2\lambda_i(\underline{x})-1) + 4\sum_{\substack{i,j=1\\ i<j}}^{n+1} v(\underline{x}^{ij})\lambda_i(\underline{x})\lambda_j(\underline{x})$$

$$\underline{x} \in e, \quad \text{for all } e \in E_h. \tag{12.3.25}$$

Theorem (12.3.26)

The family $\{V_h, \pi_h, \rho_h\}_{h \in H}$ is a stable, convergent, internal approximation of $H_0^1(\Omega)$; and

$$||v-\rho_h v||_{H_0^1} \leq c_3(v)\, \sigma(h)\, h^2 \quad \text{for all } v \in C_0^3(\bar{\Omega}) \qquad (12.3.27)$$

where $c_3(v)$ depends on third derivatives of v.

12.3.5 Linear finite element approximation of $H^1(\Omega)$

$$V_h = \{v_h \in C(\bar{\Omega}) \mid v_h|_e \in P_1(e)\}$$

$$\pi_h : V_h \to H^1(\Omega), \quad \pi_h v_h = v_h$$

$$\rho_h : C^2(\bar{\Omega}) \to H^1(\Omega), \quad \rho_h v(\underline{x}) = \sum_{i=1}^{n+1} v(\underline{x}^i)\lambda_i(\underline{x}),$$

$$\underline{x} \in e, \quad \text{for all } e \in E_h$$

Theorem (12.3.28)

$\{V_h, \pi_h, \rho_h\}_{h \in H}$ is a stable, convergent, internal approximation of $H^1(\Omega)$. Moreover:

$$||v-\rho_h v||_{H^1} \leq c_2(v)\, \sigma(h)\, h \quad \text{for all } v \in C^2(\bar{\Omega}) \qquad (12.3.29)$$

12.3.6 Finite element approximation of $\underline{V}$

In this section we shall discuss a finite element approximation of the space $\underline{V}$, where the treatment of the incompressibility constraint $\nabla \cdot \underline{u} = 0$ is of course the main difficulty (see part II). Although it is possible in 2D to construct an internal finite element approximation of $\underline{V}$ using polynomials of degree

4, we shall introduce here an n-dimensional external finite element approximation using the piecewise linear finite element of (12.3.8) or (12.3.10).
Let us first introduce the notion of external approximation. Such an approximation is used when a direct relation between the spaces V and V_h is not possible. The relation is then defined using a third space F.

Definition (cf. definition (12.1.1)) (12.3.30)

A set $\{F,\omega,\{V_h,\pi_h,\rho_h\}_{h\in H}\}$ is called an external approximation of a Hilbert space V if

(i) F is a Hilbert space and ω an isomorphism (i.e. a linear one-to-one mapping) from V into F

(ii) for each $h \in H$: V_h is a Hilbert space (with norm $||\cdot||_h$)
π_h a continuous linear operator from V_h into F
ρ_h an operator from V into V_h.

Notice that an internal approximation is a particular case of an external approximation: ω = identity, F = V.

Definition (cf. definition (12.1.2)) (12.3.31)

An external approximation of V is stable if (12.1.3) is satisfied, and is called convergent if

(i) $\lim\limits_{h\to 0} ||\pi_h\rho_h v - \omega v||_F = 0$ for all $v \in \mathcal{V}$ dense in V.

(ii) for each sequence $\{v_h\}$, $h \to 0$, in V_h, the property

there exists $\phi \in F$ such that

$$\lim_{h\to 0} (\pi_h\rho_h v - \phi, z)_F = 0 \quad , \quad \text{for all } z \in F$$

implies $\phi \in \omega(V)$, which means $\phi = \omega v$ for some $v \in V$.

In order to formulate the discrete variational problem, we give, for each $h \in H$:

(i) a continuous bilinear V_h-elliptic form $a_h(\cdot,\cdot)$ on $V_h \times V_h$ which satisfies the following consistency property:

If, for $h \to 0$, the sequences $\{v_h\}$ and $\{w_h\}$ in V_h are such that

$$\lim_{h\to 0} (\pi_h v_h - \omega v, z)_F = 0 \ , \quad v \in V \ , \quad \text{for all } z \in F$$

$$\lim_{h\to 0} ||\pi_h w_h - \omega w||_F = 0 \ , \quad w \in V \tag{12.3.32}$$

then

$$\lim_{h\to 0} a_h(v_h, w_h) = a(v,w)$$

$$\lim_{h\to 0} a_h(w_h, v_h) = a(w,v)$$

(ii) a continuous linear form $L_h(\cdot)$ on V_h which satisfies the following consistency property:

If, for $h \to 0$, the sequence $\{v_h\}$ in V_h is such that

$$\lim_{h\to 0} (\pi_h v_h - \omega v, z)_F = 0 \ , \quad v \in V \ , \quad \text{for all } z \in F \tag{12.3.33}$$

then:

$$\lim_{h\to 0} L_h(v_h) = L(v)$$

We associate now to problem (12.2.1) the following family of approximate problems:

Find $u_h \in V_h$ such that

$$a_h(u_h, v_h) = L_h(v_h) \qquad \text{for all } v_h \in V_h \tag{12.3.34}$$

Theorem (cf. theorem (12.2.3)) (12.3.35)

Let $\{F,\omega,\{V_h,\pi_h,\rho_h\}_{h\in\mathcal{H}}\}$ be an external stable convergent approximation of V. Let $a_h(\cdot,\cdot)$ and $L_h(\cdot)$ be as defined above. Then (12.3.34) has a unique solution u_h which converges to the solution u of (12.2.1) in the sense of

$$\lim_{h\to 0} ||\pi_h u_h - \omega u||_F = 0 \qquad (12.3.36)$$

In this context it is worthwhile to mention the patch-test, which is a useful practical tool for testing external finite element approximations (non-conforming elements, see Section 5.5). In its original form the patch-test is due to Irons and Razzaque (1972), but has been further refined and analyzed by Strang (1972). The idea is that if a second order linear BVP has a linear solution, the FEM must approximate this solution exactly. Although this test is useful for engineers, it is neither a necessary nor a sufficient condition for convergence of the finite element approximation (see Stummel (1980)). A generalized patch-test which yields a both necessary and sufficient condition for an important class of non-conforming elements has been established by Stummel (1979).

We shall now construct an external approximation of $\underline{V}$ using the linear finite element with reduced continuity of (12.3.8) or (12.3.10).

Let us denote by G the set of barycentres of the (n-1)-faces of the n-simplices $e \in \mathcal{E}_h$. Then we define $\underline{V}_h$ as the space consisting of all n-component (n=2,3) vector-functions on $\bar{\Omega}$ which are polynomials of degree ≤ 1 on each n-simplex $e \in \mathcal{E}_h$, which are continuous in the points belonging to G, which are solenoidel on the interior of each $e \in \mathcal{E}_h$ and which vanish on the set $G \cap \Gamma$.

In $\underline{V}_h$ we define the Hilbert structure

$$(\underline{u}_h, \underline{v}_h)_h = \sum_{e \in E_h} \sum_{i,j=1}^{n} \int_e \frac{\partial u_{hi}}{\partial x_j} \frac{\partial v_{hi}}{\partial x_j} d\Omega ,$$

$$||\underline{v}_h||_h = (\underline{v}_h, \underline{v}_h)_h^{\frac{1}{2}} \tag{12.3.37}$$

Next we define $\underline{F} = (\underline{L}_2(\Omega))^{n+1}$ and an isomorphism ω: $\underline{V} \to \underline{F}$ by

$$\omega\underline{v} = \{\underline{v}, \frac{\partial \underline{v}}{\partial x_1}, \ldots, \frac{\partial \underline{v}}{\partial x_n}\} \tag{12.3.38}$$

The prolongation operator $\pi_h : \underline{V}_h \to \underline{F}$ is given by

$$\pi_h \underline{v}_h = \{\underline{v}_h, \frac{\partial \underline{v}_h}{\partial x_1}, \ldots, \frac{\partial \underline{v}_h}{\partial x_n}\}, \tag{12.3.39}$$

$\frac{\partial}{\partial x_i}$ defined in the interior of each $e \in E_h$

The restriction operator ρ_h is defined in the following way. Since the space

$$\underline{V} = \{\underline{v} \in \underline{C}_0^{\infty}(\bar{\Omega}) | \nabla \cdot \underline{v} = 0\} \tag{12.3.40}$$

is a dense subspace of $\underline{V}$, it is sufficient to define ρ_h on $\underline{V}$. The image $\underline{v}_h = \rho_h \underline{v}$ of $\underline{v} \in \underline{V}$ is completely determined by its values in the barycentres of the (n-1)-faces of the n-simplices. Let S be an (n-1)-face with barycentre $\underline{G}$, then we define $\underline{v}_h(\underline{G})$ as the mean-value of $\underline{v}$ on S

$$\underline{v}_h(\underline{G}) = \frac{1}{|S|} \int_S \underline{v} \, dS \tag{12.3.41}$$

where $|S|$ denotes the area of S.
It is easy to verify that $\underline{v}_h = \rho_h \underline{v} \in \underline{V}_h$ for $\underline{v} \in \underline{V}$, in fact, since $\underline{v}_h$ is linear on e it is also linear on each (n-1)-face S_i of e, $i = 1,2,\ldots,n+1$. We have

$$\text{div } \underline{v}_h = \frac{1}{|e|} \int_e \text{div } \underline{v}_h \, d\Omega = \frac{1}{|e|} \sum_{i=1}^{n+1} \int_{S_i} \underline{v}_h \cdot \underline{n}_{S_i} \, dS =$$

$$= \frac{1}{|e|} \sum_{i=1}^{n+1} \underline{v}_h(\underline{G}_i)\cdot\underline{n}_{S_i} |S_i| = \frac{1}{|e|} \sum_{i=1}^{n+1} \int_{S_i} \underline{v}\cdot\underline{n}_{S_i}\, dS =$$

$$= \frac{1}{|e|} \int_e \operatorname{div} \underline{v}\, d\Omega = 0 \qquad \text{for} \quad \underline{v} \in \underline{V}$$

where $\underline{n}_{S_i}$ denotes the unit outer normal on S_i and $\underline{G}_i$ its barycentre. We can now give the following theorem:

Theorem (12.3.42)

The set $\{\underline{F},\omega,\{\underline{V}_h,\pi_h,\rho_h\}_{h\in H}\}$ is a stable, convergent, external approximation of $\underline{V}$.

12.4 Interpretation of the discretized variational problem

The aim of this section is to formulate some discretized variational problems and to give the convergence results. For the actual interpretation and the assembly of the system of linear algebraic equations we refer to the appropriate sections of part I and part II.

12.4.1 Dirichlet-Neumann problem for the Laplace operator

We shall summarize the results of two examples.

(i) Dirichlet problem for the Laplace operator.
The continuous problem is formulated by (11.1.8). A discretized formulation is given by (12.2.2) with $\{V_h,\pi_h,\rho_h\}$ the internal linear finite element approximation defined in section 12.3.3 and $L(\cdot)$ given by

$$L(v) = (f,v)_{L_2} \qquad f \in L_2(\Omega). \qquad (12.4.1)$$

The convergence result of theorem (12.2.3) states that

$$\lim_{h\to 0} ||u_h - u||_{L_2} = 0$$

$$\lim_{h\to 0} ||\frac{\partial u_h}{\partial x_i} - \frac{\partial u}{\partial x_i}||_{L_2} = 0 \qquad i = 1,2,\dots,n \tag{12.4.2}$$

When the linear approximation is replaced by the quadratic finite element approximation of section 12.3.4, we obtain the same convergence result (12.4.2).

(ii) Neumann problem for the Laplace operator.
The continuous problem is formulated by (11.2.4). A discretized formulation is given by (12.2.2) with $\{V_h, \pi_h, \rho_h\}$ the internal linear finite element approximation of section 12.3.5. L(.) defined by (12.4.1). The convergence theorem (12.2.3) implies (12.4.2).

12.4.2 Stokes problem

The continuous problem is formulated by (11.5.4). A discretized formulation is given by (11.5.2) with $\{\underline{F}, \omega, \{\underline{V}_h, \pi_h, \rho_h\}_{h\in H}\}$ the in section 12.3.6 defined external finite element approximation of $\underline{V}$ (cf. theorem 12.3.42). The forms $a_h(.,.)$ and $L_h(.)$ are defined as follows:

$$a_h(\underline{u}_h, \underline{v}_h) = \mu \sum_{e\in E_h} \sum_{i,j=1}^{n} \int_e \frac{\partial u_{hi}}{\partial x_j} \frac{\partial v_{hi}}{\partial x_j} d\Omega \quad \text{(cf. (12.3.37))}$$

$$L_h(\underline{v}_h) = \int_\Omega \underline{f}\cdot\underline{v}_h \, d\Omega$$

The convergence of $\underline{u}_h$ to $\underline{u}$ is established by theorem (12.3.35):

$$\lim_{h\to 0} ||\underline{u}_h - \underline{u}||_{\underline{L}_2} = 0$$

$$\lim_{h\to 0} ||\frac{\partial u_{hi}}{\partial x_j} - \frac{\partial u_i}{\partial x_j}||_{L_2(e)} = 0 \quad i,j=1,2,\dots,n, \quad \text{for all } e \in E_h \tag{12.4.3}$$

For the construction of a basis in the space $\underline{V}_h$ we refer to Section 12.3.6. Concerning the interpretation we shall introduce here the discrete pressure. Let $\underline{W}_h$ be the space of all n-component vector-functions on $\bar{\Omega}$ which are polynomials of degree ≤ 1 on each $e \in E_h$, which are continous in the points belonging to G and which vanish on $G \cap \Gamma$ (G is the set of barycentres defined in Section 12.3.6). Then

$$\underline{V}_h = \{\underline{v}_h \in \underline{W}_h \mid \nabla \cdot \underline{v}_h = 0 \quad \text{on each } e \in E_h\}$$

The form

$$\underline{v}_h \to a_h(\underline{u}_h, \underline{v}_h) - L_h(\underline{v}_h)$$

is linear on $\underline{W}_h$ and vanishes on $\underline{V}_h$. The space $\underline{V}_h$ is characterized by the set of linear constraints $\nabla \cdot \underline{u}_h = 0$ on each $e \in E_h$.

Consequently there exists a collection of Lagrange multipliers $\lambda_e \in \mathbb{R}$, $e \in E_h$ such that

$$a_h(\underline{u}_h, \underline{v}_h) - L_h(\underline{v}_h) = \sum_{e \in E_h} \lambda_e \nabla \cdot \underline{v}_h , \qquad \text{for all } \underline{v}_h \in \underline{W}_h$$

Let us introduce now the discrete pressure p_h by

$$p_h = \sum_{e \in E_h} \frac{\lambda_e}{|e|} \chi_e \quad (\chi_e = \text{characteristic function of } e) \quad (12.4.4)$$

then

$$a_h(\underline{u}_h, \underline{v}_h) - (p_h, \nabla \cdot \underline{v}_h)_{L_2} = L_h(\underline{v}_h) \quad \text{for all } \underline{v}_h \in \underline{W}_h \quad (12.4.5)$$

and we see indeed that p_h can be interpreted as the discrete pressure.

Chapter 13

Error Analysis of the FEM

The purpose of this chapter is to state the theorems on error estimates, i.e. estimates of the difference between the solution u_h of the approximate problem and the solution u of the continuous problem. Both simplicial as well as quadrilateral and hexahedral elements will be discussed. The influence of numerical integration on the error estimates will be examined.

13.1 H^1 and L_2 error estimates

In this section we shall give an error analysis of the FEM. We restrict ourselves to second order BVPs. Second order BVPs lead, by applying Green's formula, to variational formulations in which only derivatives of order less than or equal to one occur. This implies that the spcae V in the general formulation (11.3.1) satisfies $H_0^1(\Omega) \subset V \subset H^1(\Omega)$. A second restriction in this error analysis is that we only consider finite elements $\{e,P(e),\Sigma_e\}$ of class C^0, which are defined by the following two conditions:

(i) $P(e) \subset C^0(e)$ = class of continuous functions on e

(ii) for each face S of e, the set $\Sigma_S = \Sigma_e \cap S$ is P(S)-unisolvent where $P(S) = \{p|_S \mid p \in P(e)\}$

In a remark we shall extend the error estimates to elements which are not of class C^0.

The aim of this section is thus to compare the solution u of problem (11.3.1), where V, a(.,.) and L(.) satisfy the conditions of theorem (11.3.3) with the solution u_h of (12.2.2), where it is assumed that $\{V_h, \pi_h, \rho_h\}_{h \in H}$ is a stable convergent internal approximation of V.

A first result can be obtained immediately by the remark that, since $V_h \subset V$, the difference $u_h - u$ satisfies

$$a(u_h - u, w_h) = 0 \qquad \text{for all } w_h \in V_h$$

Writing $w_h = u_h - v_h$ and making use of the continuity and the ellipticity of a(.,.), one easily deduces the following theorem:

Theorem (Céa's lemma) (13.1.1)

The solution u of (11.3.1) and u_h of (12.2.2) satisfy

$$||u_h - u||_V \leq \frac{M}{\alpha} \inf_{v_h \in V_h} ||u - v_h||_V \qquad (13.1.2)$$

with M and α defined in theorem (11.3.3).

This result establishes that the error measured in the V-norm is bounded above by the distance in V of the exact solution u from any member v_h in the finite element space V_h. The problem of finding error estimates is now reduced to one of estimating the distance $||u - v_h||_V$, $v_h \in V_h$, which is often phrased as the problem of finite element interpolation. From estimate (13.1.2) we deduce

$$||u_h - u||_V \leq \frac{M}{\alpha} \inf_{v_h \in V_h} ||u - v_h||_V = \frac{M}{\alpha} \inf_{v \in V} ||u - r_h v||_V \leq$$

$$\leq \frac{M}{\alpha} \{||u - v||_V + ||v - r_h v||_V\} \text{ for all } v \in V. \qquad (13.1.3)$$

Using now inequality (12.3.22) for the Dirichlet problem with linear finite elements, or (12.3.27) for the Dirichlet problem with quadratic finite elements, or (12.3.29) for the Neumann problem with linear finite elements, it is possible to estimate $||u_h-u||_V$. We shall give here the following general result:

Theorem (13.1.4)

Assume $n \leq 3$. Let the family $\{e,P(e),\Sigma_e\}$ be a class C^0 with Σ_e a P(e)-unisolvent set. Let V_h be defined by

$$V_h = \{v_h \in C^0(\bar{\Omega}) \mid v_h|_e \in P_e\} \cap V$$

Assume that there exists an integer $k \geq 1$ such that

$$P_k(\hat{e}) \subset P(\hat{e}) \subset H^1(\hat{e}) \qquad (13.1.5)$$

where $\hat{e}$ is the reference element defined in section 5.4.2. Then, if the solution $u \in V$ of (11.3.1) belongs to $H^{k+1}(\Omega)$, there exists a constant c independent of k such that

$$||u-u_h||_{H^1} \leq c\, h^k ||u||_{H^{k+1}} \qquad (13.1.6)$$

where u_h is the solution of (12.2.2).

Proof See Ciarlet (1978), Raviart, Thomas (1983).

As a consequence of this theorem we see that the linear finite element (see proposition (12.3.3)) which corresponds to $k = 1$ is first order accurate. The quadratic finite element of proposition (12.3.5) corresponds to $k = 2$ and is second order accurate. The cubic element of (12.3.7) is third order accurate: $k = 3$. In all these cases $P(e) = P_k(e)$. In the case of the extended quadratic finite elements defined in (12.3.9) we have $P_2(e) \subset P(e) = P_2^+(e) \subset H^1(e)$, consequently the error estimate (13.1.6) holds for $k = 2$.
The result of theorem (13.1.4) also holds for quadrilateral ($n = 2$) and hexahedral ($n = 3$) finite elements as defined in sections 4.4 and 4.5. For the $Q_1(e)$ finite element we have in

two and three dimensions $P_1(\hat{e}) \subset P(\hat{e}) = Q_1(\hat{e}) \subset H^1(\hat{e})$, so that the error estimate (13.1.6) holds for $k = 1$. The $Q_2(e)$ finite element satisfies $P_2(\hat{e}) \subset P(\hat{e}) = Q_2(\hat{e}) \subset H^1(\hat{e})$ which implies second order accuracy: $k = 2$. Theorem (5.3.4) can not be applied to the linear finite elements with reduced continuity (12.3.8), (12.3.10), because the space V_h is not contained in the space of continuous functions. Moreover, the corresponding discrete problem is not of the form (12.2.2), because $V_h \not\subset V$.

Error estimate (13.1.6) implies also that the $L_2(\Omega)$-norm of the difference $u_h - u$ is at least of the same order:

$$||u-u_h||_{L_2} \leq c\, h^k ||u||_{H^{k+1}} \tag{13.1.7}$$

The next theorem shows that, under additional assumptions, one has in fact $||u-u_h||_{L_2} = O(h^{k+1})$. We make the following assumption:

There exists a constant c such that for any $g \in L_2(\Omega)$ the solution w of the following problem

$$\begin{array}{l} \text{Find } w \in V \text{ such that} \\ a(v,w) = (g,v)_{L_2} \quad \text{for all } v \in V \end{array} \tag{13.1.8}$$

belongs to the space $H^2(\Omega) \cap V$ and satisfies $||w||_{H^2} \leq c||g||_{L_2}$

Theorem (13.1.9)

In addition to the assumptions of theorem (13.1.4) we assume that (13.1.8) holds. Then, if the solution $u \in V$ of (11.3.1) belongs to $H^{k+1}(\Omega)$, there exists a constant c independent of h such that

$$||u-u_h||_{L_2} \leq c\, h^{k+1} ||u||_{H^{k+1}} \tag{13.1.10}$$

with u_h solution of (12.2.2).

Proof See Ciarlet (1978), Raviart, Thomas (1983).

Let us give an example for which assumption (13.1.8) is satisfied (see Grisvard (1976)). Let a(.,.) be defined by (11.4.1), then (13.1.8) is true when

(i) $a_{ij} \in C^1(\bar{\Omega})$, $a_0 \in C^0(\bar{\Omega})$

(ii) Ω is polyhedral and convex

(iii) $V = H_0^1(\Omega)$ or $V = H^1(\Omega)$

Let us remark that for the mixed Dirichlet-Neumann problem (11.2.5), (11.2.6), condition (13.1.8) is not satisfied.

The following theorem summarizes for a special case the error estimates for the linear finite elements with reduced continuity: (12.3.8) for n=2, (12.3.10) for n=3.
We consider problem (11.3.1) with

$$a(u,v) = \sum_{i,j=1}^{n} \int_\Omega a_{ij} \frac{\partial u}{\partial x_i} \frac{\partial v}{\partial x_j} d\Omega + \int_\Omega a_0 u \ v \ d\Omega \ , \quad a_{ij} = a_{ji}$$

$$L(v) = \int_\Omega f \ v \ d\Omega$$

The discrete problem is defined by (12.3.34) with

$$a_h(u_h,v_h) = \sum_{e\in E_h} \sum_{i,j=1}^{n} \int_e a_{ij} \frac{\partial u_h}{\partial x_i} \frac{\partial v_h}{\partial x_j} d\Omega + \sum_{e\in E_h} \int_e a_0 u_h v_h d\Omega$$

$$L_h(v_h) = \sum_{e\in E_h} \int_e f \ v \ d\Omega$$

V_h generated by finite element (12.3.8) or (12.3.10) with Hilbert structure

$$(u_h,v_h)_h = \sum_{e\in E_h} \sum_{i=1}^{n} \int_e \frac{\partial u_h}{\partial x_i}\frac{\partial v_h}{\partial x_i}\, d\Omega + \sum_{e\in E_h} \int_e u_h v_h d\Omega$$

Theorem (13.1.11)

Assume that the solution u belongs to the space $H^2(\Omega)$. Then there exists a constant c independent of h such that

$$||u-u_h||_h \equiv \Big(\sum_{e\in E_h} ||u-u_h||^2_{H^1(e)}\Big)^{\frac{1}{2}} \leq c\, h||u||_{H^2} \qquad (13.1.12)$$

With the supplementary condition (13.1.8) one can show that

$$||u-u_h||_{L_2} \leq c\, h^2||u||_{H^2} \qquad (13.1.13)$$

Error estimates for the Stokes equations using the linear finite element with reduced continuity can be found in Crouzeix, Raviart (1973) and Hecht (1981).

Finally, we mention that it is sometimes possible to measure the error in the infinity norm of $u-u_h$, for which estimates of the following type hold:

$$||u-u_h||_{0,\infty} \leq c\, h^2|\log h|\ ||u||_{2,\infty} \quad \text{for } k = 1 \qquad (13.1.14)$$

$$||u-u_h||_{0,\infty} \leq c\, h^{k+1}||u||_{k+1,\infty} \quad \text{for } k \geq 2 \qquad (13.1.15)$$

where generally:

$$||v||_{m,\infty} = \max_{0\leq j\leq m} \{\operatorname{ess\,sup}_{\underline{x}\in\Omega} |D^j v(\underline{x})|\}$$

with D^j all derivatives of order j. For details the reader is referred to Ciarlet (1978), Nitsche (1976), Natterer (1975), Ciarlet, Raviart (1973).

13.2 Numerical integration

In this section we shall analyse the error of the FEM when numerical integration is used. The results will be given on the basis of the following homogeneous Dirichlet/Neumann problem

$$\left|\begin{array}{ll} A\,u = f & \text{in } \Omega \\ u = 0 \quad \text{or} \quad \dfrac{\partial u}{\partial n_a} = 0 & \text{on } \Gamma \end{array}\right. \tag{13.2.1}$$

where A is given by (11.6.1) satisfying (11.6.2) and $f \in C^{\infty}(\bar{\Omega})$. We consider finite elements $\{e, P(e), \Sigma_e\}$ of class C^0, with Σ_e a P(e)-unisolvent set. The space V_h is defined as in theorem (13.1.4). The discrete problem is formulated as follows:

$$\left|\begin{array}{l} \text{Find } u_h \in V_h \text{ such that} \\ \displaystyle\sum_{e\in E_h} \sum_{i,j=1}^{n} \int_e a_{ij} \frac{\partial u_h}{\partial x_i} \frac{\partial v_h}{\partial x_j} d\Omega + \sum_{e\in E_h} \int_e a_0 u_h v_h d\Omega = \\ \quad = \displaystyle\sum_{e\in E_h} \int_e f\, v_h\, d\Omega \qquad \text{for all } v_h \in V_h \end{array}\right. \tag{13.2.2}$$

If we denote by $\{\phi_1, \phi_2, \ldots, \phi_L\}$ a basis for the space V_h, then (13.2.2) is equivalent to

$$\left|\begin{array}{l} \text{Find } u_h = \displaystyle\sum_{\ell=1}^{L} u_{h\ell}\phi_\ell\,, \quad u_{h\ell} \in \mathrm{IR}\,, \quad \text{such that} \\ \displaystyle\sum_{e\in E_h} \sum_{\ell=1}^{L} \sum_{i,j=1}^{n} u_{h\ell} \int_e a_{ij} \frac{\partial \phi_\ell}{\partial x_i} \frac{\partial \phi_m}{\partial x_j} d\Omega + \end{array}\right. \tag{13.2.3}$$

$$+ \sum_{e\in E_h} \sum_{\ell=1}^{L} u_{h\ell} \int_e a_0 \phi_\ell \phi_m d\Omega = \sum_{e\in E_h} \int_e f\, \phi_m\, d\Omega$$

$$m = 1, 2, \ldots, L$$

In general, but especially when the functions a_{ij}, a_0 and f have no simple analytical expression, the integrals

$$\int_e a_{ij} \frac{\partial \phi_\ell}{\partial x_i} \frac{\partial \phi_m}{\partial x_j} d\Omega \ , \quad \int_e a_0 \phi_\ell \phi_m d\Omega \ , \quad \int_e f \ \phi_m d\Omega$$

are not computed exactly. They are approximated through a process of numerical integration. This means that we replace, for any $\psi \in C^0(e)$, the integral $\int_e \psi(\underline{x}) d\Omega$ by a finite sum of the form $\sum_{r=1}^{R} w_r^e \ \psi(\underline{b}_r^e)$. The real numbers $w_r^e > 0$ are called the weights and the points $\underline{b}_r^e \in e$ are called the integration nodes. R is an integer independent of e. For examples of integration rules and their degree of accuracy we refer to section 5.4. Thus when we use numerical integration for the computation of the integrals we are, instead of (13.2.2) or (13.2.3), solving the following problem:

$$\left| \begin{array}{ll} \text{Find} \quad u_h^* \in V_h \quad \text{such that} & \\ a_h(u_h^*, v_h) = L_h(v_h) \qquad \text{for all} \quad v_h \in V_h & \end{array} \right. \tag{13.2.4}$$

where

$$a_h(u_h^*, v_h) = \sum_{e \in E_h} \sum_{i,j=1}^{n} \sum_{r=1}^{R} w_r^e (a_{ij} \frac{\partial u_h^*}{\partial x_i} \frac{\partial v_h}{\partial x_j})(\underline{b}_r^e) +$$

$$+ \sum_{e \in E_h} \sum_{r=1}^{R} w_r^e (a_0 u_h^* v_h)(\underline{b}_r^e)$$

$$L_h(v_h) = \sum_{e \in E_h} \sum_{r=1}^{R} w_r^e (f v_h)(\underline{b}_r^e)$$

In fact one needs only an integration rule over the reference element $\hat{e}$ defined in sections 4.4 and 5.4. Indeed, let F_e be an invertible mapping from $\hat{e}$ onto e then

$$\int_e \psi(\underline{x}) \ d\Omega = \int_{\hat{e}} \hat{\psi}(\hat{\underline{x}}) \ d\hat{\Omega}$$

with

$$\hat{\psi}(\hat{\underline{x}}) = \psi(F_e(\hat{x}))J_{F_e}(\hat{\underline{x}})$$

$$J_{F_e}(\hat{\underline{x}}) = \text{Jacobian of } F_e \text{ at } \hat{\underline{x}}$$

and

$$\int_{\hat{e}} \hat{\psi}(\hat{\underline{x}})\, d\hat{\Omega} \approx \sum_{r=1}^{R} \hat{w}_r \hat{\psi}(\hat{\underline{b}}_r) = \sum_{r=1}^{R} w_r^e \psi(\underline{b}_r^e)$$

where $\hat{w}_r$ are the weights and $\hat{\underline{b}}_r$ the integration nodes on the reference element, which are related to w_r^e and $\underline{b}_r^e$ by

$$w_r^e = J_{F_e}(\hat{\underline{b}}_r)\hat{w}_r \qquad \underline{b}_r^e = F_e(\hat{\underline{b}}_r)$$

We now have $u \in V$, solution of the variational formulation of (13.2.1), we have $u_h \in V_h$ solution of (13.2.2) and u_h^* solution of (13.2.4). In the next theorems we answer the following questions: For which integration rules do we have (i) existence and uniqueness of u_h^*, (ii) $||u-u_h^*||$ of the same order as $||u-u_h||$. The proofs of these theorems can be found in Ciarlet (1978), Raviart, Thomas (1983). First we consider the case $P(\hat{e}) = P_k(\hat{e})$.

Theorem (13.2.5)

Let $n \leq 3$. Assume that there exists an integer $k \geq 1$ such that

(i) $P(\hat{e}) = P_k(\hat{e})$

(ii) the integration rule is exact for all $\hat{\psi} \in P_{2k-2}(\hat{e})$

Then problem (13.2.4) has a unique solution.
If moreover the solution $u \in V$ belongs to $H^{k+1}(\Omega)$, then there exists a constant c independent of h such that

$$||u-u_h^*||_{H^1} \leq c\, h^k(||u||_{H^{k+1}}+1) \qquad (13.2.6)$$

This theorem applies to the linear, quadratic and cubic finite element, for instance.

The next theorem considers the case when $P_k(\hat{e}) \subset P(\hat{e}) \subset P_{k'}(\hat{e})$.

Theorem (13.2.7)

Let $n \leq 3$. Assume that for two integers $k \geq 1$, $k' \geq 1$

(i) $P_k(\hat{e}) \subset P(\hat{e}) \subset P_{k'}(\hat{e})$

(ii) the integration rule is exact for all $\hat{\psi} \in P_{k+k'-2}(\hat{e})$

(iii) the set $\{\hat{\underline{b}}_r \mid r=1,2,\ldots,R\}$ contains a $P_{k'-1}(\hat{e})$-unisolvent subset

Then one has the same conclusion as in theorem (13.2.5).

This theorem is applicable to the extended quadratic case $P_2^+(e)$ in 2D (see Section 4.2.5) in conjunction with integration rule (5.4.10). $P_2(\hat{e}) \subset P_2^+(\hat{e}) \subset P_3(\hat{e})$, which implies $k=2$, $k'=3$. The set $\{\hat{\underline{x}}^1, \hat{\underline{x}}^2, \hat{\underline{x}}^3, \hat{\underline{x}}^{12}, \hat{\underline{x}}^{13}, \hat{\underline{x}}^{23}, \hat{\underline{x}}^{123}\}$ contains the set $\{\hat{\underline{x}}^1, \hat{\underline{x}}^2, \hat{\underline{x}}^3, \hat{\underline{x}}^{12}, \hat{\underline{x}}^{13}, \hat{\underline{x}}^{23}\}$ which is $P_2(\hat{e})$-unisolvent.

The following theorem extends the result to the quadrilateral and hexahedral finite elements of sections 4.4 and 4.5 .

Theorem (13.2.8)

Let $n \leq 3$. Assume that for some integer $k \geq 1$

(i) $P_k(\hat{e}) \subset P(\hat{e}) \subset Q_k(\hat{e})$

(ii) the integration rule is exact for all $\hat{\psi} \in Q_{2k-1}(\hat{e})$

(iii) the set $\{\hat{\underline{b}}_r \mid r=1,2,\ldots,R\}$ contains a $Q_k(\hat{e}) \cap P_{nk-1}(\hat{e})$ unisolvent subset. (This implies $R \geq (k+1)^n-1$).

Then one has the same conclusion as in theorem (13.2.5).

From this theorem it can be deduced which integration rule must be used at least for the cases n = 2,3 and k = 1,2:

(i) n = 2, k = 1

$$Q_1(\hat{e}) \cap P_1(\hat{e}) = P_1(\hat{e}) = sp\{1,\hat{x}_1,\hat{x}_2\}$$

Integration rule: 4-point Newton-Cotes (5.4.24) / 4-point Gauss (5.4.26)

(ii) n = 2, k = 2

$$Q_2(\hat{e}) \cap P_3(\hat{e}) = sp\{1,\hat{x}_1,\hat{x}_2,\hat{x}_1^2,\hat{x}_1\hat{x}_2,\hat{x}_2^2,\hat{x}_1^2\hat{x}_2,\hat{x}_1\hat{x}_2^2\}$$

Integration rule: 9-point Newton-Cotes (5.4.25) / 9-point Gauss (5.4.27)

(iii) n = 3, k = 1

$$Q_1(\hat{e}) \cap P_2(\hat{e}) = sp\{1,\hat{x}_1,\hat{x}_2,\hat{x}_3,\hat{x}_1\hat{x}_2,\hat{x}_1\hat{x}_3,\hat{x}_2\hat{x}_3\}$$

Integration rule: 8-point Newton-Cotes (5.4.28) / 8-point Gauss (5.4.30)

(iv) n = 3, k = 2

$$Q_2(\hat{e}) \cap P_5(\hat{e}) = Q_2(\hat{e}) \setminus \{\hat{x}_1^2\ \hat{x}_2^2\ \hat{x}_3^2\}$$

Integration rule: 27-point Newton-Cotes (5.4.29) / 27-point Gauss (5.4.31)

Chapter 14

Mixed Finite Element Methods

Let us consider the Stokes problem the formulation of which is repeated here for convenience:

$$\left|\begin{array}{lll} \text{(i)} & -\mu\Delta\underline{u} + \nabla p = \underline{f} & \text{in } \Omega \\ \text{(ii)} & \nabla.\underline{u} = \underline{0} & \text{in } \Omega \\ \text{(iii)} & \underline{u} = \underline{0} & \text{on } \Gamma \end{array}\right. \qquad (14.1)$$

For simplicity we consider on Γ the no-slip boundary conditions, but the theory of mixed finite elements also applies to other type of boundary conditions.

Let us multiply (14.1) (i) by a function $\underline{v}$ and integrate over Ω, next we use Green's formula to reduce the second order term and we apply relation (11.5.3).

When $\underline{v}$ is taken such that $\underline{v}|_{\Gamma} = \underline{0}$, we obtain

$$a(\underline{u},\underline{v}) + b(p,\underline{v}) = \int_{\Omega} \underline{f}.\underline{v}\, d\Omega$$

where

$$a(\underline{u},\underline{v}) = \mu \sum_{i,j=1}^{n} \int_{\Omega} \frac{\partial u_i}{\partial x_j} \frac{\partial v_i}{\partial x_j}\, d\Omega \qquad (14.2)$$

$$b(\underline{v},p) = - \int_{\Omega} p(\nabla.\underline{v})\, d\Omega \qquad (14.3)$$

We multiply (14.1) (ii) by a function q and we integrate over Ω to obtain

$$- b(\underline{u},q) = 0 \qquad (14.4)$$

Choosing now $\underline{u}$, $\underline{v}$, p and q in function spaces which make the forms a(.,.) and b(.,.) finite, we can give the following formulation:

$$\left| \begin{array}{l} \text{Find } \{\underline{u},p\} \in \underline{H}_0^1(\Omega) \times \tilde{L}_2(\Omega) \\ \\ a(\underline{u},\underline{v}) + b(\underline{v},p) = (\underline{f},\underline{v})_{\underline{L}_2} \quad \text{for all } \underline{v} \in \underline{H}_0^1(\Omega) \\ \\ b(\underline{u},q) = 0 \quad \text{for all } q \in \tilde{L}_2(\Omega) \end{array} \right. \qquad (14.5)$$

where $\tilde{L}_2(\Omega) = \{q \in L_2(\Omega) | \int_\Omega q \, d\Omega = 0\}$

This formulation of the Stokes equations has been used already in chapters 7 and 8 in discrete form.
With this formulation in mind, let us formulate the following general abstract problem:

Let V and H be two Hilbert spaces such that $V \subset H$, which means $V \subset H$ and $||v||_H \leq c||v||_V$ for all $v \in V$. Then the problem is

$$\left| \begin{array}{ll} & \text{Find } \{u,p\} \in V \times H \quad \text{such that} \\ (i) & a(u,v) + b(v,p) = L(v) \quad \text{for all } v \in V \\ (ii) & b(u,q) = (g,q)_H \quad \text{for all } q \in H \end{array} \right. \qquad (14.6)$$

where $g \in H$, a(.,.) is a bilinear, $V \times V$-continuous, V-elliptic form and L(.) a linear continuous form on V:

$$|a(u,v)| \leq M||u||_V||v||_V$$

$$a(v,v) \geq \alpha||v||_V^2$$

$$|L(v)| \leq \ell||v||_V$$

and where b(.,.) is a bilinear $V \times H$-continuous form which satisfies the following condition, called the Babuška-Brezzi condition (B.B.-condition): there exists a constant $\beta > 0$ such that

$$\sup_{v \in V} \frac{b(v,q)}{||v||_V} \geq \beta ||q||_H \quad \text{for all } q \in H . \tag{14.7}$$

This condition was first derived by Ladyzhenskaya (1969) in the context of the (Navier-) Stokes equations, later on this condition was proposed in a general context by Babuška (1973a, 1973b), Babuška, Aziz (1972), Brezzi (1974), Babuška, Oden, Lee (1977, 1978).
Existence and uniqueness is guaranteed by the following theorem which we shall prove because the idea of the proof has been employed in Chapter 8 to solve the (Navier-) Stokes equations.

Theorem (14.8)

When the above mentioned assumptions on V, H, $a(\cdot,\cdot)$, $b(\cdot,\cdot)$, L and g are satisfied, then problem (14.6) has a unique solution $\{u,p\} \in V \times H$.

Proof

Consider the following family of regularized problems ($\varepsilon > 0$):

Find $\{u_\varepsilon, p_\varepsilon\} \in V \times H$ such that

(i) $a(u_\varepsilon, v) + b(v, p_\varepsilon) = L(v) \quad$ for all $v \in V$ (14.9)

(ii) $\varepsilon(p_\varepsilon, q)_H - b(u_\varepsilon, q) = -(g,q)_H \quad$ for all $q \in H$

We define for $\Phi_\varepsilon = \{u_\varepsilon, p_\varepsilon\}$ and $\Psi = \{v,q\}$ in $V \times H \equiv \mathcal{W}$

$$\mathring{A}_\varepsilon(\Phi_\varepsilon, \Psi) = a(u_\varepsilon, v) + b(v, p_\varepsilon) + \varepsilon(p_\varepsilon, q)_H - b(u_\varepsilon, q)$$

$$\mathcal{L}(\Psi) = L(v) - (g,q)_H$$

The bilinear form $\mathring{A}_\varepsilon(\Phi_\varepsilon, \Psi)$ is continuous on $\mathcal{W} \times \mathcal{W}$ and $\mathcal{W}$-elliptic, the linear form $\mathcal{L}(\Psi)$ is continuous on $\mathcal{W}$. By virtue of theorem (11.3.3), the following problem:

$$\left| \begin{array}{l} \text{Find } \Phi_\varepsilon \in W \text{ such that} \\ \mathcal{A}_\varepsilon(\Phi_\varepsilon,\Psi) = \mathcal{L}(\Psi) \quad \text{for all } \Psi \in W \end{array} \right. \tag{14.10}$$

which is equivalent to (14.9) admits a unique solution $\Phi_\varepsilon = \{u_\varepsilon, p_\varepsilon\}$.
Next we take $v = u_\varepsilon$ and $q = p_\varepsilon$ in (14.9). Adding the two equalities we get, using the V-ellipticity of $a(.,.)$ and $V \subset H$:

$$\begin{aligned} \alpha||u_\varepsilon||_V^2 + \varepsilon||p_\varepsilon||_H^2 &= L(u_\varepsilon) - (g,p_\varepsilon)_H \\ &\leq c(||u_\varepsilon||_V + ||p_\varepsilon||_H) \end{aligned} \tag{14.11}$$

Since

$$b(v,p_\varepsilon) = L(v) - a(u_\varepsilon,v) \leq ||v||_V \{\ell + M||u_\varepsilon||_V\}$$

we obtain, using the B.B.-condition

$$\beta||p_\varepsilon||_H \leq \sup_{v \in V} \frac{b(v,p_\varepsilon)}{||v||_V} \leq \{\ell + M||u_\varepsilon||_V\} \tag{14.12}$$

from which we deduce

$$||p_\varepsilon||_H \leq c(||u_\varepsilon||_V + 1) \tag{14.13}$$

Combining now (14.11) and (14.13) we obtain

$$\begin{array}{ll} \text{(i)} & \left| \begin{array}{ll} ||u_\varepsilon||_V \leq c & c \text{ independent of } \varepsilon. \\ ||p_\varepsilon||_H \leq c & \end{array} \right. \\ \text{(ii)} & \end{array} \tag{14.14}$$

This implies the existence of $\bar{u} \in V$, $\bar{p} \in H$ and a subsequence $\varepsilon' \to 0$ such that

$$(u_{\varepsilon'},v)_V \to (\bar{u},v)_V \quad \text{for all } v \in V$$

$$(p_{\varepsilon'},q)_H \to (\bar{p},q)_H \quad \text{for all } q \in H$$

Passing now to the limit ($\varepsilon' \to 0$) in (14.9) we find that $\{\bar{u},\bar{p}\} = \{u,p\}$ is a solution of (14.6).
Concerning the uniqueness of $\{u,p\}$, suppose that $\{u_1,p_1\}$ and $\{u_2,p_2\}$ are solutions of (14.6). Then

$$\begin{array}{ll} \text{(i)} & a(u_1-u_2,v) + b(v,p_1-p_2) = 0 \quad \text{for all } v \in V \\ \text{(ii)} & b(u_1-u_2,q) = 0 \quad \text{for all } q \in H \end{array} \qquad (14.15)$$

Taking $v = u_1-u_2$, $q = p_1-p_2$ we deduce from (14.15) (i), (ii)

$$\alpha||u_1-u_2||_V^2 \leq a(u_1-u_2,u_1-u_2) = 0$$

hence $u_1=u_2$. Next we use the B.B.-condition and (14.15) (i) to obtain

$$\beta||p_1-p_2||_H \leq \sup_{v\in V} \frac{b(v,p_1-p_2)}{||v||_V} = 0$$

which implies $p_1 = p_2$.
Hence the solution $\{u,p\}$ of (14.5) is unique.

In the following theorem an estimate of the difference $u-u_\varepsilon$ and $p-p_\varepsilon$ is given.

Theorem (14.16)

The assumptions are the same as for theorem (14.8).
Let $\{u,p\}$ be the solution of problem (14.6) and $\{u_\varepsilon,p_\varepsilon\}$ of (14.9) then

$$||u-u_\varepsilon||_V = O(\varepsilon) \qquad ||p-p_\varepsilon||_H = O(\varepsilon)$$

Proof

Subtraction of (14.9) (i) from (14.6) (i) gives

$$a(u-u_\varepsilon,v) + b(v,p-p_\varepsilon) = 0 \quad \text{for all } v \in V \qquad (14.17)$$

Using the B.B.-condition we find

$$\beta ||p-p_\varepsilon||_H \leq \sup_{v\in V} \frac{b(v,p-p_\varepsilon)}{||v||} \leq M||u-u_\varepsilon||_V \qquad (14.18)$$

From (14.6) (ii) and (14.9) (ii) we deduce

$$\varepsilon(p_\varepsilon,q)_H + b(u-u_\varepsilon,q) = 0 \qquad \text{for all } q \in H \qquad (14.19)$$

Next we choose $v = u-u_\varepsilon$ in (14.17) and $q = p-p_\varepsilon$ in (14.19). We get

$$\alpha||u-u_\varepsilon||_V^2 \leq a(u-u_\varepsilon,u-u_\varepsilon) = \varepsilon(p_\varepsilon,p-p_\varepsilon)_H \leq \varepsilon||p_\varepsilon||_H||p-p_\varepsilon||_H$$

Using now (14.14) (ii) and (14.18), we obtain

$$||u-u_\varepsilon||_V \leq c\ \varepsilon$$

and from (14.18) we get

$$||p-p_\varepsilon||_H \leq c\ \varepsilon$$

Remark (14.20)

Let the operator $B : V \to H$ be defined by

$$(Bv,q)_H = b(v,q) \qquad \text{for all } q \in H \qquad (14.21)$$

Then for the regularized problem (14.9) one has

$$\left|\begin{array}{l} p_\varepsilon = \frac{1}{\varepsilon}(Bu_\varepsilon-g) \\ a(u_\varepsilon,v) + \frac{1}{\varepsilon}(Bv,Bu_\varepsilon-g)_H = L(v) \qquad \text{for all } v \in V \end{array}\right. \qquad (14.22)$$

which corresponds to a penalization of the "constraint" (14.6) (ii).

Let us show that the general theorem (14.8) is applicable to the Stokes problem (14.5). We have $a(.,.)$ defined by (14.2), $b(.,.)$ defined by (14.3), $\underline{V} = \underline{H}_0^1(\Omega)$, $h = \tilde{L}_2(\Omega)$, $L(\underline{v}) = (\underline{f},\underline{v})_{\underline{L}_2}$, $\underline{f} \in \underline{L}_2(\Omega)$ and $g = 0$. The only point is to verify the B.B.-condition:

$$\sup_{\underline{v}\in \underline{H}_0^1(\Omega)} \frac{b(\underline{v},q)}{||\underline{v}||_{\underline{H}_0^1}} = \sup_{\underline{v}\in \underline{H}_0^1(\Omega)} \frac{-\int_\Omega q(\nabla\cdot\underline{v})d\Omega}{||\underline{v}||_{\underline{H}_0^1}} \geq c||q||_{\tilde{L}_2(\Omega)}$$

The last inequality is proved in Temam (1979). Consequently the solution $\{\underline{u},p\}$ of the Stokes problem exists uniquely.
Since

$$b(\underline{v},q) = (-\text{div } \underline{v},q)_H \quad \text{for all } q \in \tilde{L}_2(\Omega)$$

with

$$B = -\text{div} : \underline{H}_0^1(\Omega) \to \tilde{L}_2(\Omega)$$

the penalized Stokes equations take the following form (see Section 8.2.2):

$$\left|\begin{array}{l} \text{Find } \underline{u}_\varepsilon \in \underline{H}_0^1(\Omega) \text{ such that} \\[2ex] a(\underline{u}_\varepsilon,\underline{v}) + \frac{1}{\varepsilon}(\text{div } \underline{v},\text{div } \underline{u}_\varepsilon)_{\tilde{L}_2} = (\underline{f},\underline{v})_{\tilde{L}_2} \\[2ex] \qquad\qquad\qquad\qquad \text{for all } \underline{v} \in \underline{H}_0^1(\Omega) \\[2ex] p_\varepsilon = -\frac{1}{\varepsilon}\text{div } \underline{u}_\varepsilon \end{array}\right. \tag{14.23}$$

which is the variational formulation of the following BVP:

$$\left|\begin{array}{l} -\Delta\underline{u}_\varepsilon - \frac{1}{\varepsilon}\text{grad div } \underline{u}_\varepsilon = \underline{f} \quad \text{in } \Omega \\[2ex] \underline{u}_\varepsilon = \underline{0} \quad \text{on } \Gamma \\[2ex] p_\varepsilon = -\frac{1}{\varepsilon}\text{div } \underline{u}_\varepsilon \end{array}\right. \tag{14.24}$$

Let us finally study the finite element approximation of problem (14.6). We assume that $\{V_h,\pi_h^1,\rho_h^1\}$ and $\{H_h,\pi_h^2,\rho_h^2\}$ are stable, convergent, internal finite element approximations of V and H respectively. We formulate the approximate problem as follows:

$$\left|\begin{array}{l} \text{Find } \{u_h, p_h\} \in V_h \times H_h \text{ such that} \\ a(u_h, v_h) + b(v_h, p_h) = L(v_h) \quad \text{for all } v_h \in V_h \\ b(u_h, q_h) = (g, q_h)_H \quad \text{for all } q_h \in H_h \end{array}\right. \tag{14.25}$$

Assume that the discrete B.B.-condition

$$\left|\begin{array}{l} \text{there exists } \gamma > 0 \text{ independent of } h \text{ such that} \\ \sup\limits_{v_h \in V_h} \dfrac{b(v_h, q_h)}{||v_h||_V} \geq \gamma ||q_h||_H \quad \text{for all } q_h \in H_h \end{array}\right. \tag{14.26}$$

holds, then existence and uniqueness of a solution $\{u_h, p_h\}$ of (14.25) is guarenteed by theorem (14.8).
We remark that the continuous B.B.-condition (14.7) is not preserved in passing to the discrete problem. That is why we have to impose the discrete B.B.-condition for problem (14.25).

Finite element approximations of the type (14.25) are called mixed FEMs since two different unknown functions are concerned, i.e. u_h and p_h lying in two different spaces, i.e. V_h and H_h.

A practical way of verifying the discrete B.B.-condition is given by the following lemma.

Lemma (14.27)

If the form b(.,.) satisfies the continuous B.B.-condition (14.7) and if to any $v \in V$ there corresponds a $v_h \in V_h$ such that

$$\left|\begin{array}{l} b(v - v_h, q_h) = 0 \quad \text{for all } q_h \in H_h \\ ||v_h||_V \leq c||v||_V \end{array}\right. \tag{14.28}$$

then (14.26) holds.

Proof See Fortin (1976).

This lemma was used in section 7.2.2 to check whether a finite element approximation $\{\underline{V}_h, H_h\}$ for the Stokes equations satisfies (14.26) or not.

In order to give an estimate of the difference $\{u,p\} - \{u_h, p_h\}$ we introduce the following subsets of V and V_h

$$V(g) = \{v \in V \mid b(v,q) = (g,q)_H \quad \text{for all } q \in H\} \tag{14.29}$$

$$V_h(g) = \{v_h \in V_h \mid b(v_h,q_h) = (g,q_h)_H \quad \text{for all } q_h \in H_h\} \tag{14.30}$$

Theorem (14.31)

We make the above mentioned assumptions. If, moreover, one of the following conditions is satisfied

(i) $g \neq 0$, $V_h(0) \subset V(0)$

(ii) $g = 0$

then one has the error estimates:

$$||u-u_h||_V \leq c \inf_{v_h \in V_h} ||u-v_h||_V \tag{14.32}$$

$$||p-p_h||_H \leq c\{||u-u_h||_V + \inf_{q_h \in H_h} ||p-q_h||_H\} \tag{14.33}$$

where c is a constant independent of h; $\{u,p\}$ the solution of (14.6) and $\{u_h, p_h\}$ the solution of (14.25)

For a general theory on mixed formulations of elliptic BVPs and mixed FEMs we refer to Babuška, Aziz (1972), Babuška (1973a, 1973b), Babuška, Oden, Lee (1977, 1978), Brezzi (1974), Ciarlet (1978), Fortin (1977), Bercovier (1978). See also Glowinski, Lions, Trémolières (1976), Oden, Reddy (1976). Applications to the Stokes equations have been introduced in Chapter 7.

PART IV

CURRENT RESEARCH TOPICS

Introduction

At the end of this book some current fields of research will be overviewed to describe the state of the art. Especially capillary free boundaries (Chapter 15), non-isothermal flow (Chapter 16), turbulence (Chapter 17) and non-Newtonian effects will be treated. It is not the aim of these chapters to be complete, but to give a brief introduction to the mentioned fields with reference to textbooks for further reading. Therefore, each chapter contains the basic mathematical models, describing the flow problem concerned, and a brief indication of their numerical treatment elucidated by some practical calculations.

Chapter 15

Capillary Free Boundaries governed by the Navier-Stokes Equations

Introduction

As treated in Chapter 1, problems in mechanics and physics can oftenly be described by partial differential equations for the unknown functions. When there are additional geometrical unknowns one speaks of free boundary problems (see Cryer (1976), (1977)). A special case of them is formed by the problems with capillary free boundaries. They are defined as free boundaries on which one must take into account capillary forces, being forces due to intermolecular attractions having a non-vanishing resultant at the boundary of the liquids. There are many important technological and engineering-science applications in which capillary free boundaries play a dominant part. One needs to look only as far as lubrication (cf. Pinkus, Sternlight (1961), Cameron (1966), Coyne, Elrod (1970), Cuvelier (1978, 1979)), electrochemical plating, corrosion, coating (cf. Orr, Scriven (1978), Silliman, Scriven (1980), Saito, Scriven (1981), Kistler, Scriven (1984)), polymer technology, separation processes, metal and glass forming processes, aerospace technology (cf. Guibert, Huynh, Marce et al (1976), Ousset (1977), Greenspan (1968), Moisseev, Rumyantsev (1968)) and crystal growth (cf. Brice (1965), Hurle, Jakeman (1981), Pimputkar, Ostrach (1981)) to find abundant technologically important visous fluid flows with capillary free boundaries. For free boundaries in multi-fluid flows we refer to Dervieux, Thomasset (1981) and Dervieux (1981a,1981b).

In this chapter the governing equations will be treated in Section 15.1. As model problem the stationary motion of a liquid in an open vessel placed in the field of gravity is considered. Next two iterative methods for the numerical solution of a stationary free boundary problem will be discussed in Section 15.2, 15.3.

15.1 Mathematical model

Consider the stationary 2D motion of a liquid in an open vessel V placed in the field of gravity. When the fluid is in stationary motion the position of the fluid surface S, which is represented by a function ξ (see Fig. 15.1) is not known a priori and is called a free boundary. The object is to find the shape of this free boundary.

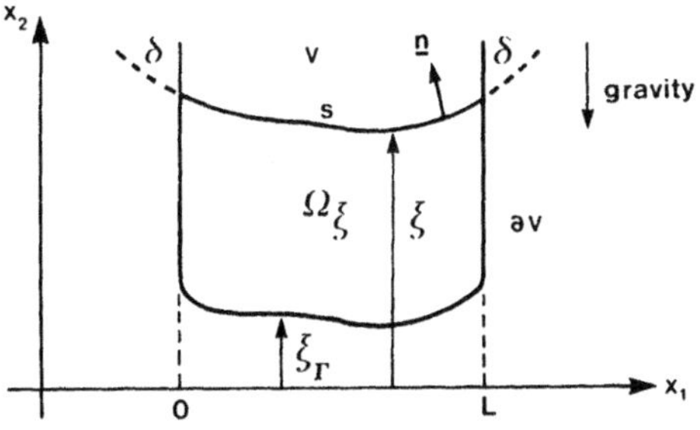

Fig. 15.1 Liquid contained in a vessel V.

To that end the fluid is assumed to be isothermal, incompressible, newtonian with constant density ρ, viscosity μ and surface tension α. With these assumptions the equations of stationary motion, which are the equations of conservation of momentum and mass, reduce to the Navier-Stokes equations for an incompressible fluid as given in Section 1.6.

The boundary conditions are as follows. On the vessel wall we prescribe Dirichlet boundary conditions $\underline{u} = \tilde{\underline{h}}$ on Γ. According to the basic principles of hydrodynamics the conditions of a balance of forces must be fulfilled on the free boundary. These conditions are the so-called traction conditions on S (cf. Batchelor (1967), Landau, Lifchitz (1963))

$$\sigma_n = \frac{\alpha}{R} - p_a \qquad (p_a = 0) \tag{15.1.1}$$

$$\sigma_t = 0 \tag{15.1.2}$$

with σ_n and σ_t the normal and tangential component of the stress tensor respectively. p_a denotes the outside (atmospheric) pressure which we put equal to zero. R denotes the radius of curvature on S and is defined by

$$\frac{1}{R} = \left(\frac{\xi'}{\{1+|\xi'|^2\}^{\frac{1}{2}}}\right)' \tag{15.1.3}$$

where the accent denotes differentiation with respect to x_1. The radius R is reckoned positive when the corresponding centre of curvature lies outside the liquid.
In the stationary case, the free boundary S is a streamline, which implies:

$$u_n = \underline{u}\cdot\underline{n} = 0 \quad \text{on } S.$$

Moreover the free boundary must satisfy the condition that the contact angle δ between S and the container wall (measured within the fluid) is prescribed.
This condition reads:

$$\frac{\xi'}{\{1+|\xi'|^2\}^{\frac{1}{2}}} = \begin{cases} \cos\delta & \text{at } x_1 = L \\ -\cos\delta & \text{at } x_1 = 0 \end{cases} \tag{15.1.4}$$

Finally the quantity of fluid $\tilde{V}_c$ in the vessel V is specified by

$$\int_{\Omega_\xi} dx = \tilde{V}_c \quad \text{given}. \tag{15.1.5}$$

The equations will now be written in dimensionless form. We introduce a length scale L, a velocity scale $U = \max |\underline{\tilde{h}}|$ which lead to the following dimensionless quantities:

$$\hat{\underline{x}} = \frac{\underline{x}}{L} , \quad \hat{\underline{u}} = \frac{\underline{u}}{U} , \quad \hat{p} = \frac{1}{\rho U^2} (p+\rho g x_2) .$$

In dimensionless form the problem is formulated as follows (where the $\hat{\cdot}$ is dropped).

Find a free boundary ξ, a velocity vector $\underline{u}$ and a pressure field p such that

$$-\frac{1}{Re} \Delta\underline{u} + (\underline{u}.\nabla)\underline{u} + \nabla p = \underline{0} \quad \text{in } \Omega_\xi \qquad (15.1.6)$$

$$\nabla.\underline{u} = 0 \quad \text{in } \Omega_\xi \qquad (15.1.7)$$

$$\underline{u} = \underline{h} \quad \text{on } \Gamma \qquad (15.1.8)$$

$$u_n = 0 \quad \text{on } S \qquad (15.1.9)$$

$$\sigma_t = 0 \quad \text{on } S \qquad (15.1.10)$$

$$(OhRe)^2 \sigma_n = \frac{1}{R} - Bo\xi \quad \text{on } S \qquad (15.1.11)$$

$$\xi'(0) = -\operatorname{cotan}\delta \qquad \xi'(1) = \operatorname{cotan}\delta \qquad (15.1.12)$$

$$\int_{\Omega_\xi} dx = V_c \qquad (15.1.13)$$

where

$$\Omega_\xi = \{\underline{x} \in V |\ x_2 < \xi(x_1)\}$$

$$V = \{\underline{x} \in IR^2 |\ 0 < x_1 < 1, \quad x_2 > \xi_\Gamma(x_1)\}$$

$$Re = \frac{UL\rho}{\mu} \qquad \text{Reynolds number}$$

$$Oh = \frac{\mu}{\sqrt{\rho\alpha L}} \qquad \text{Ohnesorge number}$$

$$Bo = \frac{g\rho L^2}{\alpha} \qquad \text{Bond number}$$

$$h = \frac{\tilde{h}}{U}$$

$$\sigma_{ij} = -p\delta_{ij} + \frac{1}{Re}\left(\frac{\partial u_i}{\partial x_j} + \frac{\partial u_j}{\partial x_i}\right)$$

$$V_c = \frac{\tilde{V}_c}{L^2}$$

The Ohnesorge number is a measure for the relative importance of the viscous forces to the capillary forces, while the Bond number relates the gravitational forces to the capillary forces. Concerning the choice of the characteristic velocity U, we remark that when $\underline{\tilde{h}} \not\equiv \underline{0}$ it is obvious to take $U = \max_{\Gamma} |\underline{\tilde{h}}|$. When $\underline{\tilde{h}} \equiv \underline{0}$, we can take $U = \frac{\mu}{L\rho_0}$ which amounts to taking Re = 1 in the system of equations.

In the static equilibrium with $\underline{\tilde{h}} = 0$, the pressure p_0 and the free boundary ξ_0 are completely determined by the equations (15.1.6), (15.1.11), (15.1.12) and (15.1.13). One easily verifies that, for V_c sufficiently large, ξ_0 is determined by the following boundary value problem:

$$-\left(\frac{\xi_0'}{\{1+|\xi_0'|^2\}^{\frac{1}{2}}}\right)' + Bo\,\xi_0 = Bo\,\bar{V}_c - 2\cos\delta \quad \text{on } (0,1) \qquad (15.1.14)$$

$$\xi_0'(0) = -\cot\!an\,\delta \qquad \xi_0'(1) = \cot\!an\,\delta \qquad (15.1.15)$$

$$\bar{V}_c \equiv V_c + \int_0^1 \xi_\Gamma(x_1)dx_1 = \int_0^1 \xi(x_1)dx_1 \qquad (15.1.16)$$

The corresponding static pressure p_0 is given by

$$p_0 = \frac{1}{(Re\ Oh)^2}(Bo\bar{V}_c - 2\cos\delta)$$

Theoretical results for the stationary capillary free boundary problem governed by the Navier-Stokes equations have been obtained by Pukhnachev (1972, 1974). Free boundary problems are considered for which the free boundary of the fluid does not intersect the container; theorems concerning the solvability of the corresponding boundary value problems are established. Solonnikov (1979) obtained results for existence and uniqueness of capillary free boundaries which intersect the container. Solonnikov's work is based on the earlier results of Pukhnachev; it constitutes a constructive existence proof which is useful for a computational approach (cf. Cuvelier (1982)). Time-dependent (moving) boundaries governed by the Navier-Stokes equations have been studied in Solonnikov (1977) using a Lagrangian formulation; the unique solvability of a solution is proved under the condition that surface (or capillary) forces are negligible. This result has been extended to capillary moving free boundaries by Beale (1984) and Allain (1984). In Cuvelier (1984, 1985) an unsteady capillary free boundary problem is studied using small perturbation techniques. Existence and uniqueness of a solution of the linearized problem is proved. Stable and convergent finite element - finite difference schemes are considered for the computation of free boundaries, eigenfrequencies, damping factors and vibration modes.

15.2 Normal stress iterative method

The normal stress iterative method is based on the following principle. A shape of the free boundary is assigned and the flow within that shape is calculated after disregarding one of the boundary conditions on the free boundary. Next, a new shape of the free boundary is computed which satisfies as closely as possible the boundary condition that was relaxed. This procedure is repeated until convergence is attained.

The boundary condition on the free surface that will be relaxed is condition (15.1.11). Let ξ be a fixed function satisfying conditions (15.1.13), (15.1.14). Then an auxiliary problem is defined by finding the solution $\{\underline{u},p\}$ of problem (15.1.6),...,(15.1.10). This means that condition (15.1.11) is relaxed and it is this boundary condition that will be used to adjust the function ξ. It can be proved (see Solonnikov (1979)) that for small values of Re, V_c sufficiently large and $\delta \in (0,\pi)$ the auxiliary problem has a unique solution $\{\underline{u},p_0\}$ where the pressure p_0 is normalized by

$$\int_{\Omega_\xi} p_0(x)dx = 0 \tag{15.2.1}$$

Next, using (15.1.12) and (15.1.13), we deduce from (15.1.11) that

$$(\mathrm{OhRe})^2 \int_0^1 p(x_\xi)dx_1 = 2\,\mathrm{Oh}^2\mathrm{Re} \int_0^1 \frac{\partial \underline{u}}{\partial n}\cdot\underline{n}(x_\xi)dx_1 - 2\cos\delta + \mathrm{Bo}\,\bar{V}_c \tag{15.2.2}$$

$$x_\xi = \{x_1, \xi(x_1)\}$$

When $\{\xi,\underline{u},p\}$ denotes the solution of the full system (15.1.6),...(15.1.13), the pressure p and the pressure p_0 corresponding to the auxiliary problem differ only a constant, which gives, using (15.2.1) and (15.2.2), the following expression for $p(x_\xi)$

$$\begin{aligned} p(x_\xi) &= p_0(x_\xi) + \int_0^1 (p(x_\xi) - p_0(x_\xi))dx_1 = \\ &= p_0(x_\xi) - \int_0^1 p_0(x_\xi)dx_1 + \frac{2}{\mathrm{Re}} \int_0^1 \frac{\partial \underline{u}}{\partial n}\cdot\underline{n}(x_\xi)dx_1 - \frac{2\cos\delta}{(\mathrm{OhRe})^2} + \\ &\quad + \frac{\mathrm{Bo}\bar{V}_c}{(\mathrm{OhRe})^2} \\ &= p_0(x_\xi) + \int_0^1 \sigma_n(p_0,\underline{u})(x_\xi)dx_1 - \frac{2\cos\delta}{(\mathrm{OhRe})^2} + \frac{\mathrm{Bo}\bar{V}_c}{(\mathrm{OhRe})^2} \end{aligned} \tag{15.2.3}$$

with $\sigma_n(p_0,\underline{u}) = -p_0 + \frac{2}{Re}\frac{\partial \underline{u}}{\partial n}\cdot\underline{n}$

Substitution of (15.2.3) into boundary condition (15.1.11) gives:

$$-\frac{1}{R(\xi)} + Bo\xi = -(OhRe)^2(\sigma_n(p_0,\underline{u}) - \int_0^1 \sigma_n(p_0,\underline{u})dx_1) + \\ - 2\cos\delta + Bo\overline{V}_c \qquad (15.2.4)$$

for $x_1 \in (0,1)$.

The function ξ will now be compared with the free boundary ξ_0 of the static problem defined by (15.1.14),...,(15.1.16). Introducing $\psi = \xi - \xi_0$, it can be deduced that ψ satisfies the following non-linear ordinary differential equation with Neumann type boundary conditions:

$$\left| \begin{array}{l} -\left(\frac{\psi'}{\{1+|\xi_0'|^2\}^{\frac{1}{2}}}\right)' + Bo\psi = G \quad \text{on} \quad (0,1) \\ \psi'(0) = \psi'(1) = 0 \end{array} \right. \qquad (15.2.5)$$

with G given by

$$G = -(OhRe)^2(\sigma_n(p_0,\underline{u}) - \int_0^1 \sigma_n(p_0,\underline{u})dx_1) + \frac{1}{R(\xi)} - \frac{1}{R(\xi_0)} + \\ -\left(\frac{\xi'-\xi_0'}{\{1+|\xi_0'|^2\}^{\frac{1}{2}}}\right)' = G(\xi,\underline{u},p_0) \qquad (15.2.6)$$

The algorithm based on these results is the following: The free boundary ξ is iteratively approximated by a sequence $\xi^0,\xi^1,\ldots,\xi^\ell$ defined by

(i) $\xi^0 = \xi_0$ static solution

(ii) assume $\xi^1,\xi^2,\ldots,\xi^\ell$ are known

(iii) solve the auxiliary problem with ξ^{ℓ}; the solution is denoted by $\{\underline{u}^{\ell}, p_0^{\ell}\}$

(iv) solve problem (15.2.5) with $G^{\ell} = G(\xi^{\ell}, \underline{u}^{\ell}, p_0^{\ell}\}$; the solution is denoted by ψ^{ℓ}.

(v) $\xi^{\ell+1} = \xi_0 + \psi^{\ell}$

Using techniques of Solonnikov (1979) it can be established that for small values of Re, V_c sufficiently large and $\delta \in (0,\pi)$ the sequence $\{\xi^{\ell}, \underline{u}^{\ell}, p_0^{\ell}\}$ converges, which proves existence and uniqueness of the solution $\{\xi, \underline{u}, p_0\}$ of problem (15.1.6),...,(15.1.13). The pressure p is calculated from

$$p(x) = p_0(x) + \int_0^1 (-\, p_0(x_\xi) + \frac{2}{Re}\frac{\partial \underline{u}}{\partial n}\cdot\underline{n}(x_\xi))dx_1 + $$

$$+ \frac{1}{(OhRe)^2}(Bo\bar{V}_c - 2\cos\delta) \qquad (15.2.7)$$

A typical example is given for the previous described model situation in Figs. 15.2, 15.3 and 15.4. Here a plot is made of the free boundaries for $Oh^2 = 1.0, 3.0, 5.0$, Bo = 0.0, 5.0 and $\delta = \frac{\pi}{2}, \frac{\pi}{4}$ and Re = 1.0.

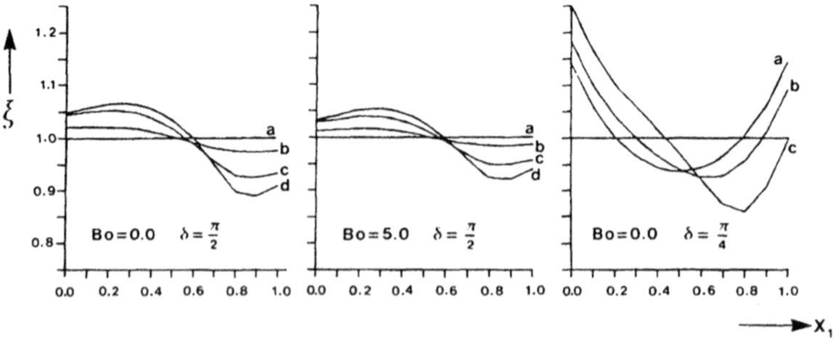

Fig. 15.2 Fig. 15.3 Fig. 15.4

(a) ξ_0 , (b) $Oh^2=1.0$, (c) $Oh^2=3.0$, (d) $Oh^2=5.0$.

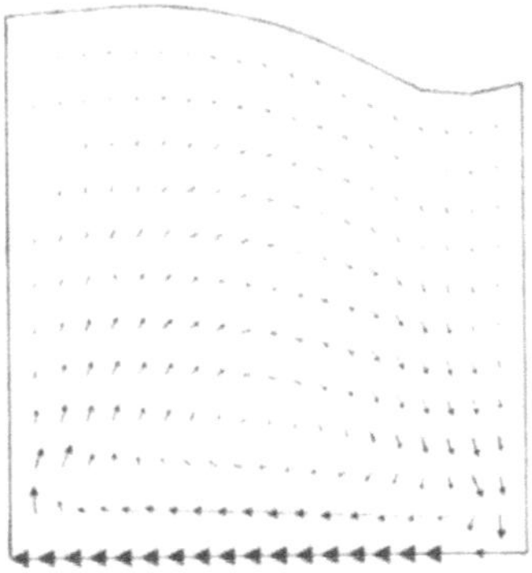

Fig. 15.5 Velocity vectors $Oh^2=5.0$, $Bo=0.0$, $\delta=\frac{\pi}{2}$

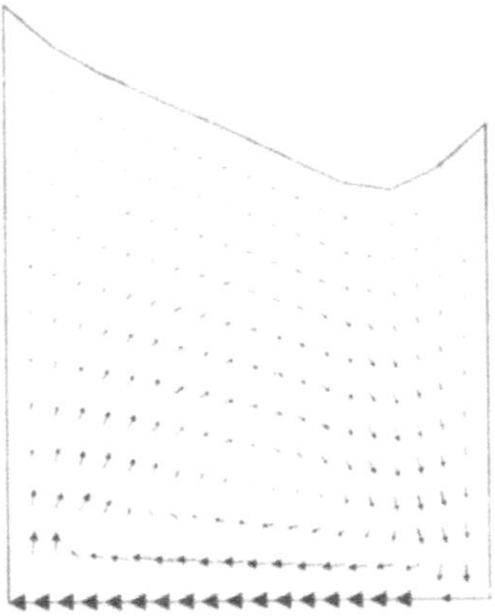

Fig. 15.6 Velocity vectors $Oh^2=3.0$, $Bo=0.0$, $\delta=\frac{\pi}{4}$

Velocity vector field plots are given in Fig. 15.5 ($Oh^2 = 5.0$, $Bo = 0.0$, $\delta = \frac{\pi}{2}$) and Fig. 15.6 ($Oh^2 = 3.0$, $Bo = 0.0$, $\delta = \frac{\pi}{4}$).

For a discussion of the numerical experiments we refer to Cuvelier (1982).

15.3 Newton's method for free boundaries

The method we discuss in this section is based on the simultaneous calculation of the position of the free surface and the field variables. The procedure is as follows. Using the FEM, a large system of (non-linear) equations is constructed where the field variables (i.e. velocity and pressure) as well as the position of the nodes of the free surface are considered as degrees of freedom. This large system is then solved by Newton's method or a quasi-Newton method using a Broyden update for the inverse of the Jacobian of the system.

To illustrate this procedure, we construct the large system of equations, which are in fact the Galerkin finite element equations. First we write down the not yet discretized Galerkin equations, which are obtained by multiplying equations (15.1.6), (15.1.7) and (15.1.11) by appropriate test functions, taking the integral over the corresponding domain and substitution of boundary conditions. To that end the momentum equation (15.1.6) is multiplied by a test (vector-) function $\underline{v} = \{v_1, v_2\}$ satisfying $\underline{v} = \underline{0}$ on Γ and $v_n = 0$ on S. Integration over Ω_ξ, application of Green's formula (7.2.4) and substitution of the boundary conditions (15.1.9) and (15.1.10) lead to

$$\frac{1}{2Re} \sum_{i,j=1}^{2} \int_{\Omega_\xi} \left(\frac{\partial u_i}{\partial x_j} + \frac{\partial u_j}{\partial x_i}\right)\left(\frac{\partial v_i}{\partial x_j} + \frac{\partial v_j}{\partial x_i}\right) d\Omega + \sum_{i,j=1}^{2} \int_{\Omega_\xi} u_i \frac{\partial u_j}{\partial x_i} v_j d\Omega -$$

$$- \int_{\Omega_\xi} p_0 \operatorname{div} \underline{v} \, d\Omega = 0 \qquad (15.3.1)$$

The incompressibility constraint (15.1.7) is multiplied by a test function q, which gives, after integration over Ω_ξ:

$$\int_{\Omega_\xi} q \operatorname{div} \underline{u} \, d\Omega = 0 \qquad (15.3.2)$$

Finally equation (15.2.4), which is equivalent to (15.1.11), is multiplied by a test function χ. Integration from 0 to 1, application of Green's formula and substitution of the boundary conditions (15.1.12) yield

$$\int_0^1 \frac{\xi'}{\{1+|\xi'|^2\}^{\frac{1}{2}}} \chi' \, dx_1 + Bo \int_0^1 \xi \, \chi \, dx_1 =$$

$$= - (OhRe)^2 \int_0^1 \{\sigma_n(p_0,\underline{u})\chi - \chi \int_0^1 \sigma_n(p_0,\underline{u}) dx_1\} dx_1 + \qquad (15.3.3)$$

$$+ (Bo\bar{V}_c - 2 \cos \delta) \int_0^1 \chi \, dx_1 + (\chi(0) + \chi(1)) \cos \delta \, .$$

Next, the equations (15.3.1), (15.3.2), (15.3.3) are discretized using, for instance, a triangular FEM. The velocity (i.e. $\underline{u}$ and $\underline{v}$) is approximated by continuous piecewise extended quadratic polynomials ($\tilde{u}$ and $\tilde{v}$). The pressure (i.e. p_0 and q) is approximated by piecewise linear polynomials ($\tilde{p}$ and $\tilde{q}$) not necessarily continous in Ω_ξ. Finally, we approximate the free boundary (i.e. ξ and χ) by continuous piecewise quadratic one-dimensional polynomials ($\tilde{\xi}$).

This finite element approximation in combination with the penalty method discussed in Chapter 8 results in a matrix system of nonlinear algebraic equations of the form

$$A\underline{u} + B(\underline{u})\underline{u} + C\underline{p}_0 = \underline{F} \tag{15.3.4}$$

$$\varepsilon M\underline{p}_0 + C^T\underline{u} = \underline{0} \tag{15.3.5}$$

$$N\underline{\xi} + D\underline{u} + E\underline{p}_0 = \underline{G} \tag{15.3.6}$$

where $\underline{u}$, $\underline{p}_0$ and $\underline{\xi}$ are the vector representations of $\tilde{u}$, $\tilde{p}_0$ and $\tilde{\xi}$ respectively. If the vector $\underline{p}_0$ is eliminated from system (15.3.4),...,(15.3.6), we obtain the reduced system

$$\begin{vmatrix} A+B(\underline{u})-\frac{1}{\varepsilon}\,CM^{-1}C^T & 0 \\ D-\frac{1}{\varepsilon}\,EM^{-1}C^T & N \end{vmatrix} \begin{vmatrix} \underline{u} \\ \underline{\xi} \end{vmatrix} = \begin{vmatrix} \underline{F} \\ \underline{G} \end{vmatrix} . \tag{15.3.7}$$

This system of (nonlinear) equations can be solved by Newton's method (cf. Section 5.6.3.2). When we write system (5.3.10) shortly as

$$K(\underline{w})\underline{w} = \underline{k}(\underline{w}) \qquad \underline{w} = \{\underline{u},\underline{\xi}\} \tag{15.3.8}$$

Newton's method reads as follows:

(i) Let $\underline{w}^0$ be an approximation of the solution $\underline{w}$

(ii) For i=0,1,2,... solve

$$J(\underline{w}^i)\underline{w}^{i+1} = J(\underline{w}^i)\underline{w}^i - R(\underline{w}^i)$$

where $R(\underline{w}) = K(\underline{w})\underline{w} - \underline{k}(\underline{w})$

and $J(\underline{w}) = \dfrac{\partial R(\underline{w})}{\partial \underline{w}}$.

Notice that the computation of the Jacobian matrix $J(\underline{w})$ is complicated since the matrices A,B,C,M,D,E,N, and the vectors $\underline{F}$ and $\underline{G}$ all depend on $\underline{\xi}$ through the integration over Ω_{ξ}. Formally we can write

$$J(\underline{w}) = \begin{vmatrix} A+\dfrac{\partial B}{\partial \underline{u}}(\underline{u})\underline{u}+B(\underline{u})-\dfrac{1}{\varepsilon}\,CM^{-1}C^{T} & \left\{\dfrac{\partial A}{\partial \underline{\xi}}+\dfrac{\partial B}{\partial \underline{\xi}}(\underline{u})-\dfrac{1}{\varepsilon}\,\dfrac{\partial(CM^{-1}C^{T})}{\partial \underline{\xi}}\right\}\underline{u}-\dfrac{\partial \underline{F}}{\partial \underline{\xi}} \\ & \\ D-\dfrac{1}{\varepsilon}\,EM^{-1}C^{T} & N+\left\{\dfrac{\partial D}{\partial \underline{\xi}}-\dfrac{1}{\varepsilon}\,\dfrac{\partial(EM^{-1}C^{T})}{\partial \underline{\xi}}\right\}\underline{u}+\dfrac{\partial N}{\partial \underline{\xi}}\underline{\xi}-\dfrac{\partial \underline{G}}{\partial \underline{\xi}} \end{vmatrix}$$

with, for instance,

$$\left(\frac{\partial A}{\partial \underline{\xi}}\,\underline{u}\right)_{jk} = \sum_{\ell} \frac{\partial (A)_{j\ell}}{\partial(\underline{\xi})_k}\,(\underline{u})_{\ell}$$

$$\left(\frac{\partial \underline{F}}{\partial \underline{\xi}}\right)_{jk} = \frac{\partial(\underline{F})_j}{\partial(\underline{\xi})_k} .$$

In order to reduce the cost of the construction of the Jacobian matrix at each iteration level, a quasi-Newton method can also be used to solve (15.3.8). This method is described in Section 5.6.3.3.

Whereas the normal stress iterative method, discussed in section 15.2, has a linear rate of convergence, it is expected that the Newton-type iterative method, studied in this section, is a quadratic process (or at least superlinear). For applications to (isothermal) free boundary problems we refer to Kistler, Scriven (1984), Ruschak (1980) and Saito, Scriven

(1981). Non-isothermal capillary free boundaries are considered in Cuvelier, Driessen (1985a, 1985b).
The study of superlinear iterative processes for the computation of capillary free boundaries governed by the Navier-Stokes equations forms a subject of our current research.

Chapter 16

Non-Isothermal Flows

Introduction

In this Chapter we focus our attention to solution methods for the system of equations describing a non-isothermal flow. This means that, besides the equations for conservation of mass and momentum, also the equation for conservation of energy must be taken into account. The interest in heat-transfer processes in fluids may be exemplified by problems of thermal insulation, thermosyphons, cooling of fluids in channels surrounding a nuclear reactor core, double-glazed windows and so on. Heat transfer plays also an important role in materials processing where the existence of temperature gradients is an inherent characteristic of the system. Examples are the manufacture of glass in reverbatory furnaces, the circulation of liquid metals in solidifying ingots and the manufacture of (large) crystals in space.

In Section 16.1 we shall write down the equations describing a non-isothermal flow and we shall make some simplifications. Section 16.2 is dedicated to two methods for solving the Navier-Stokes equations coupled with the heat equation. Finally we present some numerical experiments and discuss some of the characteristics of the two methods.

16.1 Mathematical model

Let us turn back to the equations (1.6.1), (1.6.8) and (1.6.9) which express the conservation of mass, momentum and energy, respectively. These basic equations can be simplified by making the following assumptions (cf. Batchelor (1967), Landau, Lifchitz (1963)):

(i) The term corresponding to the rate of dissipation of mechanical energy in the energy equation is negligible.

(ii) We assume that the Boussinesq approximation is valid, which means that variations in the density ρ are negligible except in the body force term $\rho\underline{f}$ where the variation induced by the temperature gives rise to a body force $\rho\underline{g}$, where $\underline{g}$ is the gravitational acceleration.

Hence, except for the latter effect, the fluid will be regarded as incompressible.

(iii) The density ρ in the term $\rho\underline{g}$, mentioned in (ii), satisfies the following equation of state

$$\rho = \rho_0(1 - \beta(T-T_0)) \tag{16.1.1}$$

where β denotes the volume expansion coefficient and where the subscript 0 refers to a reference state.

(iv) The problem is time-independent.

(v) The viscosity is constant (i.e. independent of $\underline{x}$ and T).

With these assumptions the equations (1.6.1), (1.6.8) and (1.6.9) reduce to the following system, called the Boussinesq equations:

$$\nabla.\underline{u} = 0 \tag{16.1.2}$$

$$-\mu\Delta\underline{u} + \rho_0(\underline{u}.\nabla)\underline{u} + \nabla p = \rho_0\underline{f} + \rho_0\underline{g}(1-\beta(T-T_0)) \tag{16.1.3}$$

$$-\nabla.(k\nabla T) + \rho_0 c_p(\underline{u}.\nabla)T = \rho_0 f_T \tag{16.1.4}$$

We assume moreover that k and c_p are constant and that $\underline{f} = \underline{0}$ and $f_T = 0$. To this system boundary conditions for the velocity and the temperature need to be added. As example we consider the square $\Omega : 0 < x_1 < L,\ 0 < x_2 < L$ (see Fig. 16.1).

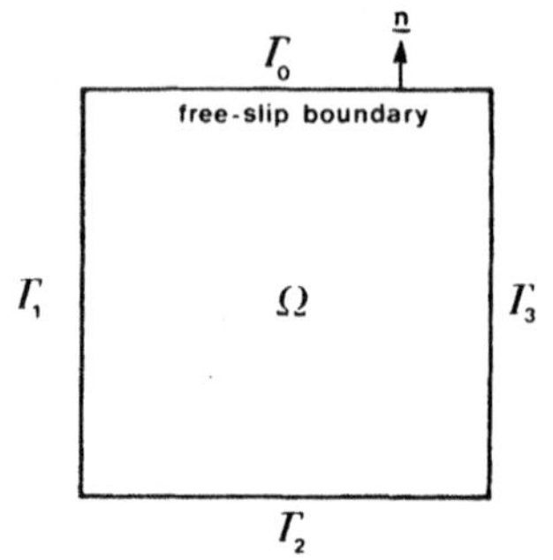

Fig. 16.1 Region Ω and boundary conditions.

The boundary conditions here assumed are as follows

Γ_0 : Thermally insulated: $\dfrac{\partial T}{\partial \underline{n}} = \dfrac{\partial T}{\partial x_2} = 0$

Free-slip boundary: $u_n = u_2 = 0, \ \sigma_t = 0$

Γ_1 : $T = T_+$ (constant)

$\underline{u} = 0$

Γ_2 : $T = T_+ - \dfrac{x_1}{L}(T_+ - T_-)$

$u_1 = \tilde{h} \qquad u_2 = 0$

Γ_3 : $T = T_-$ (constant)

$\underline{u} = 0$

The equations (16.1.2)... (16.1.4) will now be written in dimensionless form. We introduce a length scale L, a velocity scale U and form the following dimensionless quantities, using $\underline{g} = -\text{grad}\, gx_2$,

$$\hat{\underline{x}} = \frac{\underline{x}}{L}, \quad \hat{\underline{u}} = \frac{\underline{u}}{U}, \quad \hat{p} = \frac{1}{\rho U^2}(p + \rho_0 g x_2), \quad \hat{T} = \frac{T - T_0}{T_+ - T_-}$$

where $T_0 = \frac{1}{2}(T_+ + T_-)$
In dimensionless form the problem is formulated as follows (where the $\hat{}$ has been dropped)

Find a velocity vector $\underline{u} = \{u_1, u_2\}$, a pressure field p and a temperature distribution T such that

$$-\frac{1}{Re}\Delta\underline{u} + (\underline{u}.\nabla)\underline{u} + \nabla p = \frac{Gr}{Re^2} T \,\underline{g} \quad \text{with } \underline{g} = (0,1) \quad \text{in } \Omega \tag{16.1.5}$$

$$\nabla.\underline{u} = 0 \quad \text{in } \Omega \tag{16.1.6}$$

$$-\frac{1}{RePr}\Delta T + (\underline{u}.\nabla)T = 0 \quad \text{in } \Omega \tag{16.1.7}$$

$$\frac{\partial T}{\partial x_2} = 0, \quad u_2 = 0, \quad \sigma_t = 0 \qquad \text{on } \Gamma_0 \tag{16.1.8}$$

$$T = 0.5, \quad \underline{u} = 0 \qquad \text{on } \Gamma_1 \tag{16.1.9}$$

$$T = 0.5 - x_1, \quad u_1 = h, \quad u_2 = 0 \qquad \text{on } \Gamma_2 \tag{16.1.10}$$

$$T = -0.5, \quad \underline{u} = \underline{0} \qquad \text{on } \Gamma_3 \tag{16.1.11}$$

where

$$\Omega = \{\underline{x} \in IR^2 \mid 0 < x_1 < 1, \ 0 < x_2 < 1\}$$

$$Re = \frac{LU\rho_0}{\mu} \qquad \text{Reynolds number}$$

$$Gr = \frac{\beta \, g \, \rho_0^2 L^3 (T_+ - T_-)}{\mu^2} \qquad \text{Grashof number}$$

$$Pr = \frac{c_p \mu}{k} \qquad \text{Prandtl number}$$

$$h = \frac{\tilde{h}}{U}$$

$$\sigma_t = -\frac{1}{Re}\frac{\partial u_1}{\partial x_2} \qquad \text{tangential stress on } \Gamma_0$$

The Grashof number is a measure for the relative importance of the buoyancy forces to the viscous forces, while the Prandtl number is a measure for the similarity of the transport of heat and momentum. For the characteristic velocity U we can take $U = \max_{\Gamma_2} |\tilde{h}|$ when $\tilde{h} \equiv 0$. If $\tilde{h} \equiv 0$ we choose $U = \frac{\mu}{L\rho_0}$ so that $Re = 1$. The case for which $\tilde{h} \equiv 0$ (i.e. Re = 1) is known as the problem of natural (or free) convection. When there are driving forces, such as $\tilde{h} \equiv 0$ or $\underline{f} \equiv 0$, one speaks of forced convection.

Instead of working with the Grashof number, the Rayleigh number $Ra = GrPr$ is sometimes introduced.

Surveying the literature we see that problems of existence, uniqueness and regularity of solutions of the Navier-Stokes equations coupled with the heat-conduction equation have been studied by, for instance, Zenouda (1970, 1971), Lions (1969), Cuvelier (1976a, 1976b, 1978b). For references to numerical studies of non-isothermal flows governed by the Navier-Stokes equations and the heat equation, we cite Jones, Thomson (1981) where a generally accepted model problem is formulated that is used to compare and to test computer codes for the solution of the Boussinesq equations.

In the model problem discussed so far, the boundary Γ_0 is a free-slip fixed boundary In material processes, like for instance crystal growth, the temperature dependence of the surface tension coefficients α must be taken into account. This implies that the boundary condition for the tangential stress on Γ_0 changes into

$$\sigma_t = \frac{Ma}{Pr} \frac{\partial T}{\partial x_1} \quad \text{on} \quad \Gamma_0 \qquad (16.1.12)$$

where

$$Ma = \left(-\frac{d\alpha}{dT}\right)_{T_0} \frac{c_p \rho_0 L (T_+ - T_-)}{\mu\, k}$$

denotes the Marangoni number. When Ma does not vanish the shear stress (16.1.12) generates a surface flow from the region of low surface tension to that of high surface tension, which usually means from hot to cold (because usually $\frac{d\alpha}{dT} < 0$). Due to the viscosity of the fluid this surface motion penetrates into the fluid and induces a bulk flow. This type of convection is called Marangoni or thermo-capillary convection. For reviews on surface tension phenomena we refer to Ostrach (1977, 1979).

16.2 Numerical treatment

In this section two iterative methods will be discussed for solving problem (16.1.5),...,(16.1.11) with h = 0, so that Re = 1. The first one, called the coupled method, is to define a sequence $\{\underline{u}^{(k)},p^{(k)},T^{(k)}\}$, k = 0,1,2,... where $\underline{u}^{(k)}$, $p^{(k)}$ and $T^{(k)}$, $k \geq 1$, are computed simultaneously using the quantities $\underline{u}^{(k-1)}$ and $T^{(k-1)}$ of the previous iteration. When the non-linear terms are linearized by the Newton-Raphson technique (explained in Section 5.6.3.2), we find the following scheme:

(i) Start with an initial approximation, for example $T^{(0)} = 0.5 - x_1$, $\underline{u}^{(0)} = \underline{0}$

(ii) Assume that $\{\underline{u}^{(1)},p^{(1)},T^{(1)}\},\ldots,\{\underline{u}^{(k-1)},p^{(k-1)},T^{(k-1)}\}$ are known.
The next iterand is computed from

$$-\Delta\underline{u}^{(k)} + (\underline{u}^{(k)}.\nabla)\underline{u}^{(k-1)}+(\underline{u}^{(k-1)}.\nabla)\underline{u}^{(k)}+\nabla p^{(k)}$$

$$-GrT^{(k)}\underline{g} = (\underline{u}^{(k-1)}.\nabla)\underline{u}^{(k-1)} \qquad (16.2.1)$$

$$\nabla.\underline{u}^{(k)} = 0 \qquad (16.2.2)$$

$$-\frac{1}{Pr}\Delta T^{(k)}+(\underline{u}^{(k)}.\nabla)T^{(k-1)}+(\underline{u}^{(k-1)}.\nabla)T^{(k)} =$$

$$= (\underline{u}^{(k-1)}.\nabla)T^{(k-1)} \qquad (16.2.3)$$

and the boundary conditions (16.1.8),...,(16.1.11)

The second method, called the uncoupled method, defines also a sequence $\{\underline{u}^{(k)},p^{(k)},T^{(k)}\}$, $k = 0,1,2,\dots$. But now $\underline{u}^{(k)}$ and $p^{(k)}$ are computed using $\underline{u}^{(k-1)}$ and $T^{(k-1)}$ with Newton-Raphson linearization for the non-linear velocity terms. Next, $\underline{u}^{(k)}$ is used for the computation of $T^{(k)}$. We have the following scheme:

(i) Start with an initial approximation, for example

$$T^{(0)} = 0.5 - x_1 , \quad \underline{u}^{(0)} = \underline{0}$$

(ii) Assume that $\{\underline{u}^{(1)},p^{(1)},T^{(1)}\},\dots,\{\underline{u}^{(k-1)},p^{(k-1)},T^{(k-1)}\}$ are known. Then $\underline{u}^{(k)}$ is computed from:

$$-\Delta\underline{u}^{(k)} + (\underline{u}^{(k)}.\nabla)\underline{u}^{(k-1)} + (\underline{u}^{(k-1)}.\nabla)\underline{u}^{(k)} + \nabla p^{(k)} = \tag{16.2.4}$$

$$= Gr\ T^{(k-1)}\underline{g} + (\underline{u}^{(k-1)}.\nabla)\underline{u}^{(k-1)}$$

$$\nabla.u^{(k)} = 0 \tag{16.2.5}$$

with boundary conditions for $\underline{u}^{(k)}$.

Next, we compute $T^{(k)}$ from:

$$-\frac{1}{Pr}\Delta T^{(k)} + (\underline{u}^{(k)}.\nabla)T^{(k)} = 0 \tag{16.2.6}$$

with boundary conditions for $T^{(k)}$.

For both schemes, we define

$$\{\underline{u},p,T\} = \lim_{k\to\infty} \{\underline{u}^{(k)},p^{(k)},T^{(k)}\}$$

The following figures show some results of FEM computations for Ma = 0, Pr = 0.73 and for various values of the Grashof number: Gr = 2.0, 2.10^2, 2.10^4 and 10^5. For the velocity we have chosen continuous piecewise extended quadratic basis functions together with piecewise linear non-continuous basis functions for the pressure. The temperature is approximated by a full quadratic polynomial in each element. In each figure the isotherms are shown together with the velocity vectors in the nodal points of the triangulation. The velocity vectors are scaled; the maximum velocity $|\underline{u}|$ is indicated for each figure. The isotherms are given for T = 0.5 to T = -0.5 with increments of 0.1

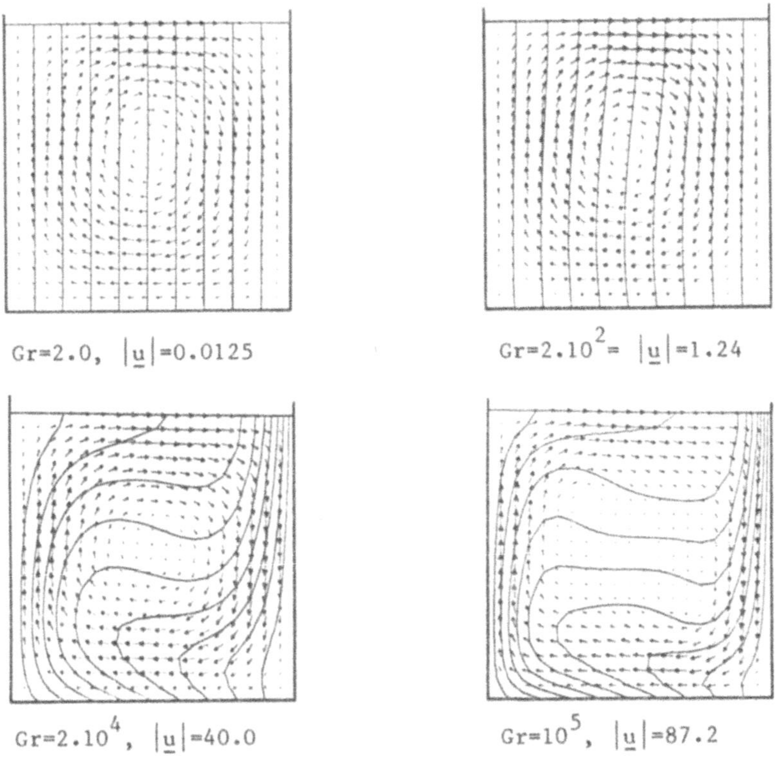

Fig. 16.2 Isotherms and velocity vectors for Pr = 0.73 and various values of Gr.

Concerning the convergence of the two iterative methods the following can be observed. The coupled method appeared to have a quadratic convergence rate. The uncoupled method has a quadratic convergence rate for low Prandtl numbers, otherwise it is a linear process. Notice that for low Prandtl numbers, the temperature distribution is almost independent of the flow, because heat is mainly transported by conduction. For both methods it is essential that a starting value for velocity and temperature is given in a neighbourhood of the final solution. Computations have been performed for various grid sizes. The results of Figures 16.2 correspond to a 21×21 grid. For values of Gr larger than 10^5 boundary layers appear and mesh refinement along the boundaries is necessary in order to get convergence. The rate of convergence for Gr > 10^5 can be improved by taking the solution for a smaller Grashof number as starting value. We also remark that the uncoupled method requires less computer memory and more computer time than the coupled method. For more details on numerical results, see Jones, Thomson (1981), de Vahl Davis (1982). We also refer to Segal (1982a). Mathematical results on the convergence of both methods can be found in Driessen (1984).

In Figures 16.3 some numerical results are given for Gr = 0, Pr = 0.73 and Ma = 50, 400.

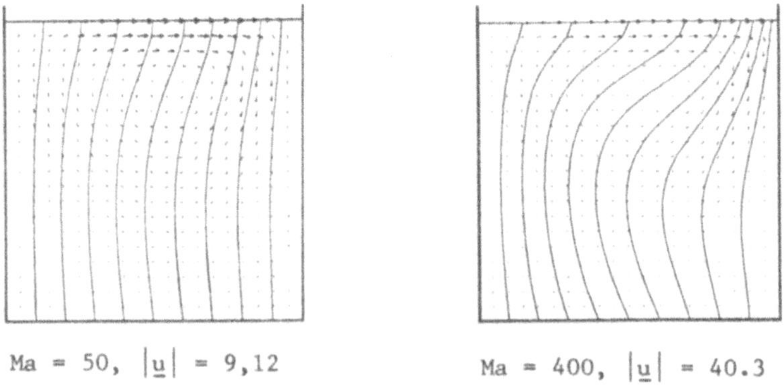

Fig. 16.3 Isotherms and velocity vectors for Gr = 0, Pr = 0.73.

We refer to Cuvelier, Driessen (1985a, 1985b) for a (numerical) study of thermocapillary free boundaries (i.e. Grashof-Marangoni convection in fluids with free boundaries) in the context of crystal growth processes.

Chapter 17

Turbulence

Introduction

In the investigations of turbulent flows of Newtonian media one assumes that these flows can be described by the Navier-Stokes equations. In principle it would be possible to compute the instantaneous flow variables (pressure and velocity) by means of direct simulation. The execution of this simulation, however, is practically impossible. This is caused by the fact that the smallest length scale in turbulent flows is an order of 10^{-3} smaller than the largest length scale (Rodi (1980)). The latter length scale is determined by the extent of the solution region. For a 3D flow at least 10^9 nodal points are necessary to model the turbulent flow in this way. Even the largest computers in the next generations are not able to compute such large problems. Another important difficulty is the representation of the results. Besides, for engineering applications this amount of information is highly redundant. That is the reason why several turbulence models have been developed which describe only a part of the turbulence phenomena. In these models the mean values of velocity and pressure are the quantities of interest, from which averaged turbulence properties of the flow can be derived afterwards. They do not pretend to be universal in the sense that every arbitrary flow problem can be described adequately. The main advantage of these models is that they are simple, and well tested for a considerable number of flow situations.

Turbulence models are the permanent subject of discussion between supporters and opposers. Opposers of the turbulence models say that these models are conceptually incorrect, in the sense that their preliminaries are not valid. The fact that, for some problems, the results nicely agree with measurements, is in their opinion only because empirical parameters ensure this succes. The models thus do not describe the physical properties of the fluid flow, but describe only a fluid flow for a restricted class of geometries and for a restricted class of boundary conditions. Supporters of the turbulence models agree that may be some of the concepts are incorrect, but on the other hand, these models produce results that, for a number of flows, have shown to be in good agreement with the measurements. Furthermore there are no other models available for the computation of such flows. In this book we shall not enter this discussion. However, it will be clear that the predictability of turbulent flow computations for practical problems may be of limited value.

In this Chapter first in Section 17.1 some mathematical models to describe the behaviour of the time averaged quantities are treated. In Section 17.2 some aspects of the numerical treatment are dealt with together with a numerical example. For a thorough study of turbulence the reader is referred to Tennekes and Lumley (1972), Hinze (1975) and Launder and Spalding (1972).

17.1 Mathematical models

Reynolds equations

The classical approach which goes back to Reynolds, is to average the momentum balance equations and the continuity equation. This so-called Reynolds decomposition is carried out as follows:

Consider the steady Navier-Stokes equations for an incompressible Newtonian fluid:

$$\rho(\underline{u}\cdot\nabla)\underline{u} = -\nabla p + \mathrm{div}(2\mu\underline{\underline{e}}) + \rho\underline{f} , \quad (17.1.1)$$

$$\mathrm{div}\ \underline{u} = 0 , \quad (17.1.2)$$

with ρ the density of the fluid,

μ the dynamic viscosity,

$e_{ij} = \frac{1}{2}(\frac{\partial u_i}{\partial x_j} + \frac{\partial u_j}{\partial x_i})$ the rate-of-strain tensor,

$\underline{f}$ an external force.

Define $\underline{u} = \bar{\underline{u}} + \underline{u}'$,

$p = \bar{p} + p'$,

where $\bar{\underline{u}}$ and $\bar{p}$ denote the mean value of velocity and pressure respectively, defined by:

$$\bar{\underline{u}} = \frac{1}{\tau}\int_t^{t+\tau} \underline{u}\ dt , \quad (17.1.3)$$

$$\bar{p} = \frac{1}{\tau}\int_t^{t+\tau} p\ dt ,$$

and $\underline{u}'$ and p' denote the fluctuations. τ must be chosen such that $\tau_1 << \tau << \tau_2$ (Hinze 1975). τ_1 is the largest periodicity or main time scale of the turbulence and τ_2 the period of possible slow variations that is not included within turbulence. The mean values of $\underline{u}'$ and p' are zero.

Averaging of (17.1.1), according to the rules (17.1.3) leads to:

$$\rho(\bar{\underline{u}}\cdot\nabla)\bar{\underline{u}} = -\nabla\bar{p} + \mathrm{div}(2\mu\bar{\underline{\underline{e}}}) + \rho\underline{f} + \mathrm{div}\ \underline{\underline{\tau}} , \quad (17.1.4)$$

$$\mathrm{div}\ \bar{\underline{u}} = 0 ,$$

where $\underline{\underline{\tau}}$ denotes the so-called Reynolds stress tensor, defined by:

$$\tau_{ij} = -\rho\ \overline{u_i'u_j'}. \quad (17.1.5)$$

Equation (17.1.4) is called the Reynolds equation.
Another commonly used quantity is the so-called mean turbulent kinetic energy defined by $k = \frac{1}{2}\sum_{i=1}^{3}\overline{u_i'^2}$ (17.1.6)

Contraction of $\underline{\underline{\tau}}$ from (17.1.5) then yields:

$$\tau_{11} + \tau_{22} + \tau_{33} = -2\rho k \,. \tag{17.1.7}$$

Actually $\underline{\underline{\tau}}$ in (17.1.4) represents the influence of the turbulence on the mean flow. Since $\underline{\underline{\tau}}$ is symmetric, this introduces 6 extra unknowns, without extra equations. This phenomenon, too much unknowns compared to the number of equations, is known as the closure problem. The closure problem can be solved by modelling the tensor $\underline{\underline{\tau}}$ as function of $\underline{u}$ and/or p. The most frequently used models are known under the name:

1. Boussinesq hypothesis,
2. Mixing length theory,
3. Energy equation model (k model),
4. Energy-dissipation model (k-ε model).

Boussinesq hypothesis

The usual way to solve the closure problem, is to treat the Reynolds stresses in exactly the same way as the stresses due to viscous actions in Newtonian media. Boussinesq (1887) introduced the notion of turbulent viscosity μ_t, and defined $\underline{\underline{\tau}}$ as follows:

$$\tau_{ij} = \mu_t\left(\frac{\partial \overline{u}_i}{\partial x_j} + \frac{\partial \overline{u}_j}{\partial x_i}\right), \tag{17.1.8}$$

or in tensor notation:

$$\underline{\underline{\tau}} = 2\,\mu_t \underline{\underline{\overline{e}}}. \tag{17.1.9}$$

Furthermore, Boussinesq assumed that μ_t is a constant. Hence, effectively an eddy viscosity is added to the laminar viscosity. The original definition needs some refinement as (17.1.7) implies:

$$\tau_{11} + \tau_{22} + \tau_{33} = 2\mu_t\left(\frac{\partial \bar{u}_1}{\partial x_1} + \frac{\partial \bar{u}_2}{\partial x_2} + \frac{\partial \bar{u}_3}{\partial x_3}\right) = 2\mu_t \operatorname{div} \bar{\underline{u}} = 0. \tag{17.1.10}$$

From (17.1.9) and (17.1.7) it would follow that k equals zero. Such an assumption is impossible since then there would be no turbulence at all. Therefore $\underline{\underline{\tau}}$ is redefined by:

$$\underline{\underline{\tau}} = (2\,\mu_t \bar{\underline{\underline{e}}} - \frac{2}{3}\,\rho\, k\, \underline{\underline{I}})\ , \tag{17.1.11}$$

where $\underline{\underline{I}}$ denotes the unit tensor. Equation (17.1.11), together with the assumption that μ_t is a constant is called the Boussinesq hypothesis. If one introduces a new pressure p^* by:

$$p^* = \bar{p} + \frac{2}{3}\,\rho\, k\ , \tag{17.1.12}$$

the Boussinesq hypothesis yields:

$$\rho(\bar{\underline{u}}.\nabla\bar{\underline{u}}) = -\nabla p^* + \operatorname{div}((2\mu + 2\mu_t)\bar{\underline{\underline{e}}}) + \rho\underline{f}. \tag{17.1.13}$$

In the sequel we shall drop the bar above the velocities and the star for the corrected pressure, and assume that we are dealing with mean values for corrected pressure and velocity. Equation (17.1.13) contains in fact two unknowns: k and μ_t, where k models the turbulent pressure terms τ_{11}, τ_{22} and τ_{33}. For the closure of the Reynolds equation, it is necessary to model k and μ_t as function of the flow variables. The standard way of modelling μ_t is to assume that μ_t is the product of a characteristic velocity scale u and a characteristic length scale ℓ:

$$\mu_t = \rho\, u\, \ell\ . \tag{17.1.14}$$

In the Boussinesq hypothesis u and ℓ are regarded to be constant. It is obvious that then the numerical solution of (17.1.13) is not different from the numerical solution of laminar flow. In the next three turbulence models more accurate estimates for u and ℓ are proposed.

Mixing length model

The mixing length theory developed by Prandtl assumes that the flow has a dominant direction, hence recirculation is excluded. For a simple turbulent shear flow in the x_1 direction u and ℓ are then estimated as follows:

$$u = \ell_m \left|\frac{\partial u_1}{\partial x_1}\right| , \qquad \ell = \ell_m , \tag{17.1.15}$$

where ℓ_m denotes the mixing length, defined by a suitable algebraic relation. See for example van Driest (1956). Generalization of (17.1.15) yields:

$$\mu_t = \sqrt{2}\ \rho\ \ell_m^2 (\underline{\underline{e}} : \underline{\underline{e}})^{\frac{1}{2}} . \tag{17.1.16}$$

It is remarked that μ_t depends on the field variables, while ℓ_m is assumed to be a constant. Furthermore Prandtl (1945) and Kolmogorov (1942) suggested the next relation between turbulent viscosity and the turbulent kinetic energy and mixing length:

$$\mu_t = C_\mu \rho\ \ell_m \sqrt{k} . \tag{17.1.17}$$

The kinetic energy of turbulence k can then be approximated by:

$$k = \beta \frac{\mu_t^2}{\ell_m^2 \rho^2} , \tag{17.1.18}$$

where β is some constant, that according to Gessner and Emery (1976) should be chosen between 3.5 and 3.8. Then the closure problem is solved. The derivation of (17.1.15) can be found in Tennekes and Lumley (1972), or Hinze (1975) and is based on a type of method that is common in kinetic gasdynamics. Unfortunately the preassumptions are quite arguable, and therefore one tries to get good results by the choice of ℓ_m, which more and more plays the role of an arbitrary length scale. In fact the choice of ℓ_m can be considered as a type of curve fitting.

The energy equation model (k-model)

An improvement of the mixing length model can be found in the energy equation model. The turbulent viscosity for this model is defined by relation (17.1.17).
The kinetic energy follows from the transport equation in which k is the dependent variable:

$$\rho \sum_{j=1}^{3} \bar{u}_j \frac{\partial k}{\partial x_j} = \sum_{j=1}^{3} \left[\frac{\partial}{\partial x_j} \left\{(\mu + \frac{\mu_t}{\sigma_k})\frac{\partial k}{\partial x_j}\right\} + \right.$$

$$\left. + \mu_t \sum_{i=1}^{3} \frac{\partial \bar{u}_i}{\partial x_j} \left(\frac{\partial \bar{u}_i}{\partial x_j} + \frac{\partial \bar{u}_j}{\partial x_i}\right)\right] - C_D \rho \frac{k^{3/2}}{\ell}, \quad (17.1.19)$$

in which C_D and σ_k are empirical constants. For the derivation of (17.1.19) the reader is referred to Taylor et al (1981).
The length scale ℓ is modelled in the same way as in the mixing length model. This is one of the main disadvantages of the k-model.

The energy-dissipation model (k-ε model)

The one-equation model (17.1.19) is usually extended to a two-equation model, where both the turbulent kinetic energy k and the length scale ℓ are determined from corresponding transport equations. The k-ε model is the most commonly used two-equation model, defined by:

$$\varepsilon = \frac{k^{3/2}}{\ell}, \quad (17.1.20)$$

$$\sum_{j=1}^{3} \rho \bar{u}_j \frac{\partial \varepsilon}{\partial x_j} = -\frac{C_2}{C_\mu} \rho \frac{\varepsilon^2}{k} + \sum_{i,j=1}^{3} \left[\rho\, C_\mu C_1 k \frac{\partial \bar{u}_i}{\partial x_j} \left(\frac{\partial \bar{u}_i}{\partial x_j} + \frac{\partial \bar{u}_j}{\partial x_i}\right)\right.$$

$$\left. + \frac{\partial}{\partial x_j} \left\{(\mu + \frac{\mu_T}{\sigma_\varepsilon}) \frac{\partial \varepsilon}{\partial x_j}\right\}\right] . \quad (17.1.21)$$

The parameters C_1, C_2, σ_ε and C_μ are empirical constants. For a derivation of (17.1.21) see for example Taylor et al (1981).

Important disadvantages of the extra equations (17.1.19) and (17.1.21) are that besides of the large number of empirical constants, also boundary conditions and initial values for k and ε are necessary.

17.2 Numerical treatment

The numerical solution of the mixing length model by the FEM is straight-forward. The only complication compared to laminar flow is the appearance of a non-linear turbulent viscosity term. This term can be linearized by a Newton method or a quasi-Newton method as depicted in sections 5.6 and 7.3.3. A typical example is given in figs. 17.1 and 17.2. Other numerical results by a FEM are reported by Gerard (1974).

The solution of the Navier-Stokes equations with the two equation models is more complicated. It requires the solution of 5 (2D) or 6 (3D) coupled non-linear equations. Taylor et al (1980) mention the following algorithm for the efficient solution of the stationary two-equation model:

i) Specify boundary conditions for $\underline{u}$, p, k and ε.
ii) Estimate and impose initial values on velocity, pressure, turbulence energy and dissipation.
iii) Use the current values of k and ε to determine the point-wise distribution of turbulent viscosity.
iv) Evaluate the distribution of $\underline{u}$, p, k and ε.
v) Return to iii) until convergence has been achieved.

Their algorithm thus decouples the computation of velocity and k and ε, which may lead to convergence problems. An alternative approach is the coupled solution of all equations, with increased computation time as consequence. Numerical experiments with the two-equation models are reported by for example Hutton and Smith (1981), Morgan et al (1979), Schamber and Laroch (1980), Thomas et al (1981), Taylor et al (1980), Smith (1984) and McLaughin et al (1983).

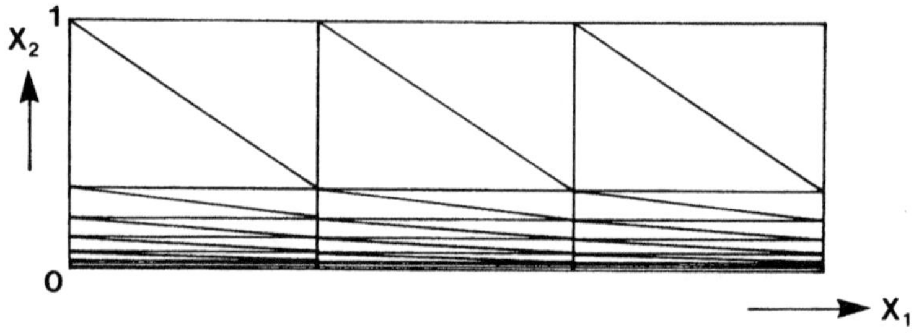

Fig. 17.1 Mesh for turbulent channel flow.

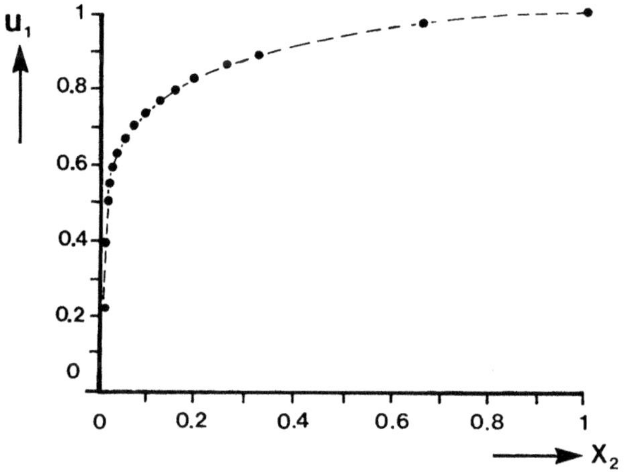

Fig. 17.2 Channel flow: velocity as function of y. Experimental values according to Hussain & Reynolds (1975). Re = 32300, mesh according to Fig. 17.1

An additional problem in handling the turbulence models in the presence of a wall (for example in channel flow) is that the velocity has a very steep gradient near the wall. The accurate solution of the velocity therefore requires a very fine grid near the wall, increasing the computation time considerably.

Many authors have solved this problem by starting the computation at a distance Δ of the wall and using boundary conditions that are found by analytical methods. This approach is known under the name "law of the wall" and a reasonable accuracy can only be found when the analytical solution resembles the actual solution. This is for example the case for channel flow. See for example Launder and Spalding (1972). It is obvious that the use of the law of the wall reduces the number of grid points and hence the computing time.

Finally a completely different approach is to model the small scale turbulence and to simulate numerically the large scale turbulence. This method is known under the name "large eddy simulation", see Peyret and Taylor (1983). The computing costs are considerable, since one calculates instantaneous pressure and velocities. For the computation of mean values of pressure and velocity it is necessary to calculate more solutions and to average them using ensemble averaging techniques. This method seems very promising but it is in its first development stage.

Chapter 18

Non-Newtonian fluids

Introduction

In section 1.6 the Navier-Stokes equations have been derived for a Newtonian fluid, i.e. a fluid with a linear constitutive relation:

$$\underline{\underline{d}} = 2\mu(\underline{\underline{e}} - \frac{1}{3}(\nabla.\underline{u})\underline{\underline{I}}) \quad \text{with } e_{ij} = \tfrac{1}{2}(\frac{\partial u_i}{\partial x_j} + \frac{\partial u_j}{\partial x_i}).$$

Such a relation is only valid for fluids without elastic behaviour, like water and air. For a number of important industrial materials it is not possible to describe the fluid with a linear relation between stress tensor and velocity gradient tensor. Non-Newtonian effects are for example that the viscosity varies with the shear rate, and the presence of normal-stress differences in shear flows. Furthermore several polymeric flows have strong memory effects, i.e. the stress tensor shows a functional dependence with respect to the strain history.
Examples of flows that do not behave Newtonian are polymer solutions and polymer melts. They are usually quite viscous; the non-Newtonian behaviour occurs at low Reynolds numbers, where the inertia terms have little effect upon the flow field.

Non-Newtonian behaviour has been the object of several books in recent years: Astarita and Marrucci (1974), Bird et al (1977), Schowalter (1978), and the numerical textbook of Crochet,

Davies and Walters (1984). For a concise review of numerical methods in non-Newtonian fluid mechanics, the reader is referred to Crochet and Walters (1983). In this section we follow their review to indicate the state of the art in this subject. In Section 18.1 first some mathematical models are discussed, followed in Section 18.2 by some aspects of numerical treatment.

18.1 Mathematical models

A non-Newtonian fluid is characterized through its constitutive relation. In the literature a number of models have been developped which approximate the relation between velocity gradients and stress tensor. Following Crochet et al (1983), we shall distinguish between the following three types of relations:

i) Flows dominated by the shear viscosity
ii) Slow flows (slightly elastic liquids)
iii) Complex flows involving highly elastic liquids.

Flows dominated by the shear viscosity

In many situations of practical importance the range of flow rates or the flow geometry is such that the influence of normal stress differences is small compared to the shear stresses. Examples are steady flow in pipes and channels and boundary-layer flows. The flow problem can then be solved by considering constitutive relations of the generalized Newtonian type. This is the simplest form of Non-Newtonian behaviour, where the constitutive law can be expressed by a simple algebraic relation, like the Power law model:

$$\mu = k|\kappa|^{n-1}, \quad k,n > 0, \tag{18.1.1}$$

with μ the dynamic viscosity,
k some constant,
κ the shear rate,
n the power in the power law model.

In Fig. 18.1 the dependence of the viscosity on the shear rate for some polymers have been plotted. When the shear rate is in the range of the power law region (straight line in Fig. 18.1), (18.1.1) gives a good description of the viscosity. When κ approaches zero and $n < 1$, μ in (18.1.1) goes to infinity, which is in contradiction with reality. A model that gives a better description of the viscosity for low shear rates is the Carreau model:

$$\mu = \mu_0(1 + (\lambda\kappa)^2)^{\frac{n-1}{2}} \quad , \quad \mu_0, \lambda > 0 \ , \ 0 < n < 1. \qquad (18.1.2)$$

The generalized Newtonian model for an incompressible fluid is a constitutive model, that is defined by:

$$\underline{\underline{d}} = 2\mu\underline{\underline{e}} \ , \qquad (18.1.3)$$

Fig. 18.1 Dependence of viscosity on shear rate for some polymers.

where $\underline{\underline{e}}$ is the rate of strain tensor (see Section 1.6), and μ is only dependent on the principle invariants of $\underline{\underline{e}}$ defined by:

$$I_e = \mathrm{tr}(\underline{\underline{e}}) = 0 \;, \tag{18.1.4}$$

$$II_e = \tfrac{1}{2}(I_e^2 - \underline{\underline{e}} : \underline{\underline{e}}) = -\tfrac{1}{2}\,\underline{\underline{e}} : \underline{\underline{e}} \;,$$

$$III_e = \det(\underline{\underline{e}}) \;.$$

The first invariant I_e is always equal to zero for incompressible flows. For a simple shear flow III_e equals zero as well. In that case the viscosity only depends on the second principle invariant. For a two-dimensional flow this second principle invariant reads in cartesian co-ordinates:

$$II_e = -\tfrac{1}{2}\left(\frac{\partial u_1}{\partial x_1}\right)^2 - \tfrac{1}{2}\left(\frac{\partial u_2}{\partial x_2}\right)^2 - \frac{1}{4}\left(\frac{\partial u_2}{\partial x_1} + \frac{\partial u_1}{\partial x_2}\right)^2. \tag{18.1.5}$$

For a simple shear flow with only $\frac{\partial u_2}{\partial x_1}$ not equal to zero, (18.1.5) reduces to:

$$II_e = -\frac{1}{4}\left(\frac{\partial u_2}{\partial x_1}\right)^2 \;, \tag{18.1.6}$$

hence

$$|\kappa| = \left|\frac{\partial u_2}{\partial x_1}\right| = (-4\,II_e)^{\frac{1}{2}} = (2\underline{\underline{e}} : \underline{\underline{e}})^{\frac{1}{2}} \;. \tag{18.1.7}$$

(18.1.7) can be used as a generalization of the shear rate. The generalization of the power law model (18.1.1) becomes:

$$\mu = k(-4\,II_e)^{\frac{n-1}{2}} \;, \quad k,n > 0 \tag{18.1.8}$$

and the Carreau model:

$$\mu = \mu_0(1 - 4\lambda^2 II_e)^{\frac{n-1}{2}} \;, \quad \mu_0, \lambda > 0 \;, \quad 0 < n < 1. \tag{18.1.9}$$

The models (18.1.8) and (18.1.9) are only valid for flows with a small value of the third principle invariant. In practice this means simple shear flows or flow that resemble them.

Practical computations, however, have shown that the application of these models to other flows sometimes also lead to useful results. The range $0.2 < n < 0.7$ is physically the most important one. For $n = 1$ (18.1.1) reduces to a Newtonian model ($\mu=k$) and (18.1.2) with $\mu = \mu_0$. For example polymers and polymer solutions are well described by this generalized non-Newtonian model. An extra complication is that the computations must be performed for high temperatures, and hence the parameter values have to be available for this particular situation. Moreover non-isothermal computations may be necessary. A special case of the generalized Newtonian models are the plastico viscous fluids, for example Bingham fluid, which have a yield stress. This means that the stress must be larger than some minimum value in order that the material behaves as a viscous liquid. In the mathematical literature much attention have been paid to this kind of materials. See for example Duvaut and Lions (1972), Fortin (1972), Glowinski et al (1976), and Fortin and Glowinski (1983).

Slow flows (slightly elastic liquids)

In more complicated situations the relation between stress tensor and deformation tensor must be expressed with the aid of some functional. One can show that such a relation implies the existence of a natural time λ that is proper to the fluid. Comparing λ with a characteristic time L/V of the flow, where V is a typical velocity and L a characteristic length of the geometry, defines the non-dimensional Weissenberg number W by:

$$W = \frac{\lambda V}{L} . \tag{18.1.10}$$

When the flow is slow, or when the viscoelastic fluid is only slightly elastic, the Weissenberg number is small. In that case the general functional relationship between stresses and deformation can be expanded into a series of approximations in the parameter W. This is called the "retarded motion expansion". A simple example is the so-called "second-order model" given by:

$$T_{ik} = 2\mu e_{ik} - \nu_1 \overset{\nabla}{e}_{ik} + 4\nu_2 e_{ij} e_{jk} , \qquad (18.1.11)$$

where μ, ν_1 and ν_2 are material constants and ∇ denotes the contravariant derivative defined by Oldroyd (1950):

$$\overset{\nabla}{e}_{ik} = \frac{\partial e_{ik}}{\partial t} + \sum_m \left\{ u_m \frac{\partial e_{ik}}{\partial x_m} - \frac{\partial u_i}{\partial x_m} e_{mk} - \frac{\partial u_k}{\partial x_m} e_{im} \right\} . \qquad (18.1.12)$$

These models are characterized by the fact that the constitutive relation consists of differential equations, which are explicit non-linear relations between the stress tensor and the rate of deformation tensor.
Because of the requirement that the flow must be slow (and slowly varying) the second-order model is of limited practical value. This model has mainly been used for finding analytical solutions. A more useful model can be found if analogous to the generalized Newtonian model, the coefficients μ, ν_1 and ν_2 in (18.1.11) are assumed to be dependent on II_e(Bird, 1976).

Complex flows involving highly elastic liquids

Non-Newtonian fluids in this category exhibit a finite memory (i.e. the relaxation time λ is not small) and especially when the flow geometry is complicated an adequate description of the constitutive relation is difficult to find. The number of models that has been proposed in literature is therefore enormous. Every model has its own range of application. For an overview of the great variety of constitutive models the reader is referred to the literature mentioned in the introduction.
The models obtained in this field are either of a differential character, where the stress and rate -of- deformation tensors and some of their time derivatives are related, or have an integral form, where the stress is explicitly represented by a time integral over the deformation history.
Important practical applications of this kind of models can be found in polymer processing e.g. extrusion, fiber spinning and injection moulding. These problems involve high rates of

deformation and abrupt changes in geometry (including free surfaces). A simple linear differential model of the Oldroyd type is the so-called "upper-convected Maxwell model" with constitutive relations of the form:

$$T_{ik} + \lambda \overset{\nabla}{T}_{ik} = 2 \mu e_{ik} , \qquad (18.1.13)$$

with ∇ defined as in (18.1.12), and λ and μ material constants. Comparing (18.1.13) with (18.1.11) shows that this model is implicit in the stress tensor. An integral form of the same model can be found in Crochet and Walters (1983).

18.2 Numerical treatment

Generalized Newtonian fluids

In this section we shall treat how the Navier-Stokes equations can be solved for a fluid with a generalized Newtonian constitutive law. First the equations (18.1.1) or (18.1.2) are substituted in the Navier-Stokes equations (1.6.5). These equations can be solved similarly to the way depicted in part 2 of this book. The types of boundary conditions treated there for the Navier-Stokes equations, can also be used for the generalized Newtonian models. The only complicating factor compared to the Newtonian model, is the non-linear viscosity term, which is of the shape

$$\mu \frac{\partial u_k}{\partial x_\ell} . \qquad (18.2.1)$$

This term must be linearized with a Picard iteration or Newton linearization as denoted in Section 5.6.
In case of a Picard iteration the viscosity can be computed from the solution of the preceding (m^{th}) iteration, using the solution vector $\underline{u}_m$:

$$\mu_{m+1} \frac{\partial u_k}{\partial x_\ell}\bigg|_{m+1} \approx \mu_m \frac{\partial u_k}{\partial x_\ell}\bigg|_{m+1} ,$$

where $\left.\frac{\partial u_k}{\partial x_\ell}\right|_{m+1}$ denotes the derivative of the solution in the $m+1^{th}$ iteration and μ_m the viscosity computed with the aid of $\underline{u}_m$. Using the Newton linearization, the Jacobian of expressions of the shape (18.2.1) must be computed. Practical computations for the Power-law fluid (18.1.1), as shown in fig. 18.2 have shown that in the interesting range $0.2 < n < 0.7$, Picard iteration converges faster than Newton iteration. However, for small values of n, the convergence is slow. Tanner et al (1975) propose a modification of the Picard process by posing:

$$\mu_{m+1} = \mu_m + \omega(1-n)(\mu_m - \mu_{m-1}) \ .$$

They found that $\omega = 0.4$ gave the best results in their experiments. Only 5 to 6 iterations were necessary in order to converge to the solution.
Numerical experiments with Bingham fluids have been reported by Fortin (1972), Glowinski et al (1976), Fortin and Glowinski (1983).

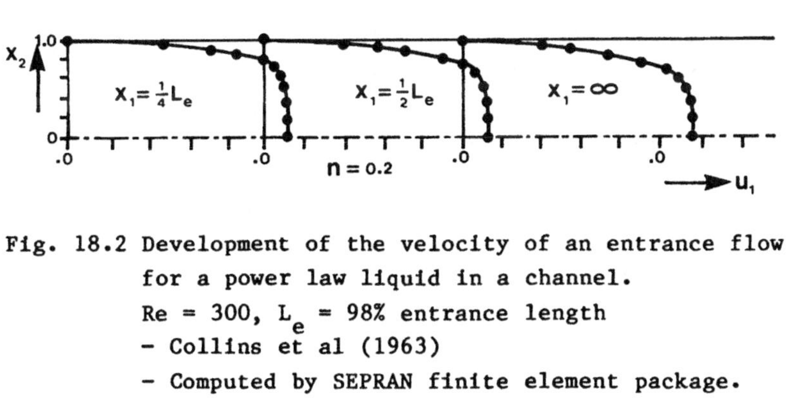

Fig. 18.2 Development of the velocity of an entrance flow for a power law liquid in a channel.
Re = 300, L_e = 98% entrance length
- Collins et al (1963)
- Computed by SEPRAN finite element package.

Second order fluids

Because of the very limited range of application only very few papers have appeared in literature which deal with numerical calculations for second-order fluid. For a review see Crochet & Walters (1983).

Viscoelastic fluids

Numerical simulation of the flow of highly elastic fluids is a very active field of research with many unresolved problems (Crochet, Davies and Walters, 1984). In this section we shall give a brief review of the methods that are used and the problems that have been encountered.

The constitutive models that are used for visco-elastic flows are either implicit in the stress tensor T_{ij}, (e.g. the upper-convected Maxwell model in (18.1.13)) or explicit but written as a single integral over the entire history of fluid particles. In both cases it is not possible to find a partial differential equation for the velocity-field similar to the Navier-Stokes equations. The various methods that are used can globally be divided into two categories according to the way the constitutive equation is treated, being mixed methods and integration along particle trajectories.

Mixed methods can only be used for differential-type models. The stresses T_{ij} are then interpolated over the elements by usual basis functions and the constitutive equation is discretized by the Galerkin method by means of weighting with the stress basis functions. The momentum equations and the continuity equation are also discretized by the Galerkin method in the usual way. The method results in non-linear algebraic equations for the nodal velocities, pressures and stresses and can be solved by Newton linearization. A great advantage of this method is that the velocities, pressure and stresses are solved simultaneously.

Integration along particle trajectories is commonly used when integral models are applied, because the integral is time-dependent over the entire history of a particle. For a given velocity field the stresses can then be found by integrating in time along the particle trajectories. The momentum equations and the continuity equation are treated by the Galerkin method. However, this method is not restricted to integral models only. Differential type models can be treated as well as they are actually ordinary differential equations in time for the stresses along particle trajectories. A disadvantage of this method is that the stresses can not be solved simultaneously with the velocities and pressure. This has a negative effect on the convergence.

This field of research has been plagued by problems in finding solutions for high Weissenberg numbers W. For example v.d. Zanden et al (1985), see Fig. 18.3, have studied the flow of an upper-convected Maxwell fluid over a slot by means of a mixed method. They found that for some rather low value of W shear stresses began to show large oscillations, generated by the corners in the geometry and no converged solution was found for slightly increased W. Mesh refinement only strengthens this effect. They concluded that the problems were due to the Galerkin method used for the constitutive equation and propose a method falling in the second category. But researchers who have used methods in the second category (mostly integral models) also report convergence problems except for Shen (1984) who uses the Leonov model, which is a differential-type model. It is believed that this kind of convergence problems have their origin in:

i) High stress gradients in the solution or even non-integrable singularities which cannot be approximated by the discretized solution.
ii) Change of type of the system of partial differential equations inherent in the system or by numerical approximation errors.
iii) Limit, bifurcation or termination points. There are

indications that many (if not all) of these points are due to the discretizing scheme and are not present in the analytical solution, see e.g. Beris et al (1984).

iv) Iteration scheme for solving the algebraic equations.

Which of the problems are present is highly dependent on the numerical discretization scheme that has been used.

a) Coarse mesh

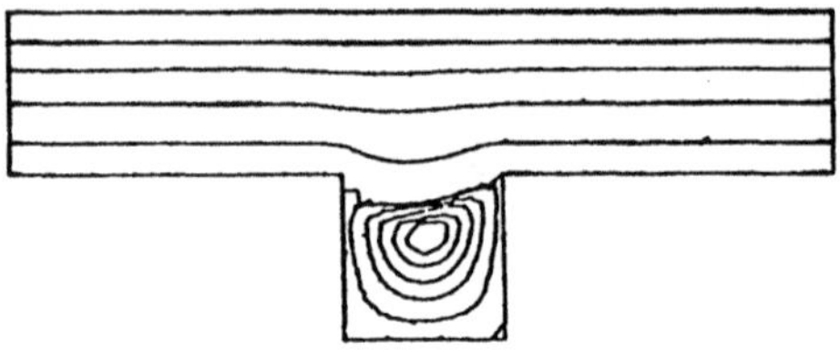

b) Finer mesh

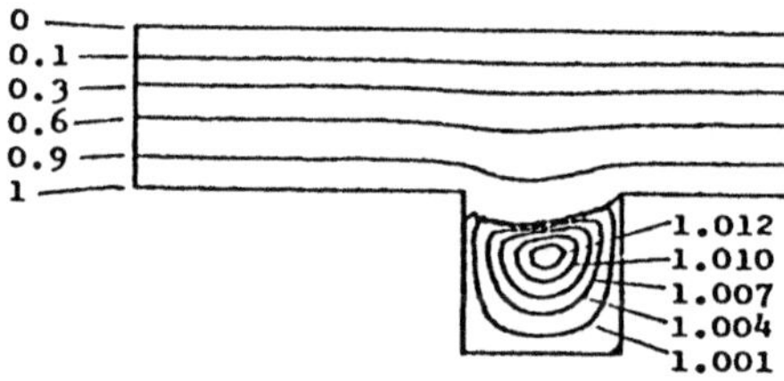

c) Fine mesh

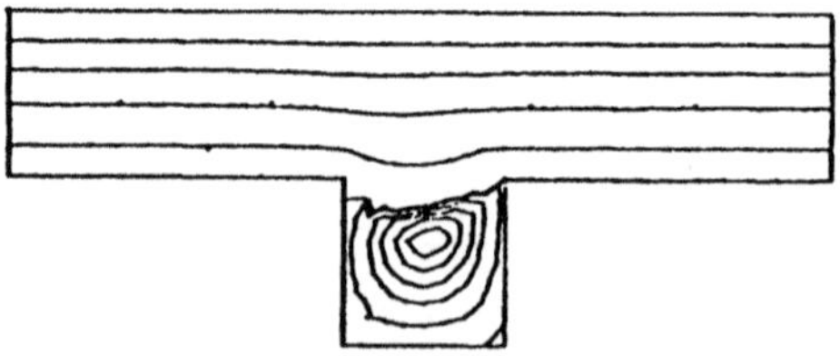

Fig. 18.3 Numerical simulation of a channel flow over a slot Maxwell-B fluid, for various meshes.
Re = 10, W = 0.1, Re = $\rho uH/\mu$, w = $\tau u/H$.

Bibliography

Adams, R.A. (1975): Sobolev spaces, Academic Press, New York.

Agmon, S. (1965): Lectures on elliptic boundary value problems, Van Nostrand, Princeton.

Allain, G. (1984): Un problème de Navier-Stokes avec surface libre et tension superficielle, Proceedings colloque International Problèmes à Frontières Libres, Maubuisson.

Astarita, G. and G. Marrucci (1974), Principles of Non-Newtonian Fluid Mechanics, London, McGraw-Hill.

Aubin, J.P. (1972): Approximation of elliptic boundary value problems, Wiley-Interscience, New York.

Aziz, A.K. (1972): The mathematical foundations of the finite element method with applications to partial differential equations, Academic Press, New York.

Babuška, I. and A.K. Aziz (1972): Survey lectures on the Mathematical foundations of the finite element method, in Aziz (1972), p. 3-359.

Babuška, I. (1973a): The finite element method with Lagrangian multipliers, Num. Math. 20, p. 179-192.

Babuška, I. (1973b): The finite element method with penalty, Math. Comp. 27, p. 221-228.

Babuška, I. and A.K. Aziz (1976): On the angle condition in the finite element method, SIAM J. Num. Anal. 13, p.214-226.

Babuška, I., J.T. Oden and J.K. Lee (1977): Mixed-Hybrid finite element approximations of second-order elliptic boundary value problems, Compt. Meth. Appl. Mech. Eng. 11, p. 175-206.

Babuška, I., J.T. Oden and J.K. Lee (1978): Mixed-Hybrid finite element approximations of second-order elliptic boundary value problems, part II: weak hybrid methods, Comp. Meth. Appl. Mech. Eng. 14, p. 1-22.

Batchelor, G.K. (1967): An introduction to fluid dynamics, Cambridge University Press, Cambridge.

Beale, J.T. (1984): Large-time regularity of viscous surface waves, Arch. Rat. Mech. Anal. 84(4), p. 307-352.

Bercovier, M. (1978): Perturbation of mixed variational problems: Application to mixed finite element methods, RAIRO, Série Rouge Anal. Num. 12 (3), p. 211-236.

Beris, A.N., R.C. Armstrong and R.A. Brown (1984): Finite element calculation of viscoelastic flow in a journal bearing, J. of Non-Newtonian Fluid Mechanics, 16, p. 141-172.

Bird, R.B., W.E. Stewart and E.N. Lightfoot (1960): Transport phenomena, John Wiley and Sons, New York.

Bird, R.B. (1976): Useful Non-Newtonian Models, Annual Review of fluid mechanics, 8, p. 13-34.

Bird, R.B., R.C. Armstrong and O. Hassager (1977): Dynamics of polymeric liquids, Vol. 1, New York, Wiley.

Blum, E.K. (1972): Numerical Analysis and Computation. Theory and Practice, Addison-Wesley, Reading.

Brandt, A. (1977): Multi-level adaptive solutions to boundary-value problems. Math. Comp., 31, p. 333-390.

Brandt, A. (1982): Guide to Multigrid Development. In: Hackbusch and Trottenberg.

Brézis, H. (1983): Analyse fonctionelle, théorie et applications, Masson, Paris.

Brezzi, F. (1974): On the existence, uniqueness and approximation of saddle-point problems arising from Lagrangian multipliers, RAIRO, Série Rouge Anal. Num. 2, p. 129-151.

Brice, J.C. (1965): The growth of crystals from the melt. North-Holland Publishing Company.

Broyden, C.G. (1965): A class of methods for solving non-linear simultaneous equations, Math. Comp. 19, p. 577-593.

Cameron, A. (1966): Principles of lubrication, Longmans,

London.

Chorin, A.J. (1968): Numerical solution of the Navier-Stokes equations, Math. Comput. 22, p. 745-762.

Christie, I., D.F. Griffiths, A.R. Mitchell and O.C. Zienkiewicz (1976): Finite element methods for second order equations with significant first derivatives, Int. J. Num. Meth. Engng, 10, p. 1389-1396.

Ciarlet, P.G. (1978): The finite element method for elliptic problems, North-Holland, Amsterdam.

Ciarlet, P.G. and P.A. Raviart (1972): General Lagrange and Hermite interpolation in IR^n with applications to finite element methods, Arch. Rat. Mech. Anal. 46, p. 177-199.

Ciarlet, P.G. and P.A. Raviart (1973): Maximum principle and uniform convergence for the finite element method, Comp. Meth. Appl. Mech. Eng. 2, p. 17-31.

Colella, P. (1985): A direct MUSCL scheme for gasdynamics, SIAM J. Sci. Stat. Comp. 6, p. 104-117.

Collatz, L. (1960): The numerical treatment of differential equations, Springer Verlag, Berlin.

Collins, M. and W.R. Schowalter (1963): Behaviour of non-Newtonian fluids in the inlet region of a channel, A.I.Ch.E.J., 9, p. 98-102.

Concus, P., G.H. Golub and G. Meurant (1982): Block Preconditioning for the Conjugate Gradient Method. Report LBL-14856, Lawrence Berkeley Laboratory, Un. of California.

Courant, R. (1936): Differential and integral calculus II, Blackie and Son, London.

Coyne, J.C. and H.G. Elrod (1970): Conditions for the rupture of a lubricating film I. J. of Lubrication Technology 92, p. 451-456.

Crochet, M.J. and K. Walters (1983): Numerical methods in Non-Newtonian fluid mechanics, Annual Rev. of Fluid Mech., 15, p. 241-260.

Crochet, M.J., A.R. Davies, K. and K. Walters (1984): Numerical Simultation of Non-Newtonian Flow, Amsterdam, Elsevier.

Crouzeix, M. and P.A. Raviart (1973): Conforming and nonconforming finite element methods for solving the stationary Stokes equations, RAIRO, Série Rouge Anal. Num. 3, p. 33-76.

Crouzeix, M. (1976): Proceedings of: Journées "éléments finis", Université de Rennes, France, 1976.

Cryer, C.W. (1976): A survey of trial free boundary methods for the numerical solution of free boundary problems M.R.C. Techn. Summ. Report 1693, University of Wisconsin.

Cryer, C.W. (1977): A bibliography of free boundary problems. M.R.C. Techn. Summ. Report 1793, University of Wisconsin.

Cuthill, E. and J. McKee (1969): Reducing the bandwidth of sparse symmetric matrices, Proc. 24th Nat. Conf. Assoc. Comput. Mech., ACM Publ. P-69, 1122 Ave. of the Americas, New York.

Cuvelier, C. (1976a): Optimal control of a system governed by the Navier-Stokes equations coupled with the heat equation. New Developments in Differential Equations (ed. W. Eckhaus), North-Holland Studies in Mathematics 21, p. 81-98.

Cuvelier, C. (1976b): Perturbation, approximation et contrôle optimal d'un système gouverné par les équations de Navier-Stokes couplées à celle de la chaleur. Thesis Delft University of Technology.

Cuvelier, C. (1978a): A free boundary problem in hydrodynamic lubrication including surface tension. Lecture Notes in Physics 90, p. 143-148.

Cuvelier, C. (1978b): Résolution numérique d'un problème de contrôle optimal d'un couplage des équations de Navier-Stokes et celle de la chaleur. Calcolo, vol XV, fasc. IV, p. 345-379.

Cuvelier, C. (1979): A free boundary problem in hydrodynamic lubrication governed by the Stokes equations. Lecture Notes in Control and Information Sciences 22, p. 375-384.

Cuvelier, C. (1982): On the solution of capillary free boundary problems governed by the Navier-Stokes equations, Report 82-09, Delft University of Technology.

Cuvelier, C. (1984): A time dependent free boundary governed by the Navier-Stokes equations. Lecture Notes in Physics 218, p. 170-174.

Cuvelier, C. (1985): A capillary free boundary problem governed by the Navier-Stokes equations. Comp. Meth. Appl. Mech. Engng. 48, p. 45-80.

Cuvelier, C., J.M. Driessen (1985a): A Navier-Stokes capillary free boundary problem in crystal growth. Proceedings Colloque International des Problèmes à Frontières Libres. Maubuisson. (Also: Report 84-34 Delft University of Technology).

Cuvelier, C., J.M. Driessen (1985b): Thermocapillary free boundaries in Crystal Growth. To appear in J. Fluid Mecanics.

Dervieux, A. (1981a): Approximation d'écoulements multifluides instationnaries. Rapports de Recherche 68, INRIA.

Dervieux, A. (1981b): Résolution de problèmes à frontière libre. Thèse université P. et M. Curie.

Dervieux, A. and F. Thomasset (1981): Sur l'approximation d'écoulements multifluides incompressibles visqueux par des éléments finis triangulaires de degré un. Rapports de Recherche 67, INRIA.

Donea, J., S. Giuliani, H. Laval and L. Quartapelle (1982): Finite element solution of the unsteady Navier-Stokes equations by a fractional step method, Comp. Meth. in Appl. Mech. and Engng, 30, p. 53-73.

Douglas, J. and T.F. Russell (1982): Numerical methods for convection-dominated diffusion problems based on combining the method of characteristics with finite element or finite difference procedures, SIAM J. Numer. Anal., Vol. 19, No 5, p. 871-885.

Driessen, J.M. (1984): Effects of non-uniform surface tension in fluid flow: Marangoni effect. Master's Thesis Delft University of Technology.

Driest, E.R. van (1956): On turbulent flow near a wall, J. Aeronaut. Sci, 23, p. 1007-1036.

Duvaut, G., J.L. Lions (1971): Les inéquations en mécanique et en physique, Dunod, Paris.

Duvaut, G. and J.L. Lions (1972): Les inéquations en mécanique et en physique. Dunod, Paris (Eng. transl: Springer (1981)).

Engelman, M.S., G. Strang and K.J. Bathe (1981): The application of quasi-Newton methods in fluid mechanics, Int. J. Num. Meth. Engng, 17, p. 707-718.

Forsythe, G.E. and W.R. Wasow (1960): Finite difference methods for partial differential equations, John Wiley and Sons, New York.

Fortin, M. (1972): Calcul numérique des écoulements des fluides de Bingham et des fluides newtonians incompressibles par la méthode des éléments finis. Thesis, Université de Paris VI.

Fortin, M. (1976): Résolution numérique des équations de Navier-Stokes par des éléments finis de type mixte, Rapport IRIA 84.

Fortin, M. (1977): Analysis of the convergence of mixed finite element methods, RAIRO, Série Rouge Anal. Num. 5, p. 341-354.

Fortin, M. (1981): Old and new finite elements for incompressible flows, Int. J. for Num. Meth. in Fluids, 1, p. 347-364.

Fortin, M. (1983): Two comments on: consistent vs reduced integration penalty methods for incompressible media using several old and new elements, Int. J. for Num. Math. in Fluids, 3, p. 93-98.

Fortin, M. and R. Glowinski (1983): Résolution numérique de problèmes aux limites par des méthodes de Lagrangian augmenté. Dunod, Paris (eng. Translation: North-Holland, Amsterdam (1983)).

Fox, L. (1961): Numerical solution of ordinary and partial differential equations, Pergamon Press, Oxford.

Gear, C.W. (1971): Numerical initial value problems in ordinary differential equations, Prentice Hall, Englewood Cliffs, New Yersey.

George, J.A. (1976): Solution of linear systems of equations: direct methods for finite element problems. In: Sparse Matrix Techniques, Eds. Dold, A. and B. Eckmann. Lecture Notes in Mathematics, Springer Verlag, Berlin, p. 52-101.

Gerard, R. (1974): Finite element solution for flow in noncircular conducts, J. hydr. div., HY3, p. 425-441.

Gessner, F.B. and A.F. Emery (1976): A Reynolds stress model for turbulent corner flow part I, J. Fluids Eng., p. 261-269.

Gibbs, N.E., W.G. Poole and P.K. Stockmeyer (1976): An algorithm for reducing the bandwidth and profile of a sparse matrix, SIAM J. Numer. Anal., 13, p. 236-250.

Gill, P.E. and W. Murray (eds.) (1974): Numerical methods for constrained optimization, Symposium, Teddington, England, Jan. 1974, Academic Press.

Ginsburg, T. (1971): The conjugate gradient method. In: J.H. Wilkinson and C. Reinsch (eds.). Linear Algebra, Handbook for automatic computation Vol. II, Springer Verlag, Berlin.

Girault, V. and P.A. Raviart (1979): Finite element approximation of the Navier-Stokes equations. Lecture notes in Mathematics, 749, Springer Verlag, Berlin.

Glowinski, R., J.L. Lions and R. Trémolières (1976): Analyse numérique des inéquations variationnelles I, II, Dunod, Paris (Eng. Translation: North-Holland, Amsterdam (1981)).

Godunov, S. and V. Riabenki (1973): Schémas aux différences, introduction à la théorie, MIR, Moscou.

Greenspan, H.P. (1968): The Theory of Rotating Fluids Cambridge University Press, London.

Gresho, P.M., S.T. Chan, R.L. Lee and C.D. Upson (1984): A modified finite element method for solving the time-dependent, incompressible Navier-Stokes equations. Part 1: Theory, Int. Journal for Num. Meth. in Fluids, 4, p. 557-598.

Griffiths, D.F. (1979): Finite elements for incompressible flow, Math. Meth. in Appl. Sci., 1, p. 16-31.

Griffiths, D.F. (1981): An approximately divergence-free 9-node velocity element (with variations) for incompressible flows, Int. J. for Num. Methods in Fluids, 1, p. 323-346.

Grisvard, P. (1976): Behavior of the solutions of an elliptic boundary value problem in a polygonal or polyhedral domain, in Hubbard (1976), p. 207-274.

Guibert, J.P., H.T. Huynh, J.L. Marce et al. (1976): Etude générale de l'influence des liquides contenus dans les satellites, Rept. ESTEC 2720.

Hackbusch, W. and U. Trottenberg (eds.) (1981): Multigrid Methods, Proceedings, Köln-Porz, Lecture notes in Mathematics 960, Springer Verlag, Berlin.

Hasbani, Y. and M. Engelman (1979): Out of core solution of linear equations with non-symmetric coefficient matrix, Computer and Fluids, 7, p. 13-31.

Hecht, F. (1981): Construction d'une base de fonctions P_1 non conforme à divergence nulle dans IR^3, RAIRO Anal. Num., 15(2), p. 119-150.

Heinrich, J.C., P.S. Huyahorn, O.C. Zienkiewicz and A.R. Mitchell (1977): An upwind finite element scheme for two-dimensional convective transport equations. Int. J. Num. Meth. Engng. 11, p. 131-143.

Hinze, J.O. (1975): Turbulence, McGraw-Hill, New York.

Hood, P. (1976): Frontal solution program for unsymmetric matrices, Int. J. Num. Meth. Engng, 10, p. 379-399.

Householder, A.S. (1964): The theory of matrices in numerical analysis, Blaisdell Publishing Company.

Hubbard, B. (1976): Numerical solution of partial differential equations, SYNSPADE 1975, Academic Press, New York.

Hughes, T.J.R., W.K. Liu and A. Brooks (1979): Finite element analysis of incompressible viscous flows by the penalty function formulation, Journal of Comp. Physics, 30, p. 1-60.

Hurle, D.T.J. and E. Jakeman (1981): Introduction to the techniques of crystal growth. PCH Physics Chemical Hydrodynamics 2, p. 237-244.

Hussain, A.K.M.F. and W.C. Reynolds (1975): Measurements in fully developed turbulent Channel flow, J. Fluids Eng., p. 568-578.

Hutton, A.G. and R.M. Smith (1981): On the finite element simulation of incompressible turbulent flow in general two-dimensional geometries. In: Num. Meth. in Laminar and Turbulent Flow. Eds: C. Taylor and B.A. Schrefler, Swansea, Pineridge Press.

Irons, B.M. (1970): A frontal solution program for finite element analysis, Int. J. Num. Meth. Engng, 2, p. 5-32.

Irons, B.M. and A. Razzaque (1972): Experience with the patch-test for convergence of finite elements, in Aziz (1972), p. 557-587.

Isaacson, E. and H.B. Keller (1966): Analysis of numerical methods, Wiley, New York.

Jamet, P. (1976): Estimations d'erreur pour des éléments finis droits presque dégénérés. RAIRO, Série Rouge Anal. Num., 10, p. 43-61.

Jones, I.P. and C.P. Thomson (eds.) (1981): Numerical solutions for a comparison problem on natural convection in an enclosed cavity. AERE-R9955, HMSO.

Kan, J. van (1985): A second-order pressure-correction method for viscous incompressible flow. SIAM J. Sci. Stat. Comp. (in press).

Kettler, R. (1982): Analysis and Comparison of Relaxation Schemes in Robust Multigrid and Preconditioned Conjugate Gradient Methods. In: Hackbusch and Trottenberg (1982).

Kistler, S.F. and L.E. Scriven (1984): Coating flow theory by finite element and asymptotic analysis of the Navier-Stokes system. Int. J. for Num. Meth. in Fluids 4, p. 207-229.

Ladyshenskaya, O.A. (1969): The mathematical theory of viscous incompressible flow, Gordon and Breach, New York.

Lambert, J.D. (1973): Computational methods in ordinary differential equations, John Wiley, London.

Landau, L.D. and E.M. Lifshitz (1967): Théorie de l'élasticité, MIR, Moscou.

Landau, L.D. and E.M. Lifshitz (1969): Electrodynamique des milieux continus, MIR, Moscou.

Landau, L.D. and E.M. Lifshitz (1971): Mécanique des fluides, MIR, Moscou.

Launder, B.E. and D.B. Spalding (1972): Mathematical models of turbulence, Ac. Press, London.

Lions, J.L. (1962): Problèmes aux limites dans les équations aux dérivées partielles, Presses de l'Université de Montréal, Montréal.

Lions, J.L. (1969): Quelques méthodes de résolution des problèmes aux limites non linéaires, Dunod, Paris.

Lions, J.L. (1973): Cours d'analyse numérique, Ecole Polytechnique, Hermann, Paris.

Malkus, D.S. and T.J.R. Hughes (1978): Mixed finite element methods-reduced and selective integration techniques: a unification of concepts, Comp. Meth. in Appl. Mech. and Engng, 15, p. 63-81.

Marchuk, G.I. (1977): Methods of numerical analysis, Springer Verlag, Berlin.

Matthies, H. and G. Strang (1979): The solution of nonlinear finite element equations, Int. J. Num. Meth. in Engng, 14, p. 1613-1626.

McLaughin, D., G. Papanicolaou and O. Pironneau (1983): Simulation numérique de la turbulence par homogénisation des structures de sous-mailles, Rapports de Recherche, no 188, INRIA.

Meijerink, J.A. and H.A. van der Vorst (1977): An iterative solution method for systems of which the coefficient matrix is a symmetric M-matrix. Math. Comp., 31, p. 148-162.

Mitchell, A.R. and D.F. Griffiths (1980): The finite difference method in partial differential equations, John Wiley and Sons, New York.

Moisseev, N.N. and V.V. Rumyantsev (1968): Dynamic stability of bodies containing fluid. Springer-Berlin.

Morgan, K., T.G. Hughes and C. Taylor (1979): The analysis of turbulent, free shear and channel flows by the finite element method. Comp. Meth. Appl. Mech. Eng., 19, 117-125.

Natterer, F. (1975): Uber die punktweise Konvergenz finiter Elemente, Num. Math. 25, p. 67-77.

Nečas, J. (1967): Les méthodes directes en théorie des équations elliptiques, Masson, Paris.

Nitsche, J.A. (1976): Uber L_∞-abschätzungen von Projektionen auf finite Elemente, Bonn Math. Schr. 89, p. 13-30.

Oden, J.T. and J.N. Reddy (1976): On mixed finite element approximations, SIAM J. Num. Anal. 13, p. 393-404.

Oldroyd, J.G. (1950): On the formulation of rheological equations of state. Proc. R. Soc. London Ser. A, 201, p. 523-541.

Orr, F.M. and L.E. Scriven (1978): Rimming flow. Numerical simulation of steady, viscous, free-surface flow with

surface tension. J. Fluid Mech. 84(1), p. 145-165.
Ostrach, S. (1977): Motion induced by capillarity. PCH Physico Chemical Hydrodynamics 2, p. 571-589.
Ostrach, S. (1979): Convection due to surface-tension gradients. Space Research 19, p. 563-570.
Ousset, Y. (1977): Etat actuel de la recherche sur la dynamique des fluides en rotation, ESA. J. p. 43-70 and p. 189-208.
Panofski, W.K.H. and M. Philips (1962): Classical electricity and magnetism, Addison-Wesley, London.
Patanker, S.V. (1980): Numerical heat transfer and fluid flow. Mc Graw Hill, New York.
Pélissier, M. (1975): Resolution numérique de quelques problèmes raides en mécanique des milieux faiblement compressibles, Estratto da Calcolo, Vol. XII, fasc. III, p. 275-314.
Peyret, R. and T.D. Taylor (1983): Computational methods for fluid flow, Springer Verlag, Berlin.
Pimputkar, S.H. and S. Ostrach (1981): Convective effects in crystals grown from melt. J. Crystal Growth 55, p. 614-646.
Pinkus, O. and B. Sternlight (1961): Theory of hydrodynamic lubrication, McGraw-Hill, New-York.
Pironneau, O. (1982): On the transport-diffusion algorithm and its applications to the Navier-Stokes equations, Num. Mathematik, 38, p. 309-332.
Pukhnachev, V.V. (1972): The plane stationary problem with a free boundary for the Navier-Stokes equation. Z. Prikl. Mekh. i. Techn. Fiz. 3, p. 91-102.
Pukhnachev, V.V. (1974): Problems with free boundary for the Navier-Stokes equations. Dissertation, Novosibirsk.
Raviart, P.A. and J.M. Thomas (1983): Introduction à l'analyse numérique des équations aux dérivées partielles, Masson, Paris.
Richtmeyer, R.D. and K.W. Morton (1967): Difference methods for initial value problems, Interscience, New York.
Rodi, W. (1980): Turbulence models and their application in Hydraulics. IAHR state of the art paper, Delft.
Ruschak, K. (1980): A method for incorporating free boundaries

with surface tension in finite element fluid flow simulations. Int. J. Num. Meth. Engng 5, p. 639-648.

Saito, M. and L.E. Scriven (1976): Study of coating flow by the finite element method. J. of Comp. Physics 42, p. 53-76.

Samarski, A. and V. Andreev (1976): Méthodes aux différences pour équations elliptiques, MIR, Moscou.

Sani, R.L., P.M. Gresho, R.L. Lee and D.F. Griffiths (1981): The cause and cure (?) of the spurious pressures generated by certain FEM solutions of the incompressible Navier-Stokes equations, Int. J. for Num. Meth. in Fluids,
Part 1: 1, p. 17-43, 1981
Part 2: 1, p. 171-204, 1981

Schamber, D.R. and B.E. Larock (1981): Numerical analysis of flow in sedimentation basins. Journal of the Hydraulics division, ASCE.

Schowalter, W.R. (1978): Mechanics of Non-Newtonian Fluids, Oxford, Pergamon.

Segal, A. (1979): On the numerical solution of the Stokes equations using the finite element method, Comp. Meth. in Appl. Mech. and Engng., 19, p. 165-185.

Segal, A. (1982a): Solution of the stationary Boussinesq equations by finite element methods. Report 82-10, Delft University of Technology.

Segal, A. (1982b): Aspects of numerical methods for elliptic singular perturbation problems, SIAM, J. Sci. Stat. Comp., 3, p. 327-349.

Segal, A. (1985): A review of some finite element methods to solve the stationary Navier-Stokes equations, Int. J. for Num. Methods in Fluids, 5, p. 269-280.

Segal, A. (1985): A short note on the implementation of boundary conditions of the type u equals unknown constant in finite element codes. Comm. in Applied Num. Methods, 1, p. 71-80.

Shen, S.F. (1984): Simulation of polymeric flows in the injection moulding process, Int. J. Num. Meth. in Fluids, 4, p. 171-183.

Silliman, W.J. and L.E. Scriven (1980): Separating flow near a static contact line: slip at a wall and shape of a free

surface. J. of Comp. Physics 34, p. 287-313.
Smith, R.M. (1984a): On the finite element calculation of turbulent flow using the k-ε model. Int. Journ. Num. Meth. Fluids, vol. 4, 303-319.
Smith, R.M. (1984b): A practical method of two-equation turbulence modelling using finite elements, Int. J. Num. Meth. Fluids, 4, p. 321-336.
Sokolnikov, I.S. (1956): Mathematical theory of elasticity, McGraw Hill, New York.
Solonnikov, V.A. ((1977): Solvability of the problem of the motion of a viscous incompressible fluid that is bounded by a free surface, Izvestia, AN.SSR.41, p. 1388-1424.
Solonnikov, V.A. (1979): Solvability of the boundary value problem describing the motion of a viscous incompressible capillary fluid in an open vessel in the two dimensional case. Izvestia, AN.SSR. 43, p. 203-236.
Sonneveld, P., P. Wesseling and P.M. de Zeeuw (1983): Multigrid and conjugate gradient methods as convergence acceleration techniques. To appear in: Holstein, H. and P.J. Paddon (eds.), Proceedings, Short Course on Multigrid Methods, Bristol, Sept. 1983.
Sonneveld, P. (1985): CGS, a fast Lanczos type solver for non-symmetric linear systems. To appear in SIAM J. Sci. Stat. Comp.
Stoer, J. and R. Bulirsch (1973): Einführung in die Numerische Mathematik II, Springer Verlag, Berlin.
Strang, G. (1972): Variational crimes in the finite element method, in Aziz (1972), p. 689-710.
Strang, G. and G.J. Fix (1973): An analysis of the finite element method, Prentice-Hall, Englewood Cliffs.
Strang, G. (1976): Linear Algebra and its Applications, Academic Press, New York.
Strang, G. (1980): Linear Algebra and its Applications, Academic Press, New York.
Stroud, A.H. (1971): Approximate calculation of multiple integrals. Prentice-Hall, Inc., New Jersey.
Stummel, F. (1979): The generalized patch-test, SIAM J. Num. Anal. 16(3), p. 449-471.

Stummel, F. (1980): The limitations of the patch-test, Int. J. Num. Meth. Eng. 15, p. 177-188.

Tanner, R.I., R.E. Nickell and R.W. Bilger (1975): Finite element methods for the solution of some incompressible non-Newtonian fluid mechanics problems with free surfaces. Comput. Methods Appl. Mech. Eng., 6, p. 155-174.

Taylor, C. and P. Hood (1973): A numerical solution of the Navier-Stokes equations using the finite element technique, Comput. Fluids, 1, p. 73-100.

Taylor, C., T.G. Hughes and K. Morgan (1980): A finite element model of one and two equation models of turbulent flow. In: Recent advances in numerical methods in fluids, eds: C. Taylor and K. Morgan. Swansea, Pineridge, p. 311-334.

Taylor, C., C.E. Thomas and K. Morgan (1980): Finite element method and the two-equation model of turbulence. In: Recent advances in numerical methods in fluids, eds: C. Taylor and K. Morgan. Swansea, Pineridge.

Taylor, C., C.E. Thomas and K. Morgan (1981): Analysis of turbulent flow with separation using the finite element method. In: Computational techniques in transient and turbulent flow, eds. C. Taylor and K. Morgan. Swansea, Pineridge. p. 283-325.

Temam, R. (1969): Sur l'approximation de la solution de Navier-Stokes par la méthode des pas fractionnaires, Archiv. Ration. Mech. Anal., 32, p. 377-385.

Temam, R. (1973): Numerical analysis, Reidel, Dordrecht.

Temam, R. (1977): Navier-Stokes equations, Theory and numerical analysis, North Holland, Amsterdam, 2nd edition.

Temam, R. (1983): Navier-Stokes equations and non-liner functional analysis, NSF/CBMS Regional Conference series in Applied Mathematics. SIAM, Philadelphia.

Tennekes, H. and J.L. Lumley (1972): A first coarse in turbulence, M.I.T. Press, Cambridge.

Thomas, C.E., K. Morgan and C. Taylor (1981): A finite element analysis of flow over a backward facing step. Computers and fluids, 9, p. 265-278.

Thomasset, F. (1981): Implementation of finite element methods for Navier-Stokes equations. Springer Verlag.

Thompson, J.F. (1982): Numerical grid generation, Proceedings Nashville Tennessee, North-Holland, Amsterdam.

Trèves, F. (1975): Basic linear partial differential equations, Academic Press, New York.

Underwood, R.R. (1976): An approximate Factorization procedure based on the block Cholesky decomposition and its use with the conjugate gradient method. Report NEDO-11386, General Electric Co., Nuclear Energy Div., San Jose, CA.

Vahl Davis, G. de (1982): Natural convection of air in a square cavity: A bench mark numerical solution. Report 1982/FMT/2, School of Mechanical and Industrial Engineering. University of New-South Wales.

Varga, R.S. (1962): Matrix iterative analysis, Prentice Hall, Englewood Cliffs.

Vosse, F.N. v.d., F.H. Vial, A.A. van Steenhoven, A. Segal and J.D. Janssen (1985): A finite element and experimental analysis of steady and pulsating flow over a two-dimensional step. Proceedings of the 4th int. conf. on num. methods in laminar and turbulent flow, Swansea, p. 515-526.

Young, D.M. (1971): Iterative solution of large linear systems, Academic Press, New York.

Zanden, J.P.P.M. van, G.D.C. Kuiken, A. Segal, W.J. Lindhout and M.A. Hulsen (1985): Numerical experiments and theoretical analysis of the flow of an elastic liquid of the Maxwell-Oldroyd type in the presence of geometrical singularities. To be published in Applied Scientific Research.

Zenouda, J.C. (1970): Résolution numérique des équations de Navier-Stokes, couplées à celle de la chaleur. Thèse de 3 cycle Paris.

Zenouda, J.C. (1971): Etude d'un schéma aux différences finies pour la résolution numérique des équations de Navier-Stokes couplées à celle de la chaleur. RAIRO, 5, R-1, p. 49-72.

Subject Index

<u>GPSR Compliance</u>

The European Union's (EU) General Product Safety Regulation (GPSR) is a set of rules that requires consumer products to be safe and our obligations to ensure this.

If you have any concerns about our products, you can contact us on ProductSafety@springernature.com

In case Publisher is established outside the EU, the EU authorized representative is:

Springer Nature Customer Service Center GmbH
Europaplatz 3
69115 Heidelberg, Germany

Batch number: 10823846

Printed by Printforce, the Netherlands